Ruediger-Marcus Flaig

**Bioinformatics Programming
in Python**

Related Titles

Hu, X., Pan, Y. (eds.)

Knowledge Discovery in Bioinformatics
approx 364 pages
2007
Hardcover
ISBN: 978-0-471-77796-0

Holmes, R.M.

A Cell Biologist's Guide to Modeling and Bioinformatics
256 pages
2007
Hardcover
ISBN: 978-0-471-16420-3

Lengauer, T.

Bioinformatics – From Genomes to Therapies
2007
1814 pages, 355 figures, 56 tables
Hardcover
ISBN: 978-3-527-31278-8

Baxevanis, A.D., Ouellette, B.F.F. (eds.)

Bioinformatics
A Practical Guide to the Analysis of Genes and Proteins
560 pages
2004
Hardcover
ISBN: 978-0-471-47878-2

Ruediger-Marcus Flaig

Bioinformatics Programming in Python

A Practical Course for Beginners

WILEY-VCH Verlag GmbH & Co. KGaA

The Author

Dr. Dr. Ruediger-Marcus Flaig
Ringstr. 22
82223 Eichenau
Germany

Cover illustration

designed by
Irén Judit Lange–Flaig

■ All books published by Wiley-VCH are carefully produced. Nevertheless, authors, editors, and publisher do not warrant the information contained in these books, including this book, to be free of errors. Readers are advised to keep in mind that statements, data, illustrations, procedural details or other items may inadvertently be inaccurate.

Library of Congress Card No.:
applied for

British Library Cataloguing-in-Publication Data:
A catalogue record for this book is available from the British Library.

Bibliographic information published by the Deutsche Nationalibliothek
Die Deutsche Nationalbibliothek lists this publication in the Deutsche Nationalbibliografie; detailed bibliographic data are available in the Internet at <http://dnb.d-nb.de>

© 2008 WILEY-VCH Verlag GmbH & Co KGaA, Weinheim

All rights reserved (including those of translation into other languages). No part of this book may be reproduced in any form – by photoprinting, microfilm, or any other means – nor transmitted or translated into a machine language without written permission from the publishers. Registered names, trademarks, etc. used in this book, even when not specifically marked as such, are not to be considered unprotected by law.

Printing: Strauss GmbH, Mörlenbach
Binding: Litges & Dopf Buchbinderei GmbH, Heppenheim

Printed in the Federal Republic of Germany
Printed on acid-free paper

ISBN: 978-3-527-32094-3

Contents

Chapter 1. Preface — 1
1.1 An axe in the house is worth a carpenter: why computers matter in the life sciences — 1
1.2 What this book is (not) about — 3
1.3 A few practical hints — 5
1.4 From the heart — 5
1.5 Structure of the book — 6
1.6 Sources — 7
1.7 Author and acknowledgements — 8
1.8 Somebody special... — 8

Part 1. EARTH — 9

Chapter 2. A Classification of Programming Languages — 11
2.1 Assembler: the language of the machine — 11
2.2 Interpreter *vs* compiler — 15
2.3 Human, all too human... — 17
2.4 Imperative languages — 19
2.5 Procedural languages — 21
2.6 Stack languages — 21
2.7 Functional languages: the λ calculus — 23
2.8 Intermission: tó mégă sýmbolon — 30
2.9 Array languages — 31
2.10 Object-oriented languages — 32
2.11 Reflective languages — 38
2.12 Declarative = logic languages: the π calculus — 39
2.13 Concurrency-oriented languages — 41
2.14 Scripting languages — 43
2.15 Markup languages — 45
2.16 Overdose kids — 47
2.17 Strong *vs* weak typing — 48
2.18 Choose wisely... — 55
2.19 Why PYTHON? — 57

Chapter 3. ◇ Propedeutics — 61

Chapter 4. Getting the Materials — 63

v

4.1	Getting the PYTHON interpreter	63
4.2	The editor	65
4.3	*sys.argv*, redirection, and pipelines: don't fear the command line	66

Chapter 5. Variables, Data Types, and Assignments   73

5.1	Relevance and purpose	73
5.2	Simple or "scalar" data types	74
5.3	Intermission: 0x07 – the hexcode never dies	82
5.4	Homogeneous groupings	84
5.5	Intermission: a note on assignment	93
5.6	Heterogeneous groupings	93
5.7	Strings	97
5.8	Assignments and unary operators	98
5.9	Overview	99
5.10	Digression: orthogonality	100

Chapter 6. Flow Control   105

6.1	Comments	105
6.2	Docstrings	106
6.3	Blocks	108
6.4	The IF...THEN...ELSE... construct	109
6.5	Modifiers	114
6.6	Multiple branching: SWITCH *vs* ELIF	115
6.7	Applying code to piles of data: FOR, MAP, FILTER and friends	116
6.8	The REDUCE function	130
6.9	What about the locals?	130
6.10	Gimme annather try: WHILE... loops	131
6.11	Take the short cut – to be continued	134
6.12	Endless pain	135
6.13	Timer-driven loops	137
6.14	One Loop To Rule Them All...	137
6.15	"Go to" statements	138
6.16	Pattern matching	140
6.17	Choosing a functional variable	141
6.18	Thou Mayst Pass	142

Chapter 7. ◇ Application: Full Impact   143

7.1	The problem	143
7.2	Approach	144
7.3	Rough and ready	145
7.4	Refinements	148

Chapter 8. Functions and Procedures   153

8.1	Subroutines – the foundation of all higher programming	153
8.2	The importance of not getting trapped between two mirrors: recursive programming	159
8.3	Iterative programming	163

8.4	Digression: the joy of stacks	164
8.5	Digression: a trick of the tail	171
8.6	The Travelling Salesman problem	173
8.7	◇ Application: cytometry (I)	177

Chapter 9. Application: Your Most Expensive Pocket Calculator — 181

Chapter 10. The Object-oriented World — 187

10.1	What happened so far	187
10.2	Class and instance	188
10.3	Constructors and destructors	190
10.4	Self *vs* non-self	191
10.5	The serpent and the crystal: PYTHON *vs* RUBY	193
10.6	Primitive types *vs* objects	194
10.7	Modules and statics	195
10.8	Inheritance	196
10.9	Ram It Down!	198
10.10	Inheritance, abstractness, and privacy	199
10.11	A different point of view: a look at SMALLTALK	200
10.12	The coffee machine: a visual example	202
10.13	Tracking and banking	205
10.14	Generatio spontanea	214
10.15	The octopus and the coffee mug: C^{++} *vs* JAVA	217
10.16	Strings as objects	224
10.17	`None` and `object` in PYTHON	232
10.18	Paster of Muppets	234
10.19	Meyer's Principles: a look at EIFFEL	238
10.20	Application: cytometry (II)	240
10.21	And then...	241

Chapter 11. ◇ Exercise: Prime Numbers — 243

Chapter 12. The Cathedral and the Bazaar: Rivalling Strategies — 247

12.1	The TOP-DOWN strategy	247
12.2	The BOTTOM-UP strategy	249
12.3	A few notes on style	250
12.4	Splitting a large project	250

Chapter 13. Ordo Ab Chao: Sorting and Searching — 261

13.1	When is a group of data ordered?	261
13.2	Bubblesort	262
13.3	Insertsort	263
13.4	Quicksort	264
13.5	Comparison	266
13.6	Others	267

Chapter 14. Welcome to the Library — 269

Part 2. WATER 275

Chapter 15. ♣ A Very Short Project: Trithemizing a File 277
- 15.1 A brief history of coding — 277
- 15.2 Johannes von Trittheim (Trithemius) and his method of steganography — 279
- 15.3 Tools of the trade — 279
- 15.4 Your task — 281
- 15.5 Professional cryptography now – random remarks — 282

Chapter 16. Some Thoughts on Compression and Checksums 287
- 16.1 Introduction — 287
- 16.2 A simple lossless compression algorithm — 290
- 16.3 The LZW compression algorithm — 301
- 16.4 Digression: prototyping and extending — 307
- 16.5 Preventing unnoticed tainting of data — 308

Chapter 17. Dealing with Errors 311
- 17.1 What are errors good for? — 311
- 17.2 When the walls come tumbling down: errors leading to program termination — 313
- 17.3 Error signalling by return value — 313
- 17.4 Error signalling by exceptions — 315
- 17.5 Assertions — 319
- 17.6 Crash & Burn — 319

Chapter 18. ♣ A Real-life Project: Generating a Restriction Map and Making Simple Predictions 321
- 18.1 Meet Micro-Willy — 321
- 18.2 What is a restriction map? — 321
- 18.3 Extracting data from a file — 324
- 18.4 Finding palindromes — 329
- 18.5 Naming the sites — 330
- 18.6 Generating the protein translations — 331
- 18.7 Making the map — 332
- 18.8 An artificial digest — 332
- 18.9 Predicting problems with star activity — 333
- 18.10 Profit motives – motive profit — 333
- 18.11 A glance at transmembrane regions — 334
- 18.12 Addressing a project — 334

Chapter 19. Advanced Techniques in Python 337
- 19.1 Escape processing and prettyprinting — 337
- 19.2 Regular expressions — 341
- 19.3 File handling — 343
- 19.4 More file handling: persistent dictionaries — 343
- 19.5 String handling — 344

19.6	Raising exceptions	344
19.7	"Curses": the functions for advanced text output	345
19.8	Module "os": the interface to all operating systems	346
19.9	Utile et iucundum	346

Chapter 20. ♣ The Third Project: Python goes PCR 347
- 20.1 Monkey business 347
- 20.2 Polymerase chain reaction: the boiling hell of Dr. Mullis 348
- 20.3 Determining the melting point of DNA 351
- 20.4 ...and still my PCR doesn't work yet! 353
- 20.5 Is there a hairpin structure? 356
- 20.6 Addressing a project 358

Chapter 21. The Wizards' Sabbath: A Gathering of Languages 359
- 21.1 Considerable considerations 359
- 21.2 FORTRAN 363
- 21.3 OBERON 364
- 21.4 LISP 366
- 21.5 ERLANG 368
- 21.6 RUBY 369
- 21.7 LUA 370
- 21.8 C++ 371
- 21.9 JAVA 374

Chapter 22. Facing up to Python-3000 377
- 22.1 Functional "print" 377
- 22.2 Standardized unicode support 380
- 22.3 Goodbye λ, it was nice to know you 381
- 22.4 Number representations, arithmetics, &c. 383
- 22.5 Function annotations 383
- 22.6 String formatting and I/O 383
- 22.7 Classes – as you've never seen them before 388
- 22.8 Class and function decorators 390
- 22.9 Miscellaneous 391

Chapter 23. Anna will Return 395

Glossary 401

Index 407

CHAPTER 1

Preface

The science of bioinformatics or computational biology is increasingly being used to improve the quality of life as we know it.

Bioinformatics has developed out of the need to understand the code of life, DNA. Massive DNA sequencing projects have evolved and added to the growth of the science of bioinformatics. DNA the basic molecule of life directly controls the fundamental biology of life. It codes for genes which code for proteins which determine the biological makeup of humans or any other living organism. It is variations and errors in the genomic DNA which ultimately define the likelihood of developing diseases or resistance to these same disorders.

The ultimate goal of bioinformatics is to uncover the wealth of biological information hidden in the mass of sequence data and to obtain a clearer insight into the fundamental biology of organisms and then to use this information to enhance the standard of life for mankind. It is being used now and in the foreseeable future in the areas of molecular medicine to help produce better and more customised medicines to prevent or cure diseases, it has environmental benefits in identifying waste cleanup bacteria and in agriculture where it can be used for producing high-yield low-maintenance crops. These are just a few of the many benefits bioinformatics will help to develop.

– taken from the EBI home page at http://www.ebi.ac.uk/2can

1.1 An axe in the house is worth a carpenter: why computers matter in the life sciences

Goethe once wrote: "May the Lord save us from the curse of a talent which we fail to develop to masterhood! For in the end we will achieve nothing and regret the waste of time and energy." I fail to agree, for my own notion, justified by everything that has occurred to me in my professional life, is that all things are intricately connected, and even a little bit of knowledge in one field may greatly help in another. Our schools make it all too easy to develop a certain gimlet mentality, encouraging people to know as many facts as possible about as few things as possible, right to the point where they are stalled for good because further progress would require input from a different field.

Computers are extremely powerful tools, or rather, can be used as such, if one only knows how to get them to do the job. Lots of tedious and costly experimental work might be avoided if people were capable of doing a few computer-based estimations beforehand. We all know the "20–80 rule": 20% of input accounts for 80% of output. That is to say, even superficial knowledge of computer science may be useful in solving real world problems. At least it will enable a life scientist to talk to a computer scientist and thus get the work done.

Let me tell you something from my own practical experience.

Not so long ago, I had to do a PCR. I selected suitable primers, used the *dan* program from the EMBOSS molecular biology suite to calculate melting temperatures (60°C), ordered the primers and did the experiment. It failed – no product was visible. My guess was that the annealing temperature, which I had set to 59°C (slightly lower than the theoretical value, according to a rule of thumb), was too high, and I reduced the annealing temperature to 57°C. Still it failed. 55°C, 53°C, 51°C, even 48°C – nothing. I began to fiddle around with Mg^{++} concentrations, various enzymes – nothing. Finally one possibility occured to me: I suspected that the primers might be able to anneal to each other. I fed them into *dan* and *supermatcher* – still nothing. At that point I hacked a program that did a somewhat refined comparison, based on the assumption that there might be "hotspots" of complementarity which might cause the primers to stick together, though they might escape a common matching program. And behold! there was indeed one such hotspot, comprising just a handful of nucleotides but amounting to a melting temperature of 62°C. After I had finished banging my head against the wall, I set the PCR machine to an annealing temperature of exactly 62°C... and this time I got so much product that the agarose gel was overloaded!

This made me wonder what I might have done if I had not, by chance, been quite skilled in programming. Probably fooled around with the PCR for several weeks more, wasted costly enzyme on reactions which I would never have known just could not work. In the end maybe, by sheer luck, might have found the correct parameters and done the cloning without ever knowing what was up. Is that smart?

Anybody should be given the possibility of learning how to solve such problems. Tackling this one will be the subject of a later book. (Please begin at the beginning, though!)

However, the main problem is that most people who know about computers do not know about researchers' needs and indulge in abstract theory where none is needed, as in the following brilliant definition of an *array*: "A function whose domains are isomorphic to contiguous subsets of integers". Intellectually certainly excellent – but who would guess from that what an array is, and what it may be used for? Now here's my own explanation: "An ordered sequence of values, each of which has a number ('index') so that the values in an array x of n elements may be referred to by its index, as $x_1, x_2, \ldots, x_{n-1}, x_n$. It can be thought of as a street with a single name but separate houses for different pieces of data, which can be addressed by their number. For example, when measuring gene expression on the single-cell level in a cytometer (p. 177), we determine the fluorescence of 10 000 cells one by one and store the values in fl_1 ... $fl_{10\,000}$."

Theory is hard to put into practice, and practice is theoretically unnecessary. Both are only inadequate substitutes for understanding.

I am only too well aware that in some academic circles a certain singlemindedness is *très chic* these days. Quite a number of scientists believe in the "pure teachings" of their own all-important field and the inviolable sanctity of the test tube. Computing, by contrast, has the image of being "nerdy" and a matter for asexual fifteen-year-olds. The truth, however, is that computers are just as important to scientists as pipettes and test tubes, since they offer a unique possibility to simulate things that are too big, too small, too fast, too slow, too dangerous or too complex for conventional experimental study; "too complex" here meaning that they involve amounts of data that would overwhelm any human "experimental genius".

Rant

1.2 What this book is (not) about

This book is about *programming in bioinformatics*.

programming in bioinformatics

My goal is not to demonstrate the intellectual beauty of information theory, nor to discuss the most advanced techniques in software development, but to teach students with little or no computing experience how to use computers to become more efficient in their scientific work, by demonstrating the way towards solving everyday problems, such as the PCR business which I just mentioned. If you are looking for an introduction to Bayesian statistics, or to neural networking, or to X-ray diffraction analysis, this book is definitely not for you; nor is this going to be about *EMBOSS*, *Husar* or other molecular biology software packages. It is also not going to be about basics such as producing smart diagrams with a spreadsheet program, writing emails or backing up the contents of your hard disk. It is about *programming in bioinformatics*. You will acquire general proficiency in algorithmic thinking, not skill in mechanically applying tools which other people have designed.

EMBOSS
Husar

programming in bioinformatics

Contrary to a widespread sentiment that details are more important than overviews, I encourage you to as wide a perspective as possible. You ought to get an idea of what is generally possible, of how many different concepts of information processing there are and of how it "feels" to develop a program. Do not waste your time with learning library functions by heart; rather toy around with the different languages touched on here. In perfect accordance with *Whorf's theorem* that language shapes minds, people who have learned only one programming language are in danger of having this language's structures forced upon their way of thinking – a thing which will seriously hamper their creativity. Of course, this taking things for granted is a general problem: being trained according to one line of thought only leads to mental inbreeding. Maybe you will find this book rather revitalizing in this respect.

Whorf's theorem

Necessarily I had to focus on one language – a choice which was far from easy. After some consideration, I selected PYTHON as the "official language" for this book because it is easy to learn, powerful and freely available[1], but my emphasis is not on PYTHON as such but on the elementary concepts of programming languages in general and the way to use them; therefore you will find quite a number of comparisons between different languages here but little about the bells and whistles inside PYTHON. Ideally, after completing this book it will take you little effort to learn *any* programming language.

The aim of this book is not to serve as a reference guide for PYTHON; it is highly recommended to have the official PYTHON documentation and tutorial at hand when working on the exercises (download them from `http://www.python.org`). This way you can also make sure that your documentation is always up date, PYTHON being a swiftly evolving language. By the way, did I happen to tell you that this book is not about PYTHON but about *programming in bioinformatics*?

<small>programming in bioinformatics</small>

Furthermore, it is designed not for going-to-be computer scientists but for future life scientists (who are also expected to have a basic understanding of molecular biology and fundamental things like transcription, translation, the gentic code and the double helix). The contents of this are purely *non scholae sed vitae* – you will not be asked for this in any exam you are ever going to take, but it may help you as soon as you assume work on your first thesis. Therefore, please take the time to read carefully, and pay attention to the ♠ sign. Be aware that not all these questions are really to be answered – some are just intended to make you think; whet your own judgement by this. Feedback is always welcome.

<small>non scholae sed vitae

♠</small>

Finally, my apologies if some parts of this book are found to be redundant. In the interest of accessibility I have repeated things where necessary and tried to present them in the respective context, rather than just using cross references to excess. On the other hand, I love to occasionally throw in things which have not been discussed yet but which may be deduced from the context, and ask you to think about that. Maybe I have exaggerated the Socratic method here and there, but it is the proper method for this task. You will also come across several chapters named "Application" describing the way to write a specific program; they are intended both to illustrate what has been described so far and to introduce what is to come next. Now and then, I could not restrain from giving you some really personal advice, which is then marked with a ♡ sign. Of course, you may always feel free to disagree.

<small>♡</small>

♡ "Computer science is no more about computers than astronomy is about telescopes", Dijkstra (one of the really great gurus of the field) once said. To be more specific: **"Computer programming is not about writing programs;**

---

[1] Please refer to "Why PYTHON?" on p. 57 for a more detailed explanation of this choice, and to "The Wizards' Sabbath" (p. 359) for a comparison of languages.

it's about understanding a problem thoroughly and devising a strategy to its solution." That given, writing the actual program is a minor task. Someone who is capable of that kind of analytical thinking will benefit in other sciences as well. Another famous quotation by Dijkstra reads: **A programming language which does not change the way you think is not worth learning.**

1.3 A few practical hints

Now and then you will find boxes comprising summaries, labelled with a capital sigma. Moreover, important "paradigms" will be noted in the margin. A large ℜ in the margin is to direct your attention to particular keywords and to the importance of the section, in particular with regard to bioinformatics. As in every field, for the beginner it is not always obvious why the path should wind through some particular thicket!

<small>Annotations</small>

In addition, there are a few sections marked with special signs. Those with a ◇ comprise exercises which are not "mainstream", but which I advise you strongly to do, since they may help you to understand (or maybe ask the correct questions). Feel free to discuss them with your friends. Those marked with a ♣, on the contrary, are the projects which you should do at all costs – and for yourself. Exemplary solutions will be found at the end of the second volume.

<small>Signs
◇
♣</small>

1.4 From the heart

Perhaps the most famous Japanese ever was 17th century Shinmen *Miyamoto Musashi*, surnamed *ken-sei*, "Enlightened Sword-Master" – a title he certainly deserved. Wandering about without home or aim for the major part of his life, he did not miss any occasion for quarreling and emerged victorious from more than 60 duels, many of them fought against renowned warriors; the *ni tō* (Two Swords) way of fencing he devised is regarded as the highest accomplishment of swordsmanship. When he finally retired, he surprised everyone by immediately becoming a notable painter and sculptor, some of his works are still considered as classics today. Upon the request of his friend and liege he also wrote a book on the subject of sword fighting, named "The Book of the Five Rings" (*Go Rin no Sho*). In this book, he gave some very interesting bits of advice:

<small>Miyamoto Musashi

Go Rin no Sho</small>

- *By following no particular School I defeated followers of all Schools:* A hint which pertains especially to going-to-be professionals. You will constantly have to face situations you are not prepared for, and coping with this is essential. This, however, can only be done when you study different approaches instead of yielding to the temptation to considering a single one as The Way.
- *The best technique is no technique, the best stance is no stance:* This was not meant as a justification for bungling or amateurishness but to express the fact that when you become really proficient in any art or ability, at a certain point formal structures will lose their helpfulness and in fact become

an impediment. To Musashi, fencing was, and ought to be, as natural as walking or talking. You do not form your words according to a conscious pattern, nor do you care about the rules of grammar or phonetics when talking in your mother language – you just talk, free from formal structures. When learning a new language, you will nevertheless first have to learn the rules, before you grow beyond them. But you do not begin to learn a language with the intention to remain glued to the grammar reference for good.

- *Masterhood in one field will entail masterhood in all:* A corollary of the former. Attaining true proficiency and mental flexibility by learning will facilitate further learning. Speaking for myself, it took me about two hours two learn PYTHON, and about a week to get into my current field of research.

The Book of the Five Rings became very popular in the West, most recently in the 1990s among managers, who grotesquely abused the "no technique" idea as an excuse for any act of sloppiness or manifestation of incompetence[2]. This is as far from Musashi's intentions as can be. He is reported to have given the following answer to a lazy student who wanted to know how much practice he still needed. "Just as much practice as will enable you to do *this*, mind you", Musashi replied. Then he had a grain of rice placed on the top of the student's head and aimed a violent *shomen-uchi* (a vertically descending blow) at him. Down fell the two halves of the grain, cut asunder cleanly without as much as scratching the student's skin.

Do yourself a favour and think about this man's advice.

1.5 Structure of the book

Beginner

It should be noted that there is no "entry level" qualification required. You can start without knowing anything about programming before. However, basic skills in handling a computer are expected, as it would be tedious to begin with these fundamentals (which are taught in more elementary books anyway). For example, you should know the the meaning of terms such as "processor", "memory", "persistent storage" and "file system". If some terms should seem unfamiliar to you, the *glossary* at the end of this book (p. 401) may be helpful.

glossary

This book is designed in two parts; according to the natural structure but also in reverence to Musashi (place smile here), it has been divided into five sections (two for the first, three for the second part):

(1) Earth – a classification of programming languages; data types; control structures; functions; objects; software project organization; iteration *vs* recursion; tips for getting started with PYTHON.

---

[2] As happens frequently to Eastern ways when brought to Euro-America: Oṁ mani padme hum; oṁ mani padme hum; wow, now I'm enlightened!

(2) Water – the three projects for the first term (cryptography, sequence analysis, oligonucleotide design); advanced techniques in Python; a comparison of languages.

(3) Fire – a gentle introduction to geometry and simulations; the Biopython package; tools for professional development; understanding data processing by building an interpreter for the programming language Lisp; two intermediate projects.

(4) Air – Graphical user interfaces; multithreading; three-dimensional graphics; bottom-up simulations; a complete virtual machine with assembler and Basic compiler included; the two highest projects.

(5) Void – Networking basics; real-life applications; algorithms in bioinformatics; summary (exemplary solutions, glossary and index).

Much care was spent on arranging "presentation" and "practice" chapters in an alternating fashion in order to enable students to switch to learning by doing as soon as possible – without getting too focused on a single language, however.

For those who are interesting in delving deeper into algorithms applied in molecular biology, the book "Genomic PERL" by Rex A. Dwyer (Cambridge University Press, 2003) is highly recommended, as it describes the approaches with exceptional clarity and furnishes examplary solutions. However, these examples are implemented in Perl, a language which is not suitable for teaching purposes, so this book requires a solid background in the basics of programming.

1.6 Sources

There is no bibliography at the end of this book. The knowledge compressed into these approximately four hundred pages was collected over the course of more than twenty years from all kinds of sources – books, journal articles, internet postings, personal discussions and, above all, the dearly-bought experience gained by lots of personal work on real-life problems which had to be solved somehow. A few selected titles can be found at the end of the second volume.

> "By all accepted standards, Musashi was not a great sword technician. Schools, styles, theories, traditions – none of these meant anything to him. His way of fighting was completely pragmatic. What he knew was only what he had learned from experience, living as an ascetic in the mountains, exposing himself to the dangers presented by nature as often as to those presented by man. He was not putting theory into practice; he fought first and theorized later."
>
> – Eiji Yoshikawa

1.7 Author and acknowledgements

Rűdiger Marcus Flaig (a.k.a. Don Rodrigo) was born in 1971 in Mannheim, Baden. In 1994, he received his Master's degree in biology from the University of Heidelberg and subsequently obtained the degrees of Dr. sci. hum. (medicine) in 1997 and of Dr. rer. nat. (pharmaceutics, with *summa cum laude*) in 2001. In the course of his adventurous life, he has worked in almost all fields of the life sciences (and also in law, languages, ... $\to \infty$). From 2003 to 2005, while in Heidelberg as a postdoctoral researcher in immunology, he taught "Programming in Bioinformatics" in cooperation with Roland Eils of the "Intelligent Bioinformatics Systems" (IBIOS) division of the German Cancer Research Centre. For his scientific and humanitarian endeavours he has received a number of decorations, including the Imperial Order of the Ethiopian Lion and the title of a Knight of the Collar of St. Agatha of Paternò.

The author gratefully acknowledges the help of many of his past and present colleagues as well as the students in his course, whose critical feedback greatly improved the quality and made him realize the unmet need for a comprehensive textbook. But above all, he would like to recognize the support he has received from his family, in particular his beloved wife, M.A. Irén Judit Lange-Flaig, without whose invaluable contributions all this would not have been possible.

1.8 Somebody special...

Meet our mascot:

Anna the hannaḥ

Part 1

EARTH

CHAPTER 2

A Classification of Programming Languages

Programming is so simple that any intelligent five-year-old can learn it. If you are more than five years old, however, you may experience some problems.

– Author unknown

I have often found it best to present material sorted in chronological order and in rising complexity, which may often be correlated. Therefore, we will begin at the very lowest end.

2.1 Assembler: the language of the machine

Relevance: Understanding the nature of data and the fundamental requirements for data processing.
Keywords: Basic design of computers; processor; RAM; registers; VM.

$\mathfrak{R}$

2.1.1 Processor architecture. A computer is a tool for data processing. Thus, it can be defined as a device capable of the following four actions:

- *Input* – accepting data from some external source (keyboard and mouse, scanner, network, file system, attached measurement device...). Input
- *Storage* – remembering data and keeping them ready. Storage
- *Processing* – manipulating data (doing any kind of arithmetical and logical operations and correlations). Processing
- *Output* – releasing data (results) to some external sink (monitor, printer, network, file system, attached manufacturing device...). Output

The development of information technology was made possible by the discovery that all four of these capacities are implemented most readily by breaking down complex information into its atomic units (e.g. a text into its characters), and greatly aided by the discovery that it is best to keep these capacities as separate as possible. Taken together, these two discoveries allow one to reduce the tasks, which the hardware must be able to perform, to a small number of basic actions.

Input and output (the *peripherals*) are self-explanatory. As for processing and storage, these are the tasks of the processor (CPU = Central Processing Unit) peripherals CPU

Bioinformatics Programming in Python. Rüdiger-Marcus Flaig
Copyright © 2008 WILEY-VCH Verlag GmbH & Co. KGaA, Weinheim
ISBN: 978-3-527-32094-3

RAM and the memory (RAM = Random Access Memory, since it may be both read and written), respectively. Some older computer designs such as the VAX actually did not discriminate clearly between storage and processing, the processor operations for manipulating data being directly applicable to data in the RAM. However, it soon became apparent that performance may be greatly improved by a design wherein all processing operations may be applied only to individual

registers snippets of data held in the processor's *registers*, and the memory is addressed by a different set of operations directed to loading values from the RAM (or, for accepting input, from keyboard, mouse, network adapter, permanent storage device...) into the processor's registers, or shoving values from the registers back into memory (or, for generating output, to the printer, network adapter, graphics card or permanent storage device...).

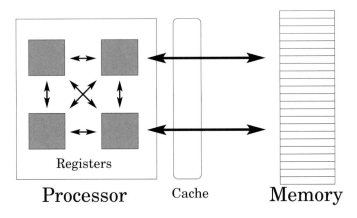

address The memory is arranged in linear fashion, with each memory cell possessing a unique *address* by which it is identified. Addresses may also be loaded into registers and used for accessing the RAM via the second set of operations mentioned above. **Think of the RAM as a huge shelf full of folders, and the registers a tiny desk. Only very few things can be placed on the desk, but they are at hand for processing.** Or if you are of the more fanciful kind, think of a many-handed Indian deity – each register is one "hand" whereby the processor may grab and manipulate one piece of data.

So to sum it up, the processor is capable only of a number[1] of primitive operations for loading data from the RAM into its registers, performing operations on data in its registers and storing back data from the registers into the RAM. These primitive operations are encoded by very short, processor-specific[2] byte

instruction codes sequences, fancifully called *"instruction codes"*, which are usually unintelligi-

---

[1] Some processors such as the VAX feature more than 300 instructions, but in fact no more than approximately 30 are required for a fully functional and highly streamlined processor. In the second volume, we will have a closer look at this.

[2] I.e.: A Pentium, a PPC, a SPARC and a VAX all have instructions for fetching and returning data, but their "genetic code" – which combinations of bytes denotes which action – is different.

ble to human beings. There is, however, a "mapping" (similar to the genetic code) where these byte sequences are represented by words, sometimes called "*mnemonics*". Taken together, the processor's instruction codes form the most basic programming language, often referred to as the "*machine language*". Any program, no matter how sophisticated the language, must and will ultimately result in a series of machine instructions, since it is only these actions which the processor can perform. **mnemonics**
machine language

Keep in mind that a processor does not have to be a hardware chip; some modern programming languages make use of a fictive processor which is emulated by a small program – the *virtual machine* or "VM" – running on actual hardware. In fact, the usage of VMs is currently increasing, as this is an easy way to standardize different hardware.

Hardware implementations usually comprise a few additional components for improving performance. As the processor is much faster than the RAM, the *cache* acts as a buffer between both. Devices – video cards, disk controllers, network adapters – occupy defined addresses; the processor has no distinct operations for, e.g., file access but instead just sends data to (or receives from) the address where the video card, disk controller or network adapter is known to be located. "Knowing" what that particular address is, and providing standardized functions for performing standard operations, is the main job of the *operating system*[3]. **cache**

operating system

Perhaps, not surprisingly, this general design – underlying all commercially available processors – is known as *register machine*. An alternative concept is represented by the **stack machine**, which will be discussed in the context of stack languages (p. 21). **register machine**

Those who desire a more thorough understanding of processor hardware and its programming are recommended to the second volume of this book, where we will design a processor of our own, an assembler and a BASIC compiler (see below). In this context, details of processor design and assembly language (see below) will also be discussed.

Today, the vast majority of computers run processors belonging to the Intel x86 family. This family is characterized by more than thirty years of continuity (starting with the 4004 in 1971), resulting in many compromises. x86 processors traditionally have few registers (originally no more than six general-purpose ones) but exceedingly high clock speeds. The AMD64 ("x64") and Itanium families are designed as successors of the x86 family and partially compatible with it. Less well known is the PPC family, jointly developed and produced by Apple, IBM and Motorola and found in large **x86**

x64 and Itanium
PPC

---

[3] For this reason, simple operating systems were often named "disk operating systems" (DOS), e.g. ProDOS™ (for the Apple II) or MS-DOS™, since before the advent of multitasking managing the file system was an operating system's prime purpose. Contemporary OSes have to take account of multi-user and multi-task activities as well and see to it that the different tasks do not cause confusion by indiscriminate access to the peripherals, but the essential requirements remain the same.

servers. They have a large number of registers (32 general-purpose and another 32 for floating-point operations), which makes them more powerful than the x86 line, but it requires adequate programming to harness this. Sun's SPARC and Silicon Graphics's MIPS are even more powerful than the PPC (the SPARC possessing no fewer than 144 registers) but found only in a few high-end machines. Acorn's ARM has an intermediate number of registers and a very smart internal design, which enables it to yield sufficient output at very low power consumption; it is no longer used in personal computers, but is used in millions of embedded devices such as cellphones. Some aspects of the differences between these processors will be discussed in the second volume.

2.1.2 Programming the hardware. This is how the expression x + 1 would be expressed in assembly language[4]:

```
POP   r51              ; get parameter value into register 51
LOAD  r52, 1           ; set register 52 to a value of 1
ADD   r51, r52, r57    ; add reg.s 51 and 52 and write the result into reg. 57
PUSH  r57              ; return the result
```

This *source text* is converted into *machine code* by means of a program named "assembler", and the language is therefore denoted "assembly language", or just plainly "Assembler". The terms "machine code" and "assembly language" are generally used interchangeably.

The characteristic trait of *assembly language* is that it is neatly tailored to the processor's architecture. This is advantageous in that it allows perfect exploitation of the processor's special capacities. At the same time, it is disadvantageous as it does not permit "porting" to a different processor architecture. Assembly code written for an x86 processor just will not run on any other processor, and vice versa. There may even be problems between different generations of the same processor line[5].

Assembly language thus enables us to write very efficient pieces of code (see p. 360 for an example), but it is not well suited for larger projects and programs comprising more complex data structures. To this end, *higher programming languages* were developed, beginning with FORTRAN, LISP and COBOL in the 1950s. In this course, we will focus exclusively on higher programming languages[6].

---

[4] The assembly language used in this little example, as in all examples relating to assembly language here, is that of the fictitious FOOBAR processor which we will build in Volume 2. It is somewhat inspired by that of an ARM processor.

[5] For example, very few systems based on the Motorola 68000 survived the upgrading to the more powerful 68020/68030; around 1990. Likewise, much 8088/8086 software had to be rewritten to run on the 80286 around 1985. Vice versa, code taking advantages of the features of a new generation simply will not run on an older processor.

[6] In the second volume, we will design a processor of our own and have a closer look at assembly language.

♠ *The C boom was sparked off in the early 1980s by a statement by Dennis Ritchie, coauthor of the programming language C and of the operating system Unix, claiming that the performance of the Unix "kernel" (i.e., the system core) rewritten in C was markedly higher than that of the assembly language original. Can you suggest any explanations for this observation?* Dennis Ritchie
Unix

♡ *Programmers' slang is difficult to get into because some of its expressions are ambiguous and context-dependent. Perhaps the ultimate sin in this respect is the expression* code, *which may mean:* code

- *a piece of source text (the "coder" being the person who actually writes the program)*
- *the machine instruction generated from the former*
- *something that deals with encryption.*

Another important slang expression is the bug, *etymologically and entomologically from the insect that got into the first ENIAC and caused the program to crash. Thus a bug is generally everything that goes wrong in a program. Hence, the process of tracking down errors is known as "debugging", and a "debugger" is usually not the poor soul condemned to such a task but a program that aids in debugging. I do not dare to say "support", because this verb is used to express the fact that a certain feature is contained in something (e.g., "*ERLANG *supports parallel computing and arbitary-precision integers"). Other terms to be remembered are:* bug

- file: *a pile of data which are packaged together and accessible under a name (file name);* file
- string: *a list of characters – what human beings call "word";* string
- user interface: *the part of the program that is visible to the user;* user interface
- command line: *either a user interface which is based on typing commands, or the commands typed in at such an interface;* command line
- shell: *the program that "encloses" the operating system and mediates its communication with the user – either a command line or a graphic user interface.* shell

> The computer's processor performs elementary actions only. A higher programming language offers more powerful expressions.

2.2 Interpreter *vs* compiler

Relevance: Appreciating the different approaches to higher programming.
Keywords: Assembler; interpreter; compiler; linker; high- and low-level languages; abstraction.

abstraction level

Higher programming languages are languages in which there is no one-to-one correlation between the instructions contained within the written program and the actions taken by the processor. That is to say, higher programming languages provide an *abstraction level*. The degree of abstraction provided, however, varies greatly, ranging from "low-level" to "high-level" languages. We will soon see what abstraction means, and what criteria make a programming language low-level or high-level. For the time being, let us content ourselves with saying that **a high-level language tries to implement human concepts on the machine, whereas a low-level language requires humans to frame their designs for the machine**.

compiler
object code file

A higher programming language may basically work in one of two ways (rarely, both have been realized simultaneously, e.g., with O'CAML). Either we have a program which does basically what the assembler does – it reads in the source file and produces a file of machine specific code. This program is named *compiler*. In general, the file generated by the compiler (for some unknown reason often referred to as *object code file*) is not executable yet but must be bound together with some auxiliary files. And to make things a little bit more complicated; some compilers, e.g., the excellent GNU C compiler (*gcc*), do not even produce object code but output an assembly language file, which is then handed to the assembler where it is converted to object code and finally thrown to the *linker*. Luckily, crosstalk between the programs is nowadays so neat that usually the programmer does not even notice that. These auxiliary files include the *runtime system* – the blue collar brigade which sets up things before, provides the catering behind the scenes during the party and does the laundry afterwards without getting or expecting any praise – this is the task of the *linker*. Complex programs may thus be divided into hundreds of source modules which may be compiled independently and linked together only when needed.

linker

runtime system

interpreter

Alternatively, the program that reads the source file(s) may not generate any output files but may run the code directly, step by step. Such a program is termed an *interpreter*. For obvious reasons, interpretation is faster in the development, but it is considerably slower (sometimes by orders of magnitude) in the execution.

 An interpreter runs a program written in a higher language by executing it line by line; a compiler translates it into a sequence of elementary processor operations which can be run independently.

♡ *From my point of view, the important thing is, and will always be, that an interpreter allows one to get straight into the show whereas a compiler needs a separate run, or rather two of them (compilation and linking). The best feature of many modern interpreters and compilers, including* PYTHON, *the Glasgow* HASKELL *compiler and* O'CAML, *is a "top level" – a mode where you can*

directly talk to the program. During this course, we will make extensive use of the PYTHON *toplevel.*

2.3 Human, all too human...

Relevance: The way from a task to a program.
Keywords: Interpreter; compiler; dealing with errors; software *vs* hardware.

Now imagine that Ruby Red and Prince Pascal[7] both have the task of calculating the square roots of the numbers from 1 to 100.

Ruby takes paper and pencil and proceeds for every number by writing down on the left the number in question and on the right the square root, which she quickly calculates on an auxiliary scrap of paper.

Pascal also grabs paper and pencil, but uses it only to formulate the little program[8]:

```
LOAD    r72, 1          ; initialize reg.72 with value "1"
LOAD    r73, 100        ; initialize reg.73 with value "100"
:label                  ; destination address for the JIFT instruction
MOVE    r72, r1         ; place content of reg.72 in reg. 1
CALL    :squareroot     ; invoke function for calculating and printing
                        ; sqrt of reg. 1
ADD     1, r72          ; increase reg.72 by one
ISGTR   r72, r73        ; is r.72 > r.73 ?
JIFT    :label          ; if so, go back to ":label" (JIFT = "Jump IF True")
```

Obviously, this is somewhat more complex than Ruby's approach, as it not only requires a particular processor to run (here a FOOBAR) and an assembler for this processor but also a previously formulated function named "squareroot" which has been designed to reliably calculate the square root of the value contained in register 1 and to print it. However, it will run much faster.

Ruby's approach is the interpreted one, Pascal's the compiled. Pascal *compiles* the algorithm – "Calculate the square roots of the numbers from 1 to 100" – into a piece of assembly language which is then *assembled* into the codes which the FOOBAR can understand and *linked* with a "library" comprising functions such as `squareroot`.

---

[7] Here the author is indebted to Samuel Delany's celebrated novel "Nova". By odd chance, RUBY is a characteristic interpreted and PASCAL a typical compiled language.

[8] Once more in FOOBAR assembly language.

Interpreter at work:

Compiler at work:

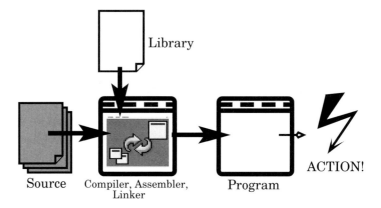

While Pascal's approach may seem to be more sophisticated, Ruby's is more straightforward, especially when it comes to identifying errors. Imagine that erroneously the task is given as "Calculate the square roots of the numbers from –1 to 100". Ruby has the alternative of either calmly writing $\sqrt{-1} = i$ or reporting the error to the boss with a request for clarification. In Pascal's case, the function squareroot will simply fail. What happens then will depend very much on circumstances, but pinpointing the source of trouble will be much more difficult.

As an interesting aside, some people may here frown upon what they perceive to be an "undue humanification of the machine". It is noteworthy, however, that unlike any other, information technology does not require any machine. Information technology is about algorithms and the processing of data, which could be done completely manually (i.e., using paper and pen). This is what underlies Dijkstra's dictum that "computer science is no more about computers than astronomy is about telescopes". A number of noteworthy pecularities of IT are based on this fact. For example, we have already learned that a processor does not require any hardware implementation but may be completely virtual. It should also be remembered that in most jurisdictions (e.g., Europe) software is explicitly exempt from patentability, as it is considered as lacking technical character. Back in 1997, the author has already pointed out in his thesis work on "pseudovirions" that indeed the capacity for information processing is what vitalists such as Uexküll considered the *vis vitalis*, which leads to the philosophical question of whether an algorithm is a living thing rather than a machine...

2.4 Imperative languages

Relevance: An introduction to some programming languages.
Keywords: The imperative paradigm.

The vast majority of contemporary languages (from COBOL and FORTRAN to C♯ and JAVA) follow the *imperative* approach, which is basically – though few people have ever realized it – an extension of the assembler concept, using abstract operations instead of processor-specific instructions. The programmer names *actions*, which are then executed by the computer. This works like the following example:

actions

```
0010 ;; Let's have some wine.
0020 While there are grapes in the vineyard:
0030     Harvest grape.
0040 Next grape.
0050 Squeeze grapes.
0060 Ferment the juice.
0070 While there is fermented juice:
0080     Fill glass.
0090     Drink.
0100 Next glass.
0110 Detoxify.
```

An imperative program is like a cooking recipe: You have a "world" – made up of objects (variables) whose states can be modified – and take actions to modify it step by step. When programs become exceedingly large, this approach becomes troublesome. Errors are hard to track down when there are thousands of variables with endless numbers of possible states, each of them potentially influencing any other.

Remember this: **A variable is a name with which a value is associated.**

The ongoing development of imperative languages is one giant attempt to keep apart what is not meant to interfere. For example, *procedures* (see p. 153) and *local variables* are the first step in this direction. Variables are restricted to where they are used and needed. They simply do not exist from the point of view of another part of the program. Other measures include restriction of possible values (see section on enum types and subtypes, p. 77), clarification of values by naming them ("PLZ_HEIDELBERG" instead of an obscure "69"), and modularization mechanisms to simplify the splitting of a large program into separate chunks.

procedures
local variables

Just to name a few:

- The Fortran languages:
  - FORTRAN (*For*mula *Tran*slator, AD 1956/1966/1977/1990/1995/2003/2008)
  - BASIC (*B*eginners' *A*ll *P*urpose *S*ymbolic *I*nstruction *C*ode, ca. AD 1967)
  - COMAL (teaching language, ca. AD 1987)
- COBOL (*Co*mmon *B*usiness *O*riented *L*anguage, ca. AD 1959)
- FORTH (announced as "the 4th generation of programming languages", ca. AD 1980)
- PL/1 (ca. AD 1966)
- The Algol languages:
  - ALGOL (*Al*gorithmic *L*anguage, AD 1958/1960/1968)
  - PASCAL (in honour of Blaise Pascal, AD 1968)
  - MODULA-2 (ca. AD 1983) and MODULA-3 (superseded by OBERON)
  - ADA (in honour of Countess Ada Lovelace, AD 1983/1995)
  - OBERON (ca. AD 1987)
  - LUA (Portuguese for "moon"; lightweighted interpreted language, ca. AD 2000)
  - DELPHI (a reincarnation of Borland's beloved TURBOPASCAL, powered up for big projects)
- The C languages[9]:
  - BCPL (antediluvian)
  - B (gracefully forgotten)
  - C (ca. AD 1971)
  - OBJECTIVE C (ca. AD 1984)
  - C^{++} (AD 1986)
  - JAVA (formerly OAK; renamed in honour of coffee, ca. AD 1991)
  - C♯ (ca. AD 1998)

branchings
loops

You will find a number of them listed again among the object-oriented languages, which is the ultimate experience in imperative programming. They all share the fact that *branchings* and *loops* are pivotal elements in them (see p. 109ff. for more details).

flow charts

Imperative programs can be represented graphically in the shape of the nefarious *flow charts*:

---

[9] Actually, the C lineage (C, C^{++}, JAVA) are also descended from ALGOL, and one cannot fail to see the similarities between C and PASCAL, but here we will use the expression "ALGOL languages" in the narrower sense.

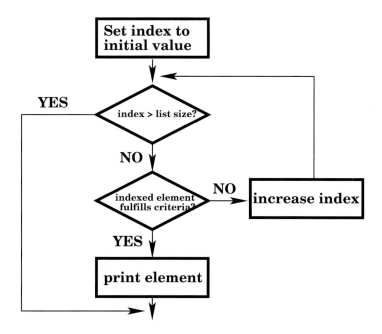

As the really interesting styles of programming are the ones to which these charts are not applicable, we will not concern ourselves with this stuff (unless you feel that it helps you to design your own programs).

 In an imperative language, actions are the basic units. The program is composed of blocks of alternating and/or repeated actions (wherein a "block" is a sequence of actions serving one common purpose and thus logically coherent).

2.5 Procedural languages

Relevance: An introduction to some programming languages.
Keywords: The procedural paradigm.

The majority of imperative languages belong here – that is to say, almost all except for the older dialects of BASIC and FORTRAN (up to FORTRAN-66 inclusive). Procedures will be described on p. 153ff. Basically, procedural languages allow extension of the "vocabulary" of a language by means of defining new "words" – indispensable for all but the most elementary tasks.

2.6 Stack languages

Relevance: An introduction to some programming languages.
Keywords: The stack language paradigm. Stacks.

These are a very small subset of imperative languages which are completely based upon the concept of the *stack* (p. 90 and p. 164). That is to say, expressions are processed by putting their elements on an imaginary pile of cards until an operator is reached, which then takes off a suitable number of operators. Thus, stack languages generally write as

```
3 3 * 4 4 * + sqrt
```

for $\sqrt{3^2 + 4^2}$ and evaluate like this:

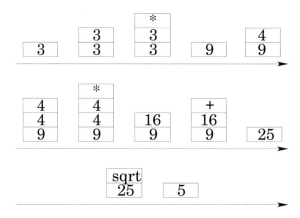

This approach was devised by a group of Polish mathematicians long before the advent of modern programming and is therefore often referred to as *reverse Polish notation*. In the 1960s and 1970, it was the standard for table calculators, but it is obviously difficult to handle and was therefore abandoned as soon as possible.

Today, there are FORTH and POSTSCRIPT to be mentioned here. POSTSCRIPT is specialized for the description of graphics, and is used extensively for the control of high-quality printers. FORTH itself has acquired a notorious reputation for being the lowest-level language imaginable (with the exception of assembly language, of course) and is used for tiny snippets of very efficient code. For example, the operating system of Apple's "Newton" nanocomputer was based upon FORTH, and had a graphical user interface with a web browser and a word processor, all fitting comfortably into four megabytes of RAM.

Among processor designers, the "stack machine" has always been considered as an alternative to the "register machine" design demonstrated at the beginning of this chapter. However, except for the hapless Inmos Transputer there never seems to have been a hardware architecture which implemented stack machinery. On the other hand, the "stack machine" is widely used not only for virtual machines but also for the internal proceedings of many interpreters and compilers. In fact, the "*bytecode*" to which PYTHON scripts are translated internally is the machine code for a virtual stack machine. When working with modules, you will notice the unforeseen appearance of files with the extension *.pyc*. They contain this "bytecode".

Stack-oriented programming is unique in that it allows programming without the need for distinct and named variables (corresponding, on the hardware level, to registers or discrete memory locations), with the values just being distinguished by their relative positions on the stack. For artificial program generators (such as printer drivers for POSTSCRIPT and compilers), this is easier to handle, whereas human programmers easily get confused by this approach. So why has the stack machine not ousted the register machine yet? Because in the end programming languages also affect processor design. Even if there is a stackish notation used as an intermediate, a variable-oriented programming language can be made to perform better on a register machine than on a stack machine. All contemporary operating systems, including Windows^{TM}, are rooted in Unix, which is – as we have learned before – written in and intricately connected with C. So all contemporary processors are optimized for programming in C, whether we like this or not. For procedural or object-oriented languages which comprise the great bulk of modern languages, from SMALLTALK and PASCAL to JAVA and RUBY, this is fine.

2.7 Functional languages: the λ calculus

Relevance: An introduction to some programming languages.
Keywords: The functional paradigm. Functions, arguments, parameters, recursion, the λ calculus.

Quite a different approach consists in giving *definitions* and making the computer construct machine code which does the right things: definitions

```
(1) Get drunk by emptying glasses filled with wine
(2)     where wine is fermented juice from grapes
(3)         where the grapes are harvested in the vineyard.
```

Perhaps not surprisingly, compilers for functional languages (ERLANG, HASKELL, CAML) are much more complex than for imperative languages – by a factor of about 10. Luckily, this does not mean that functional languages are 10 times as difficult to master as imperative ones (rather on the contrary). It is just a longer way from source to machine code, but the computer will gladly handle that for you. That is what it has been built for anyway.

Note that functional programming does away with the notion of "flow". When   No flow
beginning to learn HASKELL or ERLANG put aside all you may have heard about branchings and loops. In particular, forget the flow chart approach from the 1950s. As the name implies, functional languages rely on the mathematical concept of functions. **A *function* can be thought of as a machine that**   function
takes *something* and returns *something else*.

A very straightforward function is the root: $f(x) \to \sqrt{x}$. In the computer world, the *arguments* (what you put in) and the *results* of functions are not   arguments
restricted to numbers: any data type may be used (see p. 48 and p. 73 on the   results
data types that exist). There may be an arbitrary number (including zero) of

parameters (what the function expects; for every parameter one argument is required) but only one return value; however, there is a very elegant workaround if you should need multiple return values, the tuple. A tuple is just a group of values; we will discuss this later (p. 96).

For example, the function *length* tells you the length of a "string" (a line of text) – it takes a string and returns an integer value –, or the function *levenshtein* gives you the similarity of two strings according to the Levenshtein algorithm (see p. 230) – it takes two strings and returns a float value.

– A function is a "machine" that produces a value (result):

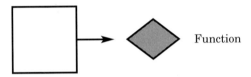

Function

– A function may accept a value (on which the result may depend directly or indirectly); of course, this value (the argument) may be the result of another function, as the result may serve as the argument to another function:

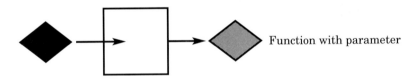

Function with parameter

– A λ is a function whose result is another function:

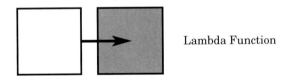

Lambda Function

Most interestingly, *arguments* and *results* can be other functions. Functions may be passed around like any other kind of data. Of course, not all operations make sense on functions, but neither do they on all types of data. For example, how would you print out a function? HASKELL will flatly refuse to. PYTHON will print something like <built-in function> or <function nn at 0x34de90>, depending on the circumstances. A λ *function* is a function which does not return a value but instead another function. E.g., $\lambda(x) \to \sqrt{x}$ does not yield the square root of any number but the function which calculates the square root.

– A function may be used as an argument to another function:

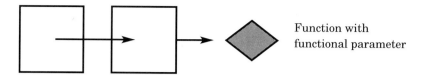
Function with functional parameter

— *Mapping is the application of a function to a plurality of "input" data to yield a plurality of results, so that each result is the application of the function to one piece of "input" data (for details, please see p. 116):*

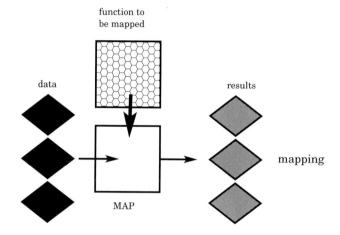

— *In recursion (p. 159), a function calls itself to yield the result:*

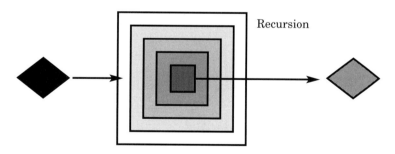

The λ *calculus* itself was proposed by Alonso Church in 1930, but the underlying theorems were developed by the mathematicians Haskell B. Curry (1900-1982) and Moses Schönfinkel (1889-1942); "*currying*", named in honour of the former, and "partial evaluation" are two related techniques which make use of λ functions to increase the modularity of programs.

λ calculus

currying

An example of how functional parameters and λ calculus can be put to rational use is given on p. 52f.

Some languages, e.g. PERL and REXX, define functions only as parameterless labels and require explicit fetching of the arguments:

```
/* Define a subroutine to print a string in a box, then call it */
call box "This is a sentence in a box."
call box "Is this a question in a box?"
exit

box: /* Print the argument in a box */
  parse arg text
  say "+--------------------------------+"
  say "|"centre(text,32)"|"      /* align the text in the box */
  say "+--------------------------------+"
  return
```

where PYTHON would have this:

```
# Define a subroutine to print a string in a box, then call it */
def box( text ): # Print the argument in a box
    print "+--------------------------------+"
    print "|%32s|" % text       # align the text in the box
    print "+--------------------------------+"

box( "This is a sentence in a box." )
box( "Is this a question in a box?" )
```

On the other hand, the Algol family have the notorious feature of "in-out" or "call-by-reference" parameters; that is to say, the function may modify the caller's variables: ADA program):

```
procedure A(X: in Integer) is
begin X := 5; end A;

procedure B(X: in out Integer) is
begin X := 5; end B;

procedure C is
  I: Integer;
begin
  I := 3;
  A( I );   Put_Line( Integer'Image(I) );
  B( I );   Put_Line( Integer'Image(I) );
end C;
```

Procedure C will generate the following output:

```
3
5
```

since A does not modify i, but B does.

Recursion (see p. 159) and mapping (projection; see p. 122) are the most mathematically precise ways of applying a function to more than one item. Actually you do not need *anything* but functions to write an operating program: A HASKELL program simply consists of a run of mutually recursive functions – and many problems are much easier to solve that way! For more examples, please see p. 359ff. where a certain problem (identification of prime numbers

2.7 FUNCTIONAL LANGUAGES: THE λ CALCULUS

by means of the "Ophidian Sieve") has been solved in a number of different languages, using either the functional or the imperative approach.

– *A function may conventionally take more than one argument:*

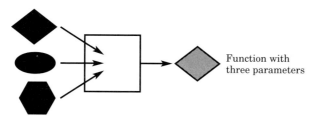

Function with three parameters

– *Alternatively, a function with a multiplicity of parameters may be broken down into a series of λ functions. For example, $f(x, y, z) = x \cdot y + z$ will, when invoked with $x = 5$, yield $f'(y, z) = 5 \cdot y + z$; and this will, when invoked with $y = 3$, in turn yield $f''(z) = 15 + z$ (e.g., when invoked with $z = 7$, 22):*

breaking down functions

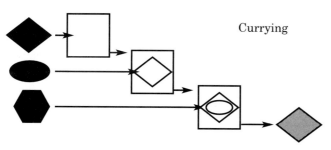

Currying

E.g.:

Argument	f	f'	f''	Result
$x = 5$	$x \cdot y + z$			$5 \cdot y + z$
$y = 3$		$5 \cdot y + z$		$15 + z$
$z = 7$			$15 + 7$	22

♡ *Now you see why functional programming, in spite of its advantages, has never really made it... it is powerful but very, very abstract (two things which are unfortunately strongly correlated). Understanding functional programming may become a brain-wracking endeavour for people who have too much experience with exclusively imperative languages. Dijkstra has been much quoted in this context: "The use of* COBOL *cripples the mind; its teaching should therefore be considered as a criminal offence."*

Crippling the mind

Of course, the ideas of functional programming have also influenced imperative languages. All modern imperative languages also possess *functions*, which are similar to procedures but return values, as the following implementations of square(x) → $x * x$ show:

PASCAL:

```
function square( n:  real ):  real;
begin square := n * n end;
```

C, C++:

```
float square( float n ) { return n * n; }
```

PYTHON:

```
def square( n ):  return n * n
```

or...

```
square = lambda( n ):  n * n
```

LISP (the ancestor of all functional languages):

```
(define square (lambda n) (* n n))
```

ERLANG (a modern functional language):

```
square n -> n * n.
```

RUBY (python-like):

```
def square( n )
    return n * n
end
```

or...

```
def square( n ) n * n end
```

However, the use of functions is rather limited in imperative languages when compared to functional languages, especially where the passing to and fro of functions is concerned. Wirth's original definition of PASCAL allowed this, but few compilers ever implemented it; C and its interpreted offspring S-LANG simulate it by the tasteless mechanism of "pointers to functions" (see Vol. 2 for the horrors of pointers). PYTHON, on the other hand, is fully λ-capable but it does not permit currying and other useful gimmicks.

♠ *A function which compares two values is easily implemented: in mathematical notation, it reads something like g(x,y) = (x>y). Thanks to the λ calculus, in a fully functional language this function may be called with one parameter only. So what is the return value of g(5)? How can this return value be used to screen a list or an array (see below) for values which are greater than 5? – you will find a hint on p. 81 under "Functions".*

2.7 FUNCTIONAL LANGUAGES: THE λ CALCULUS

♠ *From the single line of* LISP *that you have read, how would you expect this language to work? Can you write a* LISP *function which calculates the diagonal width of a rectangle whose sides are of the lengths* a *and* b*? (Hint: In* LISP*, you use* (sqrt x) *to calculate* $\sqrt{x}$*, and* (+ a b) *to get the sum of* a *and* b*.)*

♠ *Do the same for* PYTHON *and* ERLANG*.*

♠ *Try to explain the meaning of the word* return *which is found in* C *and* PYTHON*.*

♠ *Do you have any idea what the expression* square = lambda(n): n * n *means? Please explain this in clear words and complete sentences. If you have not, please turn back to p. 24.*

Just to name a few functional languages:

- The Lisp languages: (semi-functional, completely based on lists)
  - LISP (*List Processor*, AD 1956)
  - SCHEME (reduced complexity LISP, ca. AD 1990)
- FP (this first truly functional language was created in 1978 by John Backus, who had been the driving force behind the development of FORTRAN)
- The ML languages: (*Meta Language*, originally intended for abstraction only; proposed by Milner in 1978 who in turn based it on the ISWIM project of the mid-1960s)
  - STANDARD ML
  - MOSCOW ML
  - CONCURRENT ML
  - CAML (O'CAML and CAML LIGHT; reintroduction of iterative features)
  - F♯ (Microsoft's mysterious .NET adaptation of O'CAML; experimental)
- ERLANG (named in honour of Agner Krarup Erlang; a simplified, inherently concurrent language)
- The array-oriented languages (see below)
- The lazy languages:
  - SASL
  - MIRANDA (never completed)
  - CLEAN
  - HASKELL
  - GØTEBORG ML

Oh yes, sure! These ones are still missing:

HASKELL:

```
square::float->float
square n = n * n
```

O'CAML:

```
let square n = n *. n;;
```

LUA:

```
function square( n )
    return n * n
end
```

FORTRAN:

```
FUNCTION SQUARE( N )
SQUARE=N*N
RETURN
END
```

It would be boring to add more: By now you will have grasped that there are only two basic ways to do all this. Some languages have additional quirks, like PERL, but they are not really instructive.

> Σ In a functional language, relations between data are the basic units. The program is composed of series of functions defined in terms of each other. Functional programs live in a timeless world where no chronological order exists. The interpreter or compiler has to "solve" functional programs by converting them into a series of executable steps.

2.8 Intermission: tó mégǎ sýmbolon

Relevance: Understanding how fundamental abstraction works.
Keywords: Symbol tables, file systems.

At the beginning, we have learned that the processor is capable of very basic operations only, and that the entities the processor is designed to handle are (A) registers and (B) memory locations (addresses). Now in the previous chapter, functions and variables have been discussed at some length. Obviously, these are things beyond the sphere of the processor hardware. So how is this made "processor-worthy"?

identifiers
symbol table

A language's *identifiers* – i.e. the names of variables and functions – are simply *mapped* to memory addresses by means of a so-called *symbol table* which every compiler or interpreter has to keep. The interpreter uses it directly to perform the necessary actions, the compiler to design suitable machine code, but essentially it's the same in both cases: a "dictionary" (see p. 88 for more on this subject) where every identifier name is associated with the corresponding address. If it is a variable, this address is used to load the value from (into a register) or store the value at (from a register); if it is a function, to continue execution at this address. The location of the symbol table is in turn "hard-wired" into the interpreter or compiler.

A similar symbol table is well-known from everyday use: the *file system*. A storage device does not know anything about files and directories; all it is capable of handling are physical locations within its storage system. The file system is likewise essentially a symbol table that provides a mapping from file names to such physical locations.

file system

The internet also uses a symbol table to map URLs (such as www.sanctacaris.net) to IP numbers (such as 129.204.126.135). This particular symbol table is provided by a computer known as the *name server*, and the name server's IP number must be entered when the network is set up.

name server

Here it is worthwhile to recall the origin of the word *symbol*: In Greek, "syn" means "together" and "bállein" "to throw"; and the word initially alluded to the ancient Greek custom of slicing in halves little tables into which words or signs had been engraved, so that the original could be restored only by combining these halves. Such *sýmbola* were used as tokens of friendship or, presumably, authorization. In computing, the term is thus very apt, as it refers to the connection between a *key* – the name of the identifier or file, respectively – and the associated addresses holding the values.

symbol

key

Thus, the symbol tables bridge the gap between *low-level* entities, such as memory addresses, locations on disks, IP numbers, etc., and *high-level* entities, such as variables, file names, URLs, etc.

low-level
high-level

2.9 Array languages

A detailed description of what an array is will be given on p. 84. Let it suffice here to say that an array is an ordered group of data, wherein all data elements are of the same type, and that large collections of data are usually structured as arrays. As may be guessed, array languages offer special features for the efficient processing of such data (up to automatic load distribution on multi-processor systems). In general, they are based on the functional approach, but the prevalent use of arrays instead of lists justifies the separation.

Array languages are probably the most rare and exotic group described so far:

- APL ("A Programming Language" – derived from a mathematical formalism, the Iverson notation, developed in the 1960s; a language which is extraordinary in that it uses specialized glyphs in lieu of words, which does not actually make it easy to implement)
- A^+
- J
- K

- SISAL (a parallel language designed specifically for "shared memory" architectures)

2.10 Object-oriented languages

Relevance: A first glance at the concepts of object-orientation (OO), which feature prominently in PYTHON, and their general importance in IT.
Keywords: Simple and compound data structures; encapsulation of data together with the program code for handling them.

Object-oriented (OO) languages are generally considered as "the next evolutionary step" of imperative languages, and many OO languages are in fact based on conventional imperative languages (C^{++} and JAVA on C, OBERON and DELPHI on PASCAL, and others).

To explain what OO is and why it has become so popular, we will have to dig somewhat deeper first.

From the very beginning, it was clear to programming language designers that **sometimes different kinds of data may belong together to form a logical unit**. For example, an address file comprises a number of entries, each of which contains

- membership number – a positive integer number
- first name – a group of alphabetical characters
- last name – ditto
- Postal code – a number between 00000 and 99999
- City – a group of alphabetical characters
- Street – a group of alphabetical characters
- membership type – one of the following: *regular member, associate member, honorary member*.

We will have more to say about aggregates later (see p. 93). One of the resons for the success of COBOL was its rather sophisticated scheme for dealing with such aggregates, but from the 1970s on, all non-toy languages included similar "structured data types". For example, in PASCAL the above address file might be defined as follows:

```
type addressEntry = record
                    membershipNumber :  integer;
                    firstName :  string[ 20 ];
                    lastName :  string[ 30 ];
                    postalCode :  00000..99999;
                    city :  string[ 25 ];
                    membershipType :  ( regular, associate, honorary
);
                    end;
```

```
type adressfile = file of address;
var annafanclub :   addressfile;
```

You can think of the newly defined data type *addressEntry* as a "form" which provides spaces for a certain number and kind of entries (e.g., string[20] is a field for up to 20 letters). Such forms are usually referred to as "structures" or "records". Any *field* can be accessed separately:

```
var boss :   addressEntry;
boss.membershipNumber := 23;
boss.firstName := 'Dolores';
boss.lastName := 'Peinemann';
boss.postalCode := 69345;
boss.city := 'Waldwimmersbach';
boss.membershipType := regular;

if boss.postalCode <> meetingPlacePostalCode then
    write( "Please appoint a representative for the meeting area"
);
```

– *A structure or record is a compound comprising a plurality of simple data elements:*

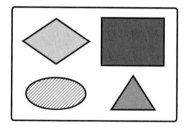
Structure or record

– *An array is a sequence of a defined number of data elements, all of which are of the same type. This type may be simple:*

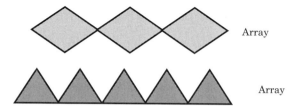
Array

Array

– *An array may also consist of structures or records, all of which must be of the same type:*

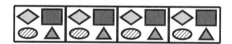
Array of structures

– *A list is an association of compound elements, wherein each list member comprises at least a reference to one other element. The number of list elements is not defined; the end of the list is indicated by the last reference being manifestly invalid. Whereas in an array any element may be accessed simply by its number (index = "house number") within the array, the list consists essentially of "this element" and "all that follows":*

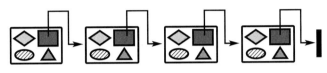

List (of structures)

Hence, lists may comprise lists, or arrays, as arrays may be arrays of lists, or arrays of arrays (the latter being effectively tantamount to a two-dimensional array).

– *A list may be doubly linked to facilitate traversing in both directions (i.e., we can move not only head→tail but also tail→head):*

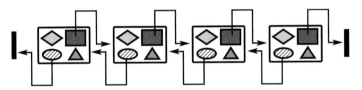

Doubly linked list (of structures)

From the above, it is obvious that the list approach provides several features which the array is not expected to consider:

- appending an element (to the end of the list)
- "consing" an element (i.e. placing it before the previous first element)
- inserting an element (at an arbitrary position)
- removing an element from the list
- "slicing": producing a sublist which contains only a portion of the original list.

Of course, when using arrays, this can be emulated by creating a new array and copying the original data. However, this is *much* slower and memory-consuming than the normal list operations. By contrast, normal access to arrays is faster (sometimes by several orders of magnitude) than to lists.

The general notion of clustering of related pieces of information is compatible with any fundamental approach. For example, in the functional language ERLANG we would express the above as follows (the obvious difference being that ERLANG does not require types):

```
-record(addressEntry, {membershipNumber, firstName, lastName,
postalCode, city, membershipType}).
```

and

```
boss = #addressEntry{membershipNumber = 23, firstName = "Dolores",
lastName = "Peinemann", postalCode = 69345, city = "Waldwimmersbach",
membershipType = Regular},

if boss#addressEntry.postalCode <> meetingPlacePostalCode
    io:write( "Please appoint a representative for the meeting area" ).
```

When dealing with a group of data items clustered as one – i.e., a "structure" (p. 93) –, we sometimes want to handle the cluster as a whole (that's what these things are good for anyway), but other times its individual components, as seen above. Nowadays virtually all languages use the ALGOL notation[10], which is *cluster.component*, as in `boss.membershipNumber`. This order is also known as the *Hungarian notation*, as in Hungarian the family name precedes the given name[11].

Hungarian notation

PASCAL and its immediate successors also allow the "dereferencing" of clusters by using the *WITH cluster DO* expression: `WITH boss DO BEGIN membershipNumber := 23; firstName := 'Dolores'; lastName := 'Peinemann'; ... END;`. This has not made it into other languages because it does not really contribute to clarity. In C, you will also find the $\rightarrow$ operator: `boss_ptr->membership_number = 23`. For an explanation of this, please see the table of pointer operations (Vol. 2).

The PASCAL token `..` is known as *ellipsis* and can be read as "from–to". Apart from PASCAL's successors, some high-level languages also feature it, including HASKELL and RUBY. In PYTHON, there is the `range` function instead.

♠ *Have a look at a GenBank file and try to imagine what fields would be necessary to store this information in the form of records.*

So far, so good. Now please remember functional programming and the λ calculus, or turn back to p. 24f. If it is basically possible to treat functions – or speaking more generally, pieces of code – like any other kind of data, why shouldn't we also include the code necessary for the processing of an aggregate data structure right into that data structure?

Let's stick to our example to show what that may be good for. You will have noted that this is rather limited: it works only for addresses in Germany[12].

---

[10] Notable exceptions: `cluster%component` in FORTRAN, and `cluster#component` in O'CAML.

[11] Just as in Linné's binominal nomenclature the genus name precedes the species name: *Lilium martagon, Paramecium caudatum, Homo sapiens*.

[12] So if you decide to collect addresses of professional boxers, tennis players or racing-drivers, you've got it made.

What are we going to do if our society extends to Great Britain? Adding a field *country* and modifying the *postalCode* field so that it can cope with foreign code systems such as the British is not really a problem (for the latter, just use a "string[10]" field or the like). Then you can write two splinters of software which work on the German and the British system, repectively, and make the choice of these depend on *country* – no problem either. But then we will have to modify our program literally everywhere an *addressField* is used! We do this (with much cursing), only to find that next we get members from Ghana, Paraguay, Indonesia...

It would be much easier if we could add our comparison function directly to the structure, so we could write something like:

```
if not boss.livesInArea( meetingPlacePostalCode ) then
    write( "Please appoint a representative for the meeting area"
);
```

Then we would only have to extend the function *livesInArea* every time a new code system is introduced.

This is the starting point of OO programming. **A structure which contains both data and code is termed an *object*, and its type is called its *class*.** Associating data and code in such a way can be used to establish *singularity of choice* (p. 239).

singularity of choice

public
private

attributes
methods

Things become exciting when we consider additional possibilities. We can define components of the object as visible to everyone and everything (*public*), or only to the members of the object itself (*private*). By doing so, we turn the loose bundle into a foolproof "black box". Hardcore OO devotees use their own terminology: they refer to the data members of an object as *attributes* and to the code members as *methods*. The program is seen as a number of "black boxes" which exchange messages (by invoking each other's methods) and are otherwise completely detached.

This not only increases stability and coherence of programs but is also very suitable for *graphical user interfaces* and similar things like that. A window or a menu bar can be represented very neatly by an object; window contents and menu entries are attributes, and user actions are just another kind of incoming messages to which the graphical object reacts by sending messages to other objects.

graphical user interfaces

In fact, the first OO language was so intricately connected with the first graphical user interface that they were considered a single system. This was Xerox's SMALLTALK, back in the 1960s, which was eventually bought by Apple and turned into the basis for their Lisa / Macintosh computers. It is indebted to SIMULA for many of its OO features.

Imagine a simple text editor. This consists of three objects: a window object, a menu object and a "core" object which keeps track of the actual text. When the window is active, it watches out for mouse clicks and keyboard events; if it receives such, it sends requests for modification to the core object. When the menu item "Save" is selected, the menu object sends a request for saving to the core object. The core object then creates a fourth object, the file selector, which takes over interactions with the user and finally sends a file name back to the core, which then deletes the file selector object and saves the file...

OO programming may also be used to simplify access to very complex structures: e.g., there is a function in PYTHON which returns the complete contents of a given directory on the disk, much like `ls` or `dir`, but as one single object. If the program wants to inspect the contents of the directory, it can use this object's methods to get information.

As for myself, I once wrote some programs which handled GenBank information[13] and found it very convenient to have a full GenBank file accessible as one object. Methods of this object included one that yielded the number of genes, and one that took an index number and returned the respective gene – again as an object, with functions for its DNA sequence, its protein sequence, its name, its annotations, its...

The OO approach becomes even more powerful by "inheritance" – new classes may be defined on top of old ones: e.g., a graphics program uses as its most fundamental item a dot, specified by (x/y) and possessing a *redraw* method. By adding a third coordinate, we get the description of a circle; by adding a fourth, an ellipse and so on. In the end, we have a list of all graphical objects we are working with, from simple dots to 3D text objects with surface textures, and each of them has its *redraw* method. So when scrolling the screen, all we have to do is to send redraw messages to all our objects – and we're done.

---

[13] These will not be presented here, for the simple reason that writing such software will constitute one of your later projects.

♡ *Please make sure that you have understood this chapter thoroughly.* Python *is an OO language, and though we will not exaggerate this, you will hardly be able to exploit the full potential of* Python *nor of* Java *or* C++ *without a proper notion of these things.*

Considering all that has been said so far, it is hardly astonishing to see that OO has by now been built into virtually any imperative language. We may roughly differentiate between four groups of OO languages:

- genuine OO languages
  - Simula, once upon a time...
  - Smalltalk and its offspring AppleScript
  - Eiffel
- languages which combine OO and non-OO elements
  - Ada
  - Java
  - Python
  - Ruby
  - Oberon
  - o'CAML, a combination not of OO and imperative but of OO and functional elements
- languages in which OO extensions are more or less optional additions to a non-OO core ("a squid obtained by nailing four wooden legs on a dog")
  - C++
  - Delphi
- languages without any structure or conscious design into which something like OO was introduced because it is fashionable
  - Perl
  - Visual Basic.

 Object-oriented (OO) languages are a subgroup of the imperative languages in which pieces of data and the peculiar blocks of executable code related to them are tethered together to form a distinct unit. This leads to a "building block" approach and thus facilitates larger projects.

2.11 Reflective languages

"Reflective programming" is bombast for object-orientation taken to its extreme. Apart from Python, Smalltalk, Eiffel and Ruby have to be mentioned here; it is the Smalltalk mujaheddin who cling to this term, vigorously defending their perceived "artificial intelligence" heritage[14].

---

[14] **Artificial intelligence** is – or rather was – a branch of computer science based on a dictum by von Neumann in 1958 (one year before the Feynman speech) that ideally a computer should be able to program itself. This stimulated the development of languages

2.12 Declarative = logic languages: the π calculus

Relevance: An introduction to some programming languages.
Keywords: The declarative paradigm, recursive descent and the π calculus.

Terminology is somewhat incoherent: some people speak of this group as "logic" languages and use the expression "declarative" to denote the group comprising both them and the functional languages, explaining that "functional languages are based on the λ calculus and logic languages on the π calculus". I dislike the term "logic" language because of its pejorative attitude towards other languages (is this meant to imply that other languages are illogic?), and I dislike explanations which do not explain things unless you know enough not to require the explanation any more. In fact, the declarative languages *sensu stricto* share many features with the functional group (up to the possibility of distributing a "non-strict" program over a multiprocessor system) but follow a quite different approach. They are based on logic predicates (the so-called Horn clauses – this is the π calculus) which define whether some attribute applies to some object, and the language has intrinsic facilities for solving dependencies. This works as follows. Facts:

- "Hannahs are reptiles without legs"
- "Serpents do not have legs, lizards have"
- "Scaly reptiles are either serpents or lizards"
- "Anna is a hannah"

The program is invoked by asking a question:

"Is Anna a lizard?"

The interpreter solves this recursively:

    Lizard(x) is true if Serpent(x) is false
       Serpent(x) is true if Has_legs(x) is false
          Has_legs(x) is false if Is_hannah(x) is true
            Is_hannah(Anna) is true
         ergo: Has_legs(Anna) is false
      ergo: Serpent(Anna) is true
   ergo: Lizard(Anna) is false

Thus, the output of the program reads:

The answer is "NO!"

---

with self-modifying code. However, during the next two decades research in this field yielded few encouraging results, and by now attempts to emulate processes of cognition and learning *in silico* have generally moved to other fields such as neuronal networks and Bayesian statistics. To put it bluntly, in 1958 nobody was able yet to fathom how dumb computers really are.

Obviously, this is tricky stuff to deal with, especially where more complex tasks such as building a compiler are concerned. Of a number of declarative languages designed so far, only PROLOG (*Programming in Logic*) has gained some importance (among other things, by being implemented as TurboProlog by Borland), but it still suffers from major conceptual weaknesses, especially where the proper treatment of runtime errors is concerned[15]. MERCURY is a declarative language which strives to be the declarative equivalent of HASKELL, but it is currently still under development.

At the core of declarative programming, there are "declarations" (who would have imagined that?) which establish relationships between things, expressed in terms of "predicates":

```
:- pred product(int, int, int).
:- mode product(in, in, out) is det.
product( X, Y, Result ) :- Result = X * Y.
```

This piece of MERCURY code declares the predicate "product" as follows.

- The values of one integer (*Result*) depends one the values of two other integers (X and Y).
- The nature of this relationship is "determinative" (the normal case).
- The exact relationship is $Result = X \star Y$.

Obviously, this is very similar to functional programming, and in fact most, if not all, algorithms applicable in functional languages can be transferred directly to declarative languages, though the declarative languages possess features not to be found in functional languages, especially the orientation towards "goals" which are resolved automatically, as described above for reptiles. Most importantly, declarative programming rejects *eo ipso* the notation of flow and actions, as the higher functional languages do. Of course, this causes the same kind of problems where temporal structuring of actions is required, that is to say, for input and output[16]. In MERCURY, a workaround has been established which serves the same purpose as the nefarious "monads" in HASKELL – and is almost as difficult to handle.

♠ *You've read "Harry Potter and the Philosophers' Stone", I guess? Well, here's a Snapish riddle for you:*

Some scholars are spending their holidays in the same hotel. This is what we know about them.

---

[15] In general, the further away from imperative style we move, the more difficult it gets to catch errors properly: if there is no user-defined flow structure it is hard to answer the question "and then". But as ERLANG demonstrates, it is possible to write absolutely foolproof systems using FP, so this should be possible with declarative programming just as well.

[16] ERLANG keeps out of trouble by insisting – unmathematically, but in perfect keeping with common sense – to process its functions when they are called, and in the very order in which they are called.

- *Prof. Shallow (UK) has room #1, Prof. Flachmann (FRG) #2, Prof. Doulophobos (Greece) #3 and Prof. Sarvahasya (India) #4.*
- *Prof. Flachmann is currently reading the "Mahabharata" (Sanskrit).*
- *Prof. Shallow does not smoke the pipe.*
- *Prof. Sarvahasya does not keep any pet.*
- *The cigarette smoker is the left neighbour of the cigar smoker.*
- *The guy who knows Usbekian has a left neighbour.*
- *The dog owner and the bird owner have two neighbours each.*
- *The cigar smoker is currently reading the "Song of Manas" (Usbekian).*
- *The guy with the cat is a non-smoker.*
- *The fellow who is reading the "Mabinogi" (Welsh) does not smoke the pipe.*
- *The right neighbour of the bird owner knows Welsh.*
- *No smoker knows Japanese.*

QUESTION: *Into which room shall we put the copy of the Japanese "Kojiki" which was accidentally left at the breakfast table?*
How would you approach this? Try to solve it if you like, and explain what you are doing. If this is too fuzzy for you, describe how you would feed it into a declarative language such as PROLOG or MERCURY.

Declarative languages are closely related to functional languages. They sport a special mechanism for dissecting and solving complex logical statements.

2.13 Concurrency-oriented languages

Relevance: An introduction to some programming languages.
Keywords: The concurrent paradigm. Parallelization, multiprocessing and Hoare's CSP (Concurring Sequential Processes) model.

At this moment, there is only one language really to be named here, and this is ERLANG, originally a functional language possessing all the extravagant features of its tribe, but following a rather different philosophy. An ERLANG program consists of a number of *processes* – independent but mutually communicating units running all at the same time (Lat. *con-currere* = to run together), as described by Hoare in his classic paper on concurrent sequential processes. We are used to seeing a handful of programs running at the same time; the UNIX command `top` reveals that on a normal workstation there may be about a hundred "threads" up and running. By contrast, it is not unsual for a larger ERLANG program to have 20 000 or more processes running. Each of them consists of a function recursing until it meets a termination condition; while it is running, it may exchange messages with other processes. Thus, in the end the overall architecture is quite similar to OO, but with the advantage of greater mutual isolation (processes may be distributed all over a multiprocessor network, and if one crashes, the others are not affected).

There are concurrent extensions of ML (CONCURRENT ML) and Haskell (EDEN), but their concurrency is not so prominent a feature; and there are also concurrent variants of declarative languages.

Of the imperative languages, ADA also incorporates concurrency at the "grammar level"[17]; the practical use of this lies in the ability to easily write machine control software. Once upon a time, there was also PEARL (not to be confused with PERL), an algolish language based on an operating system of its own (RTOS = Real Time Operating System) and designed specifically for machine control, containing the most extraordinary time-dependent looping structures ("all 5 sec do..."), but by now it has sunk into utter oblivion, whereas OCCAM is at least still remembered by some.

♠ *Parallel processing...* *In the late 1980s, Atari bought the small company Perihelion and attempted to build a computer based on the Inmos T800 processor, also known as "transputer" because it was specifically designed for working in multiprocessor networks. The Helios workstation (official name: ABAQ ATW800) was intended to accommodate up to 17 processors. However, Atari failed and went broke. For a long time, the only multiprocessor machines were huge servers which are intrinsically parallel as they have to serve many different requests at the same time. At the moment, the evolution of processor speed, exponentially so far, is about to slow down, and consequently new interest in parallel computing has arisen. Sun have inflated their servers to up to 96 processors, SGI even to 512, and Apple are routinely selling dual-processor PCs (well, a nice beginning).*
Think about it. Which kinds of program might benefit much from parallelization, and which might not? Is it easier to parallelize imperative or functional languages? When you imagine a program that works on an array of 10^6 elements, each of which has to be processed by a "pipeline" of some 10^9 subsequent operations, how could you harness the full power of a 96-processor computer so that the program finishes its task in less than two hours instead of more than four days?
Hint: There are two basic approaches, one horizontal, one vertical. Combine both.

 Concurrency-oriented languages may be based upon any paradigm (imperative, functional, or declarative); they are designed for running programs on multiprocessor systems.

---

[17] ADA was designed to incorporate *anything* – minus, of course, such weirdos as λ functions and lists which its creators had obviously never heard of.

2.14 Scripting languages

Relevance: An introduction to some programming languages.
Keywords: Scripting languages.

This is an awfully ill-defined group of rather recent languages which share the following traits.

- Most of them began as an attempt to bridge the gap between shell scripts (the "batch file" equivalent of Unix) and fully-fledged development languages.
- They refuse conventional classification, by combining imperative, functional and objective elements, allowing for a variety of styles.
- They generally provide a very high level of abstraction (especially for data of variable size), large libraries and powerful error handling mechanisms, while at the same time using a gracefully simple syntax.
- They are interpreted, not compiled – an advantage for web applications which must not be processor-dependent.
- Taken together, this allows extremely rapid development of small programs. For the same reasons, they are also well suited as teaching languages.
- They are easy to extend or embed into other projects.
- However, their "performance" is rather poor, mostly due to their weak typing, but it can be greatly improved by clever use of library functions (see below).
- Projects of more than intermediate size cannot be handled reasonably without additional effort.

The term "scripting languages" refers to the fact that these languages are usually used not to implement low-end functionality but to arrange low-end building blocks into a rational arrangement. As one bigwig at Industrial Light & Magic put it with reference to the software for the more recent *Star Wars* movies: "From rendering to crowd simulation, all things are *strung together* by PYTHON."

Therefore, it is not unexpected that extending, embedding and interfacing with lower level languages plays an important role for all scripting languages. For example, you might want to write a word processor. The core which manages the handling of the data and a few other speed-critical and not too complicated things might be written in C^{++} or o'CAML, but the remaining 90%, from the user interface to printing and project administration, would be written in a scripting language.

At the end of the second part you will see a scientific example for this: *The Thing That Should Not Be* deals with simulation of macromolecular assembly processes where the "engine" was written in C and addressed as a PYTHON object, allowing definition of highly complex scenarios.

The rationale behind this is that scripting languages, due to their higher-levelledness, permit much higher speed of development. My personal experiences are largely congruent with several studies on productivity:

Language	relative speed of development
C++	1x
JAVA	2x
PYTHON	5x–10x
ERLANG (for specialized networking applications)	9x–25x

Some authors have devised pretty graphics to show the relationship between the time required to *develop* the program and the time required to *run* the program, which is considered to be inversely proportional. However, I strongly disagree with the notion that programming languages are intrinsically "fast" or "slow" in the execution of a program. A program is slowed down by the use of inefficient algorithms, which in turn is a consequence of selecting inappropriate data structure (such as linked lists for the representation of large and fixed amounts of numeric data). When these are properly selected, intelligent use of library functions may boost the performance of a well-designed scripting language to approach that of a fully compiled language.

As noted before, scripting languages become most powerful when used in conjunction with a low-level core for heavy-duty calculations. To some degree, the same thing may be achieved by providing extensive libraries implemented in low-level language.

Currently, there are some eight scripting languages worth noting:

- PERL (with special features for lexical analysis)
- PYTHON (see below)
- RUBY (somewhat similar to PYTHON but currently little known outside Japan)
- LUA
- REXX
- PHP
- TCL (with special features for graphical user interfaces).
- S-LANG.

Of these, the former four are true general-purpose languages; REXX has always been intended as a shell component, PHP is for the net and TCL for graphical user interfaces. The odd thing about LUA is that almost nobody has ever heard of it, though LUA-based software is sold in millions: The computer gaming industry exploits the advantages of a lightweighted scripting language. Currently, many 3D games – some people claim that it is more than 50% – use a plain C core for the rendering of three-dimensional objects which is completely embedded into, and controlled by, the LUA interpreter which runs the game proper.

The sizes of the libraries are very different, with PERL > PYTHON > RUBY > LUA > S-LANG. Again, a discrimination has to be made between libraries implemented in the language itself – which areactually just "convenience packages" – and low-level languages. PERL libraries are generally written in PERL, whereas PYTHON are partly in C and partly in PYTHON, and RUBY and LUA have almost pure C libraries – the ease of interfacing the higher language with C being PERL < PYTHON < RUBY ≈ LUA.

In addition to the ones listed so far, there is the "mother language" of the shell, Unix's text-based user interface, which does not have a real name but is often referred to as SHELL. And there is JAVASCRIPT, whose very name emphasizes that it is designed to be a scripting language, but which is restricted to the ecological niche of running small programs over the Internet[18].

♠ *What do you expect to be the most time-consuming part of a programmer's work?*

> Scripting languages are designed for solving problems of arbitrary complexity in as little time as possible. They are generally system-independent but not suitable for "brute force" purposes.

2.15 Markup languages

Relevance: The concept of "markup" data.
Keywords: Markup languages; XML; HTML.

These are clearly different from all the others; sometimes they are not regarded as "real" programming languages because their primary aim is not to "do" anything but to structurize data. Nevertheless, they will be described here briefly not only because of their growing importance but also because of their relationship with "real" programming languages.

When considering a mass of data, either as a block (file) or stream (user input), one will immediately see that two different types of data exist: mass data and control data[19]. For example, in typing a text such as this, the letters are mass data to be included into the document, whereas key combinations such as *Ctrl–S* (for saving the current document) are control data. Likewise, a word processor has to save (and reload) not only the characters making up the document but also codes defining font, attributes, etc. These codes must be shaped in such a way that any confusion with mass data can be excluded.

---

[18] This niche is hyped to be expanding, but behind the spin there are unsurmountable security problems.
[19] This is the same problem as described on p. 318ff., for the handling of "exceptions", i.e., error states within a program.

Traditionally, special ("non-printable") characters were used for containing such control data within files. However, a generalized formalism based exclusively on normal characters would be advantageous in several ways: It would

- avoid potential incompatibilities between different systems
- simplify the writing of applications
- provide a very flexible tool for storing and organizing *any* kind of data
- ensure that data do not become illegible due to a change in a company's preferred data format.

This is what the markup languages do. Probably best known among them is HTML, the "HyperText Markup Language" which manages the control data required for WWW pages (mostly text formatting and links), but basically any kind of information and "meta-information" may be combined using markup languages. For example, the configuration of the widely-used Apache web server is completely done using a special markup jargon.

The structure of the markup languages is very simple: It comprises "tags" which may have additional specifiers and represent a blocklike structure enclosing the piece of data they refer to. So basically, we have

```
xxxxxxxx<TAG>yyyyyyyy</TAG>zzzzzzz
```

and

```
xxxxxxxx<TAG additional="anything">yyyyyyyy</TAG>zzzzzzz
```

with the tag referring in both cases to the *yyyyyyyy* only.

Tags may be nested like the control sequences for boldface and italic print in the following HTML example:

```
Sir: The <B>Council of <I>King</I> Cobras</B> disapproves
     of <B><U>any</U> profanity</B>.
```

which will be presented as:

Sir: The **Council of *King* Cobras** disapproves of **<u>any</u> profanity**.

If they are nested, they must be perfectly nested; that is to say, one tag pair must be completely enclosed within another. A construction like <B><I>Hey you</B></I> is not valid.

Now you will see that this is actually nothing but a slightly modified form of LISP! In LISP, this expression could be represented as

```
("Sir: The " (B "Council of " (I "King") " Cobras")
        " disapproves of "
        (B (U "any") " profanity") ".")
```

The LaTeX text formatting system is generally not considered a markup language, although it comprises the same set of features, albeit with a slightly different appearance:

\ninety{\includegraphics[scale=0.3]{schopenhauer}}

where the markup languages sensu stricto would have

<NINETY><INCLUDEGRAPHICS scale="0.3">schopenhauer</INCLUDEGRAPHICS></NINETY>

XML is the all-purpose markup language *kat' exochen*, and it can be counted upon to become more important over the years[20]. PYTHON is equipped with a rich functionality (modules xml.*) to deal with this (see documentation for details).

Markup languages serve the purpose of representing data in a system-independent, easily accessible form.

2.16 Overdose kids

Relevance: Languages with problematic design. Some aspects of good and not-so-good design.
Keywords: PL/1; ALGOL-68; ADA

When reviewing the more important programming languages in general, one will perceive that each of them is built around one certain central strength – either a design principle which was later found to be fruitful or a "killer application" where this language was peerless. It is therefore tempting to try to combine all these strengths into one and thus build the Ultimate Language.

Indeed, several attempts at this have been made before by ambitious programmers, serenely ignoring the fact that programming languages would not have branched that much if it were not that most individual strengths were incompatible with each other (beginning with the argument about recursion and iteration which caused the schism between FORTRAN-I and LISP designers in the late 1950s). In general, few of them have ever yielded working compilers, and those who have are unmanageable monsters of complexity. Just to mention a few: PL/1 (including its offsprings BRUIN, PLUM and PL/M in which the once famous CP/M operating system was written), which was designed in the 1960s in academic Britain and described by Dijkstra as a "fatal disease"; ALGOL-68, whose design committee seem to have spoilt all advantages of its

---

[20] The first large application to use the XML format is the OPENOFFICE suite, comprising a word processor, a spreadsheet and a drawing/presentation program, which exclusively uses XML to store its data. To save disk space, it employs the *zip* compression algorithm, which is also available under PYTHON (module zlib). Thus, OPENOFFICE documents can be both read and written using PYTHON.

successor ALGOL-60 by their greed for additional features; and ADA, designed by the American military as the "one-for-all" language.

In fact, good design of a general-purpose language depends not on inclusion of as many features as possible but on their sensible selection and combination. PYTHON, with its well-knit combination of procedural, object-oriented and functional features concealed behind the clarity and lucidity of its structure, is a very good example of this.

2.17 Strong *vs* weak typing

Relevance: Strength of typing
Keywords: Types of variables.

Another matter which has caused many jihads among programming language geeks is the *strength of typing*. Basically, this means whether an algebraic variable may accommodate different types of data or only one.

For a description of what types of data there are, please see the next chapter: "Variables, data types, and assigments" (p. 73). Here it is sufficient to know that there are, among other things, Booleans (true/false binary values, p. 74), integer numbers, floating point numbers, single characters, strings (sequences of characters or pieces of text) and lists consisting of any of these data types – and aggregate types (see section on object-oriented programming, p. 32) also composed of any combination of the others, as well as lists of aggregates.

For those interested in details: The compilers use recursion (see p. 159) to dissect type declarations.

Examples:

Boolean
- **Boolean:**
  TRUE

Integer
- **Integer:**
  -385

Float
- **Float:**
  3.1415927

Character
- **Character:**
  'A'

String
- **String:**
  "The quick brown fox jumps over the lazy dog"

List of integers
- **List of integers:**
  2, 9, -7

List of strings
- **List of strings:**
  "alpha", "beta", "gamma", "delta"

inference and polymorphisms

♡ *This section, especially the part on* inference and polymorphisms, *is rather* tough – as it is very abstract – and not really necessary for understanding PYTHON. If you would prefer to get into "real" things as soon as possible, you may skip this up to the point where the application of λ functions is explained (p. 52) and not feel that you have missed anything. But don't forget to mark the page; I strongly advise you to turn back some day. It may help you to cope with C^{++} or JAVA, if you should ever get in touch with them. Remember, this is not going to be a "Python, the full Python and nothing but Python" course. . .

The most extreme strength of typing is found in BASIC including COMAL and PERL, where the name of each variable must contain special characters describing the type of variable. In BASIC, the variable a$ is a string variable, a% an integer variable, a! a single-precision float, a# a double-precision float and so on[21]. In PERL, $a is a simple (scalar) variable – i.e., a single number of single string – @a an array (list) of numbers or strings[22] (p. 84ff.), %a a dictionary or "hash table" (p. 88).

Why should one thus want to restrict things? There are two answers.

- Strong typing allows the generation of *much* more efficient code. When x and y are known beforehand to be integers, the operation can be performed (under ideal circumstances) by a single processor operation, which will take place typically within about 10^{-9} sec. The same applies when they are known to be floats. If, however, the type is not defined, it must be determined *at runtime* whether an integer or a floating point addition is supposed to be performed. (From a human point of view the difference is small; from a processor's, enormous: the floating point operation uses other instructions and takes place in a different area of the processor – on older machines even on a separate processor! – typically, floats are 64 bit and integers 32 bit wide.) So the processor will have to first look up in the "symbol table" what is meant and select the correct operation[23].

---

[21] Many BASIC dialects also feature a FORTRAN-like way of declaration, with DEFSTR defining a variable as a string even without the terminal $, DEFINT declaring it as an integer, DEFDBL as a double-precision float, etc.

[22] Please do not confuse this with the usage found in RUBY. In the extremely weakly typed language RUBY, prefixes such as @a or $a do not denote data types but variable scopes, that is to say, whether the variable in question is part of an object, a class or neither. See the chapter on objects (p. 187) for more about this.

[23] Not so long ago, I decided to write a compiler for PYTHON which produced a C output that could be compiled to native code and linked to the PYTHON core serving as a runtime system. What incited me to try this was that I got vexed at C – a really disgusting language in my opinion – being about 30 to 200 times faster than the much more versatile and elegant PYTHON. I got the prototype working, only to find that it required a some three megabytes of runtime code and achieved not much more than twice the speed of normal "frozen" PYTHON code; in other words, it still required between 15 and 100 times as much processor time as native C, though it consisted of 100% machine code by now. This made me realize that the most time-consuming step during execution of PYTHON programs is, in fact, not the interpretation process but the handling of the weakly typed variables.

- Strong typing also allows for the detection of many serious errors at compile-time, especially where large projects are concerned. When several people share a project, interfaces must be defined between the chunks of code they produce. Strongly typed code takes almost complete care of this. The notorious "Saturn V incident" (see footnote on p. 58) would not have been possible with proper type checking at compile-time. The take-home message is that you cannot compare apples and oranges and should therefore not be allowed to try.

The C and Algol families, too, are strongly typed, and they achieve this by *declarations*. Any variable must be declared before being used for the first time, e.g., int a; (C'ish) or var a: integer; (Algolish). Otherwise the compiler will reject the program[24].

Among the Great Old Ones, COBOL is outstanding in that it is extremely file-oriented, as can be expected from a language that was designed mostly for administration of commercially relevant data; most data are structures, whose components are very precisely defined in the DATA DIVISION. portion of a program, including even the number of digits acceptable for a numeric value – a form of strong typing not to be found in any other language. In FORTRAN, variables are assumed to be floating-point numbers unless explicitly declared otherwise. In these languages, this was instituted for the benefit of the compiler rather than for that of the programmer.

HASKELL and the ML family are also strongly typed, but do so by *inference*. Inference means that the compiler determines the proper type of a variable by the operators which are first applied to it. This is very smart but also has some drawbacks; for example, in o'CAML there are different sets of operators for integers and floats: +, -, *, / and mod for integers, +., -., *., /. and ** for floats (modulo is not defined for floats, and power not for integers). At any rate, inference is incomplete, and it is possible to specify at least the type of parameters. In HASKELL, you are explicitly encouraged to use type signatures like this:

```
foo::Int->[Int]->(Char->Bool)->Float
foo bar blah critter = ...
```

This means that *foo* is defined as a function which takes the following parameters: *bar* as an integer, *blah* as a list of integers, *critter* as a function (remember

---

[24] *How* strongly typed? When you define two types a and b the same way, say as integers, then C and PASCAL compilers will allow mutual assignment between variables of type a and b (since their definition is equal). ADA, on the other hand, refuses this – probably due to the notion (not absolutely devoid of logic coherence) that if this behaviour had been desired and allowed by the guy who wrote the definitions, there would have been no need for two separate definitions. For this reason, Adaists like to claim that C is not a strongly typed language.

functional parameters as discussed on p. 24?) that takes a character and returns a boolean (p. 74) value – and finally produces a float. There is something more to explicit type declarations. All instances of strong typing we have seen so far were *monomorphic* – there was no variability allowed. But now imagine you have written a novel sorting algorithm – let's call it "bestsort". Do you really want to implement this separately for lists of integers, lists of floats, lists of doubles, lists of strings, ...? The solution is *polymorphism*.

monomorphic

polymorphism

There seems to be some uncertainty about the exact meaning of this word[25], which just means "being of many shapes". In JAVA, polymorphism is the multiple usage of the same name for different sets of parameters connected to different pieces of code (e.g., `int foo( int bar ) { return 2 * bar; } int foo( float bar ) { return bar / 8.54; }`). However, what we are interested in, is getting the same piece of code to work on different types of arguments.

In C^{++}, there are things called "templates" to this end. Basically, your algorithm is implemented once and for all using the headline `template<class T>class bestsort`, all relevant internal variables being defined as of the "placeholder" type T, and where you use it, you declare either a `bestsort<int>` object, a `bestsort<float>` object or whatever you like. This will cause the objects to be individually compiled in such a way that the variables of "type T" are in fact defined as `int` (for `bestsort<int>`), `float` (for `bestsort<float>`)...

HASKELL is much more elegant in this respect. Here there is a universal placeholder (in effect deliberately weakening the typing system where desired):

```
foo::a->b->a
foo bar blah = ...
```

specifies that the parameters *bar* and *blah* may be of any type, and the return value of function *foo* is of the type of the first argument. You may, for example, pass `Char` and `Boolean` (p. 74) to the function, then your output must be of type `Char`. You may, with equal legality, pass `[Float]` and `(Int->Boolean)` to the function, then the output must be a list of floats again[26]. Like Mephistopheles, the function must get out again the same way it got in. Likewise,

---

[25] I prefer to stick to the following explanation, given by INRIA's Pierre Weis, one of the authors of o'CAML, at http://caml.inria.fr/FAQ/general-eng.html: *Caml features "polymorphic typing": some types may be left unspecified, standing for "any type". That way, functions and procedures that are general purpose may be applied to any kind of data, regardless of their types (e.g., sorting routines may be applied to any kind of array).*

[26] The compiler won't accept any other type as a functional value. Things become really nasty when you try to return a result which may be of the proper type, or not. And of course, the "internals" of the function must be able to deal with such data types properly – naturally, this cannot be guaranteed by the type signature which only defines the "social life" of a function.

```
chandra::a->[a]
chandra gupta = ...
```

will take anything you throw to it and return a list of the same type: Char – a list of chars (i.e. a string), Int – a list of integers, (Float→[Float]) – a list of functions from a float to a list of floats...

Now HASKELL types are arranged in "type classes". For example, there is the type class Num for all numerical types (ints, floats & Co.), Eq for all types to which the concept of equality/inequality may be applied, Ord for all which are ordered (i.e., may be used with ⩽, ⩾, etc.). Obviously the latter is a natural choice for a sorting algorithm. So we would write:

```
bestsort::Ord a=>[a]->[a]
bestsort rawdata = ...
```

defining that *bestsort* takes any list of sortable items and returns a list of the same type. Characters, integers and floats, for example, are sortable, whereas functions are not. Therefore, the compiler will accept passing of characters, integers and floats to *bestsort* but will reject functional arguments.

Now let us reconsider functional parameters and λ calculus (see p. 24). What if we have some aggregate type or something else which we know to be suitable for ordering, but which is too complex for the built-in comparison functions? In this case it is a smart thing to pass the comparison function as a parameter:

```
bestsort::[a]->(a->a->Bool)->[a]
bestsort rawdata mustBswapped = ...
```

meaning that *bestsort* now has two parameters, the first being a list of any type and the second (*mustBswapped*) a function which takes two elements of this type and returns a Boolean value (used to determine whether they are in order or not). The compiler will be extremely squeamish here and refuse, for example, a function of type (Char→Int→Bool), or (Char→Char→Bool) when *a* is Int.

When invoking this for a list of simple things, we may shorten things by using a λ function as comparator:

```
sortedlist = bestsort [0,8,15,-1,7,5] (\ x y -> x >= y)
```

The expression (\ x y -> x>y) is read as: "λ x y → x ⩾ y" or "A small nameless function with two parameters x and y that answers whether x ⩾ y"[27].

---

[27] In HASKELL, the backslash is read as λ, assertedly because of its similarity to the Greek letter. Oh dear.

Lisp, Erlang and Python, on the contrary, are virtually type-free[28]. A proper Python definition for the "bestsort" would simply read:

```
def bestsort( rawdata ):
```

No more than that. And it would accept lists of any kind – lists of integers, lists of floats, lists of chars (strings), lists of lists (i.e., lists of strings), lists of objects, even mixed-typed lists... And this is, smart though it is, *the* problem in Python. You may quite legally invoke this with `bestsort( [0, 8, 15, "schulze"] )` – and how is the comparison between the string and the numerical values to be achieved? Python does it, though. Start your Python and try the following:

```
3>4
4>3
"foo">"bar"
"bar">"foo"
3>"foo"
"foo">3
```

As you will see, it works. But comparing *"foo"* and *3* requires conversion of the numeric value 3 into a string... and that is a multi-step procedure, as is the comparison between two strings (see footnote on p. 49 for the effects of this on speed).

The great advantage of weak typing is that it is much closer to human thinking and thus allows much higher productivity. The table on p. 44 is in accordance with other studies which suggest that a type-free language with powerful error handling facilities is the most productive one[29]. The lower speed of execution can be shrugged off lightly. By now, computers are fast enough for most purposes and spend most of their time in endless futile cycles, waiting for new input. Really speed-critical routines may be implemented in a lower-level language, as described on p. 43; in our section on simulation of molecular growth processes (Vol. 2) we will encounter an example of such a hybrid.

Occasionally it will be necessary to enforce type checking, especially where larger projects are concerned. In a more advanced chapter (Vol. 2) we will see how to do this. Basically, a variable's type is just another attribute which can be checked. If this attribute's value is not correct, then we will raise an exception (p. 344) to prevent "shit happening".

---

[28] Lisp is amazing in that it is virtually free of anything – types, grammar, whatever – but even more amazing that it accommodates features for everything which it is not supposed to have. Common Lisp does have types indeed, but they are optional and serve as mere hints for the compiler, allowing assignment checking and optimization of code, but they are optional.

[29] Whereas C is a strongly-typed language with almost nonexistent error handling facilities. Do you still wonder about the lack of reliability in most "commercial" software?

There is more to weak typing, too. Object classes are also types, of course, and in a strongly typed languages like JAVA that means having to remember hundreds of different types. By contrast, in PYTHON the "black box" model is valid. Everything is an object. When you open a file you are provided with an object containing all methods for access to the file; if opening the file fails, another object is created, an "exception" which contains all information as to what has happened. There is no need to memorize endless lists of data types or consult the reference once a minute. These objects are just objects, and they can be talked to quite easily. If nothing else works, they will be "marshalled" into spewing out a textual description of what they are[30].

Now imagine we are implementing our address file program from p. 32, where objects comprise fields like *membershipNumber*, *firstName*, *lastName*, *postalCode* and maybe others. We could get this marshalled into a string, but probably the comparison of the strings will not yield that kind of order which we desire. So once more we turn to functional parameters:

```
def addressesMustBswapped( a1, a2 ):
    ...    # return TRUE if a1 and a2 in wrong order, FALSE otherwise
def bestsort( datlist, comparator ):
    ...    # apply comparator to datlist
    return datlist;
    addressList = bestsort( addressList, addressesMustBswapped )
```

For lists of simple things, we may again resort to λ functions:

```
intList = bestsort( [ 0, 8, 15, -1, 7, -5 ], lambda(x,y): x >= y )
strList = bestsort( [ "Q","W","E","R","T" ], lambda(x,y): x >= y )
```

In PYTHON, the only disadvantage of this is that it may become somewhat difficult to read. From the expression def bestsort(datlist, comparator): ... it is not possible to infer what kind of thing is to be expected here (unlike the HASKELL version where the type declaration says a lot of things). However, this can and should be amended by generous commenting (imagine the source code itself were your lab journal):

```
def bestsort( datlist, comparator ):
   "Sorting algorithm: Itzig W & Assfogl SP, J.Irreprod.Res.35(2002):66-69."
   # Two arguments are expected: datlist must be a list of things, and
   # comparator must be a function taking two parameters and returning
   # the comparison result.
   # E.g. for datlist: ["aleph","bayt","gimal","daleth"]    or    [9,8,7]
   # and for comparator: lambda(x,y):x<y
   #
   ...
   return datlist
```

---

[30] The would-be OO language VISUAL BASIC has taken this principle to excess. When nothing else is specified, take it for granted that a "default method" is invoked. PYTHON and JAVA are less extreme, but they do have mechanisms of the family variously referred to as "pickling", "marshalling" or "serialization", the latter perhaps being the most appropriate expression, since this is what it is about: the contents of the amorphous objected are converted into a *serial* stream of data which can be printed or saved to file.

Comments will be explained on p. 105.

You will note the string immediately following the *def* clause; please see p. 106 for an explanation of this. This is known as a *docstring*.

Anyway you put it, this is certainly the best thing to do!

Sometimes you just cannot escape it though... there is `str` for converting things to strings, `int` for converting things to integers, and `float` for converting things to floats. Note that when you divide an integer by another integer, Python will use an integer division! Thus, x/y amounts to 2 if x is 5 and y is 2. Write `float(x) / float(y)` to enforce a floating point operation!

Well, there *are* tradeoffs for everything, I guess.

Certain scripting languages try to reconcile some of the advantages of strong and weak typing by causing a variable to be declared before its first use in spite of being type-free. Here JAVASCRIPT and S-LANG have to be mentioned, where you will find untyped declarations like `variable a, b, c;`. This makes more sense when the declaration is combined with an initialization, which may also define the respective variable to be an aggregate.

Variables can be of different types: e.g., 625 can be either a sequence of the three letters "6", "2" and "5", or an integer number amounting to 5^4, or a floating-point number of the same value. Strength of typing basically means the language's demands for exactness when different variables interact. It also implies restrictions which can be imposed on the values which any variable may have. Strong typing generally means more work, faster programs, fewer errors and less flexibility.

2.18 Choose wisely...

Relevance: Criteria for choosing a programming language.
Keywords: professionality, flexibility, expressiveness, POLS.

To sum it up once and for all: *there is no such thing as the perfect programming language.*

Perhaps programmers should be directed to a dictum by Emanuel Lasker, philosopher, writer and chess expert: "In chess, it is futile to seek the perfect move. There is never any 'perfect' move, there is only the move which is most troublesome *for the respective opponent under the given circumstances.*" Using this strategy, Lasker remained World Champion for more than twenty years!

Wanna be an IT macho? – A not-quite-serious evaluation of some established languages and the prejudices pertaining thereto:

Wanna be an IT macho?

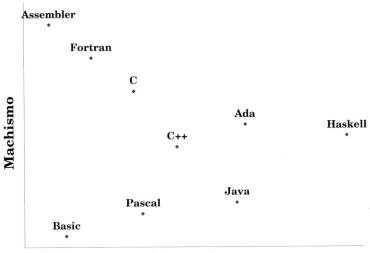

Which programming language is best to use, depends on the task at hand and the personal preferences of the programmer. A well-written C^{++} program, such as the KDE Desktop, can be superior in terms of stability, performance and ease of maintenance to a rotten HASKELL program.

Which programming languages have received general acceptance, then? That also depends on what you mean by "general acceptance".

- When judging by training programs, employment requirements, etc., one gets the impression that there are only two important languages for developing applications: C^{++} and JAVA. Looking at marketing investments, C♯ will also have to be named here – there is not one single language which has been advertised so noisily, not even JAVA.

- According to some estimates, about 35% of all the worldwide software volume is written in COBOL and another 35% in VISUAL BASIC. (The COBOL programs are mostly "legacy" software which has been in use for several decades and is constantly being maintained and extended – it just would not pay to redesign such mature programs.) That is to say that C^{++}, JAVA and all the rest of the world together have no more than 30% share!

- What about servers, probably the biggest business of all? You will find neither of the above, but PERL, PYTHON and PHP.

- So-called "embedded devices", that is to say, machines with a hidden computer system inside, are considered by many to become the most important application of all. Considering the fact that cellphones, microwave ovens and cameras are used even by people who normally would not touch a computer, even if they were paid for it, this may be true. Here again JAVA is currently dominant, but strongly challenged by PYTHON. However, ADA is far from dead yet, and this is the purpose it was originally devised for – if the functioning of the "embedded device" has any safety implications, ADA cannot be ignored. And there is also ERLANG, built for the special purpose of distributing fault-tolerant software across large networks.

- So perhaps one should have a look, not at the making of special applications, which may dictate the selection for other reasons, but at what the top shots think best? Well then, at Oxford University[31] the first language which is taught is HASKELL, and the second is OBERON!

2.19 Why Python?

PYTHON, designed in the 1990s by Guido van Rossum in the Netherlands "and a cast of thousands", currently competes with Larry Wall's PERL for being the #1 of scripting languages and is thus among the few languages which have gained some "commercial importance". It was designed with the explicit intention to have a language which is as easy to learn as BASIC, as powerful as a "professional development" language, and of clear and intelligible syntax into the bargain.

<small>commercially important
easy to learn
powerful
clear syntax</small>

It is strongly object-oriented (including even features which are lacking in JAVA!) but does not enforce the use of classes and objects upon the programmer. One may also write a plain procedural style, or alternatively, a functional program – it possesses virtually all features of a functional language, including λ functions, and complex data types. Therefore, it is ideal as a propedeutic language.

<small>OO

λ</small>

PYTHON is characterized by a very favourable learning curve[32]. There are only few things to be inculcated before you can switch on the computer, start the program and speed up your pace through "learning by doing", which is my recommendation for acquiring real proficiency anyway. By contrast, in many other programming languages you need a well-developed understanding not only of the concepts but also of many details before you can even get the compiler to accept your program. Without an intuitive grasp of the nature of pointers (see p. 80), C is bound to be a nightmare. Likewise, without distinct ideas about the complex class system, you will not enjoy JAVA overmuch. This does not lessen the power and usefulness of any of these languages, though, as professional programmers can be expected to cope with such things – but novices should be spared frustration wherever possible.

<small>easy to start</small>

---

[31] In the UK, professional boxers, tennis players and racing drivers are not yet recognized as being much more important than universities.

[32] Provided that learners *do* their homework, of course.

On the other hand, there is nothing unprofessional about the structure of PYTHON. Indeed, there are some languages around which are very popular at the moment just because of their low "entry level", but most of these click-and-go languages, legitimate successors to good old BASIC in that respect, are totally unsuitable for any larger project. And you will never be able to deal with a larger project unless you have learned to, therefore I am very much opposed to "beginner" or "teaching" languages. To quote an example which won't hurt anyone: LOGO – a thing of beauty in itself but deliberately unprofessional. PYTHON is suitable, and actually used, for as large projects as any language; see below.

suitable for large projects

Maybe even more important, the POLS – the Principle Of Least Surprise – is generally observed. There are no "miracle and wonder" things like the famous "implicit variable declarations" of FORTRAN[33] or the confusion about calling by value *vs* calling by identity caused by JAVA's "clandestine pointers". Error messages are generally clear and to the point.

no quirks

Furthermore, PYTHON is powerful not in terms of execution speed[34] but in the structure of its grammar. We have already seen that PYTHON programs are inclined to less verbosity and more clarity than other languages. Constructs like the *for... in* save a lot of typing and possible errors (there are not many things you can make wrong in PYTHON). Likewise, the availability of high-level data types allows the learner to get straight into the interesting things.

expressive grammar

As for speed, the PSYCO and PYPY projects (aiming at the construction of compilers for PYTHON which generate optimized code) hold considerable promise for the near future.

compilers available

PYTHON's main appeal, however, lies in its expressiveness. Compare the following PYTHON line

concise

```
for i in l: print i
```

with its equivalent in a certain other language, which is very fashionable at the moment:

```
for ( int i = 0; i < l.size(); i++ ) {
    System.out.print(((Integer)l.get( i )).intValue());
}
```

In spite of being very easy to learn, PYTHON is a fully professional language among whose users there are Google, Industrial Light & Magic and the NASA.

famous users

---

[33] In FORTRAN, all undeclared variables are implicitly floats with the exception of those from *i* to *n*, which are are implicitly integers (intended for looping; this is why many programmers still exhibit a preference for i and j as loop counters). This is no joke. Due to this pecularity, one of the first Saturn V rockets crashed – one programmer used a variable out of the i–n range during a floating point calculation and thus caused a significant error!

[34] Though it is possible to make programs run really fast by relying heavily on library functions!

There are also some larger Open Source applications available from which students may learn: the vector graphics program SKETCH, the text editor LEO and the integrated development environment ERIC are to be mentioned here.

This is at least partly because PYTHON's development time profile is excellent, both absolute and in relation to the execution time profile obtained for most problems (with the exception of pure number crunching). *development time profile*

The handling of the PYTHON software is plain and simple and does not require much familiarity with the underlying operating system. As for the underlying machine, PYTHON is virtually independent and can be – and has been – made to run on any "platform", providing unified access to most of the system's features, with the exception of only a few highly specialized aspects such as sound. *handling* *hardware independent* *OS independent*

There is an excellent, and well-maintained, implementation of PYTHON available for free, and the standards are also reliable (unlike, for example, BASIC and PASCAL, the LISP and to a lesser extent the ML languages, where dozens of implementations compete, all of which differ slightly from each other). This implementation also includes a really nice help facility. *freely available* *well documented*

Moreover, there is some pretty interesting bioinformatics software available for PYTHON (see the chapter on BIOPYTHON in the second volume, for details). *bioinformatics*

And last, but by no means least, programming in PYTHON is *fun*! That is to say, a well-written PYTHON program is (unlike, for example, a script written in PERL) an aesthetically pleasing experience. Even where "money talks and no-one walks, 'cos everybody's crawling"[35], personal satisfaction is important. Somebody who is continually fighting back his dislike of the language he has to use will never be productive, still less satisfied. Here PYTHON, with its clear syntax and legible expressions, virtually sets the standard. *fun*

See you at `http://www.python.org`*!*

---

[35] Ronnie James Dio: "Evilution"

CHAPTER 3

◇ Propedeutics

The first three steps of the seven are grammar, rhetoric and logic. Grammar is the science which enables us to express our ideas in correct language, but it is by rhetoric that elegance of diction is taught. Logic is the science which helps us to form clear and distinct ideas, and prevents us from being misled by analogy or similitude.

– from a Freemasons' ritual

The following exercise is taken from Schopenhauer's *Die Welt als Wille und Vorstellung*:

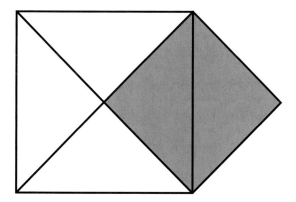

Use this figure to demonstrate that $a^2 + b^2 = c^2$.

I would like you to occupy yourselves with this smart little gadget until you have fully understood it, because this is also an example of how I have structured this course. Do not try to learn things by heart. Instead, attempt to get an intuitive grasp. For programming languages it is important to get proficient, not in terminology, but in concepts.

Bioinformatics Programming in Python. Rűdiger-Marcus Flaig
Copyright © 2008 WILEY-VCH Verlag GmbH & Co. KGaA, Weinheim
ISBN: 978-3-527-32094-3

"It is against the grain of modern education to teach children to program. What fun is there in making plans, acquiring discipline in organizing thoughts, devoting attention to detail, and learning to be self-critical?" – ALAN PERLIS

CHAPTER 4

Getting the Materials

Talking to serpents has always been considered a very Dark Art.

– Joanne Rowlings: "Harry Potter and the Chamber of Secrets"

Relevance: How to install and run PYTHON.
Keywords: Installation; editors; IDEs; command line arguments.

To enjoy PYTHON, only two programs are required:

(1) PYTHON itself.
(2) An *editor* for writing PYTHON programs.

editor

4.1 Getting the Python interpreter

PYTHON is *open-source software*; that is to say, everybody is free to inspect or modify the source code, pass on or use the source or the compiled program or do otherwise as he pleases, as long as he does not preclude others from doing the same thing. The advantage of this concept is twofold:

open-source software

- There is only one version of PYTHON[1], as everybody who is interested is invited to join (unlike the situation in earlier languages such as PASCAL, which swiftly ramified into numerous and only partially compatible dialects). New development will therefore be to the common benefit of all.
- This version can be obtained free of charge.

4.1.1 Python for Unices (including Linux, Solaris™, MacOS-X™, BSD and Cygwin).

4.1.1.1 PYTHON *pre-installed.* If you are using a unixish system, the odds are that PYTHON is already installed on your computer. In particular, many flavours of Linux (most notably those of the Red Hat – Yellow Dog – Black Lab line) employ PYTHON for set-up, configuration and administration. Apple's MacOS-X™ also comes with PYTHON pre-installed. To make sure, start a

---

[1] There *are* JPYTHON, IRONPYTHON and other implementations, but these are direct and official spin-offs of PYTHON "as we know it", rather than competitors.

"terminal" or "console" and type `which python` . If PYTHON is installed, you will see the location where it is installed.

Cygwin

Cygwin is an open-source Unix compatibility layer for Windows^{TM} computers, offering most of the advantages of unixish systems[2]. During installation, PYTHON can be (and is by default) selected as one of the packages to be installed. If in doubt, use the `setup.exe` program to review and/or install software.

package management

4.1.1.2 PYTHON *not pre-installed.* All contemporary Linux distributions support *package management* for the neat installation and removal of software; however, the details may be very different. For the most popular families, proceed as follows.

Ubuntu
- *Ubuntu* – use "Add/Remove" in the "Applications" menu, or proceed as with Debian.

Debian
- *Debian* – use the `synaptic` tool.

Gentoo
- *Gentoo* – use the `emerge` tool[3].

SuSE
- *SuSE* – use the `YaST` tool.

Red Hat
- *Red Hat* – use the `RPM` tool or any of its front ends such as `gnorpm`.

FreeBSD

FreeBSD comprises the `ports` system for building and installing packages from source code.

Solaris^{TM}

Solaris^{TM} is among the few unixish systems which do not comprise a package management system. For Solaris^{TM} in particular, a pre-compiled version may be downloaded from `http://www.sunfreeware.com` and installed according to the instructions given there.

tarball

Alternatively, you may download the PYTHON interpreter as a source *tarball* from `http://www.python.org` and proceed to install it in accordance with the instructions included.

Windows^{TM}

4.1.2 Python for Windows^{TM}. A *Windows*^{TM} port, complete with InstallShield^{TM} wizard, is also available at `http://www.python.org`. This will also create a file type and (rather ugly) icons.

PYTHON for Windows^{TM} maintains an excellent degree of compatibilty with the original Unix version. However, the two weaknesses of the port are currently[4]:

(1) sluggish memory management,

---

[2] With PYTHON, the greatest drawback of Windows^{TM} is its memory management scheme, which can tremendously slow down the system, in particular when using extensive recursion or object orientation. This is not ameliorated by Cygwin, as it is a matter of the kernel.

[3] Like the BSD systems, this will download the source and compile it, rather than a precompiled binary.

[4] N.B.: The author's personal experiences with Windows^{TM} are very limited. In particular, nothing can be said about PYTHON and Windows Vista.

(2) an incompatibility with some kinds of firewalls, which will result in spurious error messages upon start up which it is best to simply ignore.

4.2 The editor

Basically, any program capable of handling plain ASCII files is suitable, at least for making the first steps into the world of PYTHON. However, many advanced text editors provide language support features (e.g., syntax-dependent colouring and indentation) which make life a lot easier.

4.2.1 Python editors for Unices.
Traditionally, most Unix programmers are sworn devotees of either EMACS or VI, each of which is available in a number of different flavours. Both are extremely powerful and versatile tools but possess the disadvantage of having a handling that differes widely from commonly established text processor standards – VI much more so than EMACS. However, their modern graphical versions (e.g., XEMACS and GVIM) provide the remedy for this problem.

Most modern systems comprise, apart from these two venerable dinosaurs, other editors with a more conventional look'n'feel, such as NEDIT (the preferred editor on Solaris^{TM}), KEDIT and KATE (components of the KDE project), GEDIT (components of the GNOME project), SCITE or XJED. Any of these is more than adequate for professional development in PYTHON; the choice will ultimately depend on personal preferences. (The author generally prefers to work with GEDIT, both for programming and for preparing his L^AT_EX manuscripts.)

4.2.2 Python editors for Windows^{TM}.
Factory-default Windows^{TM} is poorly provided with editors, but NOTEPAD will do for a start. Alternatively, there are ports of most of the above now freely available for download, or as an integral part of Cygwin.

4.2.3 Integrated development environments for Python.
An *integrated development environment* or IDE is an umbrella program which provides a "development central" by unifying editor, language and auxiliary programs, such as a debugger, into one common interface. This is, of course, particularly useful when working with compilers or, in general, languages which are difficult to manage; however, they may also facilitate getting along with PYTHON. For example, an IDE will generally comprise a language-sensitive "browser" which keeps track of all the definitions within a particular file and thus is very helpful in navigating through a file of more than a few hundred lines.

integrated development environment

The *de facto* IDE standard for PYTHON is the simple but efficient *IDLE*, which has been included into the Windows^{TM} port and is also available for Linux. Alternatively, Linux users may use one of the more sophisticated "professional"

IDLE

IDEs such as ANJUTA or ECLIPSE. There are also a number of IDEs dedicated to PYTHON alone, such as ERIC or DR. PYTHON.

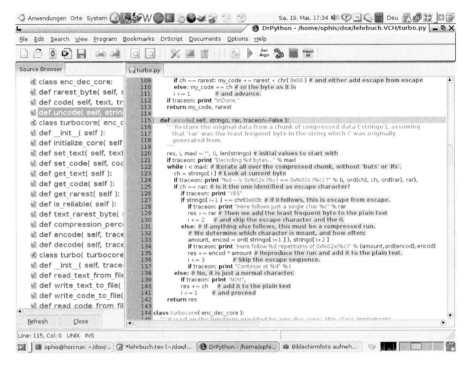

Dr. Python at work (on Ubuntu Linux)

Essentially, handling an IDE is very straightforward. It offers a text editing facility (which may be connected to a browser), a program start control and a variety of ancillary features which are best explored by "learning by doing".

4.3 *sys.argv*, redirection, and pipelines: don't fear the command line

Relevance: This section describes how to start the PYTHON interpreter with a bit more sophistication than that which a double click affords. Program arguments (which are implicitly based on the "automaton" concept of the whole program as a function), input/output redirection and pipelining – all of which can be used only when starting from a command line shell – are an easy way both for the user to control programs and for programs to interact with each other, and are particularly valuable for the tasks often encountered in scientific computing.
Keywords: Stdin, stdout, pipeline, argv.

The following section explains some fundamentals principles which will be helpful later, but which are not indispensable for the beginner. For the time being, those in a hurry may continue on p. 71, provided they promise to come back later, e.g., when beginning to work on the first project (p. 277)!

4.3.1 Stdin and Stdout: a homage to Kant and Uexküll. As mentioned above (p. 18), as early as 1997 the author had already pointed out in his thesis work on pseudovirions that indeed the capacity for information processing is what vitalists such as Uexküll considered the *vis vitalis*. Following Kant's "transcendental aesthetics", which in turn show an uncanny resemblance to Einstein's theory of relativity, Uexküll noted that any life form differs from an inanimate thing in that it has a "mind" (*Innenwelt*) of its own, which is fundamentally separate from the "environment" (*Umwelt*) of the life form[5]. The mind comprises an intangible representation of relevant elements of the environment arranged in a coordinate system belonging to the life form, not to the world it moves in (in the case of human beings, this coordinate system consists of three-dimensional space and time). Connection between the mind and the environment is twofold: the life form reads information in the form of "input" (*Merkmale*) from its surroundings and reacts by imprinting "output" (*Wirkmale*) on its surroundings.

<small>Innenwelt
Umwelt

Merkmale
Wirkmale</small>

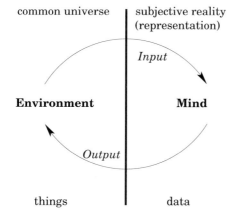

Oddly enough, basically the same concept is the basis for Unix-style programming which has also been implemented under Windows™. A program, possessing its own "mind" of data[6], possesses two fundamental channels for communicating with the outside world, namely the *standard input* channel (**stdin**) for receiving a stream of input data and the *standard output* channel (**stdout**) for disposing of output data[7].

<small>standard input
standard output</small>

---

[5] This was actually the origin of the term "environment" as it is widely used nowadays!

[6] Here mention should be made of Olaf Stapledon's philosophical masterpiece *Star Maker*, wherein he defines a "mind" as "a self-contained system of experience".

[7] In their Star Trek novel "Memory Prime", the Reeves-Stevenses described the world from the point of view of a "wayfinder", an intelligent computer, wherein software and data are "reality" and the physical universe is referred to as "datawell", which, in the wayfinder's opinion, is probably mere fiction. This is closer to reality than one would readily assume.

By default, standard input is what comes from the keyboard, and standard output what goes to the screen.

Things become much more complicated with the use of graphical user interfaces, of course, but here we will consider only the conventional "automaton" style of a program (as exemplified, e.g., in a compiler), which is essential to all scientific programming. **Scientific programming is not about glitter but about results.** If glitter is required, so-called "frontends" for automata are readily written.

redirection

It is easy to understand that sometimes it may be desirable to change this default association. For example, for producing a hardcopy, one might wish to set the standard output channel to the printer instead of the screen. This is where *redirection* comes into play. Redirection is, essentially, the technique of shifting standard input and/or standard output to sources or sinks, respectively, of data different from the default. This is done on the operating system level, without the program knowing or needing to know anything about this.

os.system

It should be noted here that a PYTHON script may call any other program, independent of whether the target is another PYTHON script or something different, using redirection and all the other advantages of the command line. This is done via the *os.system* call (see p. 346), which in fact orders the shell program to process the request.

Of course, redirection has to be established first. The shell provides two operators to this end: > and <. They are intended to represent arrows, → indicating the redirection of output and ← indicating the redirection of input. Thus, if we have a program that takes an amino acid sequence and calculates the most probable secondary structure ("alpha-helix", "beta-sheet" or "undefined"), this will typically accept standard input and write to standard output. Hence, instead of retyping the input every run, we can direct the program to accept a file's content as input:

```
ophis@horcrux:~/patente/trilobit$ protstruct
SEGGGSEGPGSEGGGSEGGGSGGGGPMKQIEDKIEEILSKIYHIENEIARIKKLIGEAPGGSGGHHHHHH
```

4.3 SYS.ARGV, REDIRECTION, AND PIPELINES

```
alpha-helix
```

Here the second line was what the user typed in, and the third the program's output.

```
ophis@horcrux:~/patente/trilobit$ protstruct <t-sequence.txt
alpha-helix
ophis@horcrux:~/patente/trilobit$ protstruct <t-sequence.txt >prediction.txt
ophis@horcrux:~/patente/trilobit$
```

Now the result has been written to file *prediction.txt*:

```
ophis@horcrux:~/patente/trilobit$ cat prediction.txt
alpha-helix
ophis@horcrux:~/patente/trilobit$
```

Moreover, programs may be concatenated so that the input of one program is used as the output of the next. This is called a *pipeline*:

pipeline

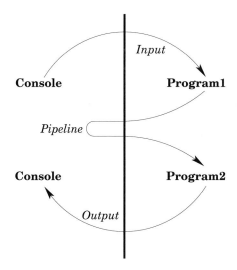

In the following example, `predictstruct` is supposed to generate a vast output of grapical data (using data supplied on standard input and dumping its output to the standard output) which can be fed to `showstruct` which then opens a window and displays what it has been fed via its own standard input (the pipeline being symbolized by the vertical bar glyph):

```
ophis@horcrux:~/patente/trilobit$ predictstruct <t-sequence.txt | showstruct
```

I/O redirection and pipelining represent a powerful instrument to easily create complex workflows from software building blocks. In scientific programming, where programs are judged by power rather than by eye candy, this is often the method of choice. Thus, the command line shell is not a "legacy application" but a sophisticated toolbox.

4.3.2 Argv: how programs are properly addressed. Apart from I/O redirection, the command line offers the possibility of invoking a program with arguments, like a function. Such arguments can be used to give directions to the program. On the command line, they are simply added to the program name:

```
ophis@horcrux:~$ just-show-arguments alef bayt gimal daleth
4 arguments
 Argument 1 = alef
 Argument 2 = bayt
 Argument 3 = gimal
 Argument 4 = daleth
ophis@horcrux:~$
```

The established custom is that objects of the program are simply listed like this, whereas options to be set are denoted by prefixing a hyphen.

Compared to the, now common, method of input via menus and dialogue boxes, this offers three distinct advantages.

- It is *much* easier and quicker to implement.
- It is system-independent.
- It offers itself to automatization and the building of workflows, as scripts may be written into which extensive parameter sets are incorporated, thereby saving the trouble of either manually configuring the program every run or sacrificing flexibility by use of the notorious "preference files". A real-life example is the following argument set for the GNU C compiler (*gcc*):   -v --enable-languages=c,c++,java,f95,objc,ada,treelang --prefix=/usr --enable-shared --with-system-zlib --libexecdir=/usr/lib --without-included-gettext --enable-threads=posix --enable-nls --program-suffix=-4.0 --enable-__cxa_atexit --enable-clocale=gnu --enable-libstdcxx-debug --enable-java-awt=gtk-default --enable-gtk-cairo --with-java-home=/usr/lib/jvm/java-1.4.2-gcj-4.0-1.4.2.0/jre --enable-mpfr --disable-werror --with-tune=pentium4. This is most conveniently incorporated into a script, which is then called in lieu of the compiler proper.

Of course, these arguments must then, unlike the redirectioning commands, be accessible from within the program. In a high-level language such as PYTHON, this is not really difficult. In the module *sys*, the variable *argv* (for ARGument Vector – you may remember this from p. 70) is a list of all the arguments passed to the PYTHON script. The blank serves as the delimiter. Thus, if the above argument set were passed to a PYTHON script, we would have *sys.argv[1]= "-v"*, *sys.argv[2]= "--enable-languages=c,c++,java,f95,objc,ada,treelang"*, ..., *sys.argv[19]= "--with-tune=pentium4"* . It is then up to the program to react in some sensible fashion.

sys
argv

For the pathologically curious: Of course, the comand line arguments have to be transferred somehow to *sys.argv*. This is done by a cooperation of the shell (which dissects the command line), the operating system core (which passes the dissected arguments) and the *runtime system* of the program, which is a piece of code that is run before the very first line of the program is executed. The runtime system does all the necessary setting up, such as obtaining the arguments from the operating system core and placing them in the *sys.argv* variable. The runtime system is an integral part of the PYTHON interpreter (or any other interpreter), and only those who are developing the PYTHON interpreter, rather than merely using it, will ever have to get into contact with it. (Actually, the PYTHON interpreter itself is written in C, so most of it just uses the C runtime system, and only those who are developing the C system, rather than merely using it, will ever have to get into contact with it...)

runtime system

4.3.3 In everyday life.
Like any other program, the PYTHON interpreter may be invoked from a graphical user interface *or* from a command line shell such as the *DOS command prompt* or any of the various Unix shells, in the latter case simply by typing `python` followed by hitting the RETURN key.

DOS command prompt

When PYTHON is started from the command line, expediently the name of the script to be run is added as an argument, e.g.,

```
python demo-script.py
```

or

```
python fortran2008-compiler-in-python.py multiarray.f -v
--processor=SPARC >f07
```

(Individual items of the command line are separated unconditionally by spaces, and by convention, hyphens introduce options, wherein a single hyphen is followed by a one-letter option such as the ubiquitous `-v` which switches programs into a "verbose" mode, and a double hyphen by a multi-letter option such as `--processor=SPARC`). In the latter case, the items following the name of the PYTHON script are passed as arguments not to the PYTHON interpreter but to this script. So the latter command will cause the PYTHON interpreter to load the script `fortran2008-compiler-in-python.py` and run it in such a way that

sys.argv

the script receives `multiarray.f, -o` and `--processor=SPARC` as arguments. As described above, these arguments will be stored in the variable *sys.argv* which has the form of a list, each of whose members comprises one of the arguments – more about this later (p. 280). Of course, the output of the entire business – which can be guessed to be the executable machine code – will go to file *f07* .

Naturally, redirectioning and pipelining commands will *not* be passed as arguments, as they are taken care of by the shell. The shell performs a "sifting" of the components of the command line, passing on only those which it does not handle at its end.

When working with an IDE, all these arguments may be set by selecting a menu entry named "Parameters..." or the like. So to run the above script, click this and enter `multiarray.f --processor=SPARC` before running your `fortran2008-compiler-in-python.py` .

Ready, dude – beat me, break me!

CHAPTER 5

Variables, Data Types, and Assignments

Too much information going through my brain,
Too much information driving me insane.

– THE POLICE: "Ghost in the Machine"

5.1 Relevance and purpose

As discussed on p. 11, computers are about the processing of data. It is therefore not astonishing that every higher programming language has representations for various kinds of data.

We have mentioned before two of the fundamental concepts upon which all modern information technology is based:

α – A computer must be capable of input, storage, processing and output.
β – These capacities are best kept separate from each other.
γ – They are readily implemented by breaking down all information into its atomic units.

To (γ), the following corollaries should be noted:

δ – All atomic units are numbers. *All information can be represented as numbers*, where a number has an intrinsic meaning or is used as a reference to something else.
ε – All numbers can be represented as associations of binary values.

<small>All information can be represented as numbers</small>

Here (δ) deserves some explanation, as *numbers with intrinsic meaning* are well-known from everyday life and in particular from scientific measurements; they directly convey meaning. Driving at 53 mph makes you liable to a fine if there is a speed limit of 30 mph. These figures have a relevance which is independent of their context.

<small>numbers with intrinsic meaning</small>

By contrast, *numbers used as references* do not convey a meaning in themselves but serve to establish a link with a piece of information stored elsewhere. For

<small>numbers used as references</small>

Bioinformatics Programming in Python. Rüdiger-Marcus Flaig
Copyright © 2008 WILEY-VCH Verlag GmbH & Co. KGaA, Weinheim
ISBN: 978-3-527-32094-3

example, the speed limit sign may be in accordance with §41 of the *Straßenverkehrsordnung*. Nothing can be directly inferred from the number 41, and in fact there is no cogent reason why 41 has been chosen, rather than any other number; but rebus sic stantibus §41 does comprise the relevant section of the code (no pun intended), and you may use it to read the full text.

The Number of the Beast

The Number of the Beast: The discovery that numbers can be used to represent things which are not numbers – ultimately going back to Descartes's insight that numbers can be used to represent geometrical objects – is pivotal to informatics. It began with the discovery that numbers can be used to represent instructions; and whenever it was found that some particular thing could be represented as a number, a new field of applications for computers was opened. When it was realized that letters can be represented as numbers, text processing was born; colours, graphics; amino acids and nucleotides, bioinformatics.

An example from computing can be found in the description of colours, for which two modes are predominantly used nowadays, 8-bit and 24-bit. In 24-bit colour, the first 8 bits give the intensity of the red component of the colour, the middle 8 bits give the green and the last 8 bits give the blue component of the colour; so we can see immediately that a colour of $708090 represents a blueish grey and that $709890 is slightly brighter and a bit more greenish. Thus, 24-bit colour values possess an intrinsic meaning. In 8-bit colour, by contrast, each colour value is used to refer to one colour in a table of 256 shades. This table comprises the actual definitions – usually in 24-bit format as described before – and the "colour" value of the individual pixel is to be used as an index to this table. For example, a pixel may be of colour 41. Again nothing can be directly inferred from the number 41, and in fact there is usually no cogent reason why 41 has been chosen, rather than any other number; but rebus sic stantibus colour table entry 41 does comprise the relevant colour values, and computers or human beings may thus look it up. So basically this is how computers may store information of *any* kind even though they can essentially handle only binary (true/false) values.

Now data types can be very different in complexity, and most languages use only of few of them. Ordered (approximately) by rising complexity, there are first the following.

5.2 Simple or "scalar" data types

ℜ

Relevance: These data types are the foundation of all programming.
Keywords: Bytes and words, integers and floats, ...

Boolean

- **Boolean**
  Named in honour of the mathematician Boole who first described binary calculations, this is a "flag" or "logical value" which can be either

TRUE or FALSE (and nothing else). Theoretically, one bit ought to be sufficient, but for technical reasons, usually a whole byte is used for the representation of a Boolean value.

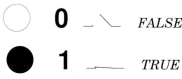

Boolean values

- **Byte**
  A block of 8 bits, allowing for $2^8 = 256$ combinations, from 0x00 to 0xFF in hexadecimal notation[1]. A byte is the smallest chunk of RAM or disk space which is normally allocated.
- **Nibble**
  Half a byte, i.e., a block of 4 bits, with $2^4 = 16$ possible combinations, and hence what may be represented by a single hexadecimal digit. In earlier times, file systems made extensive use of nibbles for storing meta-information. With storage capacities soaring and file systems becoming ever more complex, this custom has gradually faded into oblivion.
- **Character**
  Something which contains one legible glyph. Either 1 byte (ASCII) or 2 bytes (Unicode; more modern but rarely used).
- **Word**
  A block of 16 bits, allowing for $2^{16} = 65536$ combinations, from 0x0000 to 0xFFFF in hexadecimal notation.
- **Longword**
  A block of 32 bits, allowing for $2^{32} = $ ca. 4×10^9 combinations, from 0x00000000 to 0xFFFFFFFF in hexadecimal notation. Modern processors usually use registers of 32 bit size.
- **Quadword**
  Also known as "long long", a block of 64 bits, allowing for 2^{64} = ca. 2×10^{19} combinations, from 0x0000000000000000 to 0xFFFFFFFFFFFFFFFF in hexadecimal notation. Many processors are able to do 64-bit calculations, though usually with some overhead. Processors with 64-bit registers are currently emerging (PPC-G5, Athlon, SPARC64).
- **Paragraph**
  A run of 256 bits or 32 bytes, which was the basic memory management unit on 8-bit machines. Obsolete.

---

[1] Binary notation comprises 0 and 1, octal notation 0 to 7, decimal notation 0 to 9; for hexadecimal, the letters A for decimal 10, B for decimal 11, ..., F for decimal 15 were added to the decimal system. Hexadecimal letters are usually written with a suitable prefix: 0x1F or $1f for decimal 31.

Page
- **Page**
  Basic memory management unit on machines that support virtual memory, i.e. using a part of the hard disk as extension of the RAM. Capacity may vary, usually from 1024 to 4096 bytes.

Integer
- **Integer**
  An integer number. Depending on the system, this may be a word, longword, or quadword. In general, longwords are now used by programming languages unless explicitly stated otherwise – a range from $-2\,147\,483\,647$ to $+2\,147\,483\,647$ is sufficient for most purposes.

Byte: 8 bit $= -128 \,..\, +127$ ($00 .. $FF)

Byte: 16 bit $= -32\,768 \,..\, +32\,767$ ($0000 .. $FFFF)

Longword: 32 bit $= -2\,147\,483\,648 \,..\, +2\,147\,483\,647$ ($00000000 .. $FFFFFFFF)

Quadword: 64 bit $= -9\,223\,372\,036\,854\,775\,808 \,..\, +9\,223\,372\,036\,854\,775\,807$ ($0000000000000000 .. $FFFFFFFFFFFFFFFF)

Unsigned (Cardinal)
- **Unsigned (Cardinal)**
  Usually one bit is reserved for indicating negative numbers; thus, the effective range of a word is -32767 to $+32767$. For some purposes, this may be modified; thus, an unsigned word ranges from 0 to $+65535$. Only rarely used in some C programs[2].

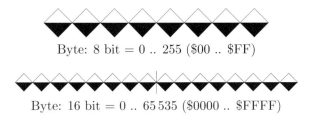

Byte: 8 bit $= 0 \,..\, 255$ ($00 .. $FF)

Byte: 16 bit $= 0 \,..\, 65\,535$ ($0000 .. $FFFF)

Longword: 32 bit $= 0 \,..\, 4\,294\,967\,295$ ($00000000 .. $FFFFFFFF)

---

[2] In theory, bytes are also signed, and characters have to be considered as "unsigned bytes".

Quadword: 64 bit = 0 .. 18 446 744 073 709 551 616
($0000000000000000 .. $FFFFFFFFFFFFFFFF)

- **"Long" and "short"**
  These are integers which are supposed to be longer or shorter, respectively, than the "standard" integer, whatever that may be. In C, the data types *long int* or *long* and *short int* or *short* are used extensively even in the absence of unambiguous definitions. In most earlier implementations, both *int* and *short int* were 16 bit, and *long int* was 32 bit; nowadays, *int* and *long int* are usually 32 bit, and an additional *long long* is often available for 64-bit values. (It need not be so[3].) All of these data types may be combined with *unsigned*, if so desired.
- **Float**
  A floating-point number. Range and precision are implementation dependent.
- **Double**
  A higher-precision floating-point number. Range and precision are also implementation dependent, but, in general, a float uses 32 bits and a double 64 bits. There are also some computers which use 128-bit floating-point arithmetics but generally 64 will be more than enough.
- **Real**
  A floating-point number, either a float or a double. With most floating-point hardware now supporting doubles, the tendency is to use doubles rather than floats.
- **Enum**
  When there are non-numeric states, these may be enumerated, as in the following example. It will be known to everyone familiar with the basics of molecular biology that DNA sequences comprise the four bases guanine (G), adenine (A), thymidine (T) and cytosine (C), while in RNA sequences thymidine is replaced by uracil (U); and that peptide sequences comprise twenty amino acids, each also represented by a single letter.

  Thus we get the very clear C definition:
```
typedef enum { G, A, T, C } dna_base;
typedef enum { G, A, U, C } rna_base;
typedef enum { A,C,D,E,F,G,H,I,K,L,M,N,P,Q,R,S,T,V,W,Y,Z } amino_acid;
```
  Likewise, PASCAL has[4]

---

[3] Luckily, the C function `sizeof` can be used to determine the size of a variable or type.

[4] You will note two differences here. The first is that in PASCAL the word "type" may introduce an arbitrary number of type definitions, whereas C's "typedef" gives only single definitions. This probably reflects the fact that in PASCAL the use of type definitions has always been extensive, whereas it was not featured in the original C standard. The second is, that in the C complex, names usually use an underscore, whereas in the PASCAL family the so-called "CamelCode" – reminiscent of a camel when seen from its side – is preferred. However, this is only a matter of convention. PYTHON, being indebted to C more than to PASCAL, usually also uses underscores.

```
type DnaBase = ( G, A, T, C );
     RnaBase = ( G, A, U, C );
     AminoAcid = ( A,C,D,E,F,G,H,I,K,L,M,N,P,Q,R,S,T,V,W,Y,Z );
```

In both cases, the compiler will issue an error message when any other value than G, A, T or C is assigned to a variable of type "dna_base" (or mutatis mutandis for the others), thus trapping a lot of errors early.

This is an important example of numbers being used as references. Here the table which is referenced is, in fact, the definition.

In theory, integers are enumerations from the lowest possible value to the highest; in practice, the compiler, once it has checked for correctness, uses 0 for the first value (in this case G), 1 for the second and so on. The rules for operations on enums are ill-defined; in C, it is quite possible to add `A+T` and obtain `C`, which is obvious nonsense. There is, however, one fix for this problem: Just don't do that.

Things become interesting when you try to assign enums of one type to those of another.

In PASCAL, you can also use the ubiquitous ellipses with enumerative data types; thus, `G..T` will comprise G, A and T. This can be rather useful in *case* structures (p. 115). You can also build sets and bags (p. 90) of enumerative data types.

- **Atom**

  Some higher-level programming languages (traditionally the weakly typed among the more functional ones, e.g., LISP and ERLANG) allow enum-like designators without an enumeration.

  An atom can be considered as a piece of explicit information without any hidden meanings; or in other words, a variable that evaluates to itself, i.e., contains its name as the value: "It does not mean anything. It simply exists."[5] Thus, in contrast to an enum, an atom does not belong to any type or group either (except that of atoms, of course). It is an entity with such a total lack of intrinsic significance that it is difficult to find any comparison. It is simply a thing that is, and that is different from something else that is. Thus, it is mostly used as a label that is placed on something else and thus conveys information about that "something else", like a flag[6] or jack, and may thus be used to identify the "something else" and determine suitable reactions for it:

Jack:	British	Swiss	Chinese	Troublemaker
Identify as:	British	Swiss	Chinese	Troublemaker
1st Action:	Play Anthem	Hail	Hail	Shoot
Next Action:	Hail	Reconfirm	Make WanTan	Report

---

[5] Stanisław Lem: "Invincible".

[6] In the conventional sense of the word, not in the IT sense of binary toggle (p. 74).

In ERLANG, the passing to and fro of messages between processes is a very important matter; usually, these messages use atoms to specify their meanings. The nice thing about an atom is that it is very straightforward in its meaning – as it has no meaning whatsoever beyond its name. So if an atom appears which has not been foreseen, we may do something suitable, whereas an unforeseen value in an enumerative data type will probably crash the program.

Very often, atoms are used in conjunction with pattern matching (p. 140):

```
-module(manowar). % name of the module
-export([react/5]). % Function "react" is made publicly available

% Function "react" takes five arguments:
%   1. the atom itself
%   2. the day when it appeared
%   3. the number of people on board
%   4.-5. the coordinates where it appeared

react( british, _, _, _, _ )          % don't care about where and when
   -> jukebox:play( 'God Save the Queen' ),
      communication:hail( 'Cheerio Miss Sophy!' );
react( swiss, Today, _, _, _ )        % use date only
   -> communication:hail( 'Gruezi vielmalen' ),
      harbor:contact( [Today,'Sighted apparently Swiss vessel;
      please reconfirm'] );
react( chinese, _, People, _, _ )     % use number of people only
   -> communication:hail( 'Nimenhao tongzhimen' ),
      kitchen:prepare( [People,'helpings of Wan-Tan'] );
react( troublemaker, Today, _, X, Y ) % use date and coordinates
   -> turret:fire( X, Y ),
      harbor:contact( [Today,'fired on pirate vessel'] );
react( Name, Date, X, Y, Z )          % simply everything
   -> harbor:contact( ['Met odd fellows today:',X,Y,Z,Date,Name] ).
```

Depending on the type of atom, the proper reaction will be selected. In the case of *L'imprevu*, this will be duly recorded.

Atoms vs *enumeration:* Armstrong et al. define atoms as "constants with names"; the example they give is the following list of atoms: Monday, Tuesday, Wednesday, Thursday, Friday, Saturday, Sunday to designate the day. This immediately recalls an enumeration. The difference is that an atom does not have to be a member of a predefined list; when using atoms, you could set the day to Sabbath without any problems – in an enumerative data type, this were simply undefined and would cause a compilation error:

```
typedef enum { monday, tuesday, wednesday, thursday,
               friday, saturday, sunday } weekday;
main() {
  weekday conference = tuesday;
  weekday holiday = sabbath;
```

}

```
[ophis@horcrux ophis]$ gcc wd.c
wd.c: In function 'main':
wd.c:6: 'sabbath' undeclared (first use in this function)
wd.c:6: (Each undeclared identifier is reported only once
wd.c:6: for each function it appears in.)
[ophis@horcrux ophis]$
```

Still, one might say that atoms are the "soft version" of enumerations. ERLANG programs, for example, simply *cannot* use enums because ERLANG is a weakly typed language and does not require declaration of variables. Thus, the abstract labels which enum types are designed to be are created "on the fly" and without any particular order, but basically for the same purpose. One might also say that C programs, for example, simply *cannot* use atoms because C is a strongly typed language and requires declaration of variables. Thus, if we want to use abstract labels, the acceptable values for any type have to be defined.

- **Newts**
  Newts are amphibia that have simply nothing to do with informatics. Nevertheless, they are obviously useful for catching a reader's attention. Quod erat demonstrandum.

- **Subtypes**
  In very complex languages such as ADA, there is the possibility of selecting sub-ranges of existing data types: e.g., we may define a type "postal_code" which is a subset of integer, restricted to the values from 10 000 to 99999. At attempt to assign anything beyond that results in an error.

- **Type classes**
  HASKELL has something to the same effect – here variables are grouped in "classes"[7], to which certain operators may be applied or not: the class *Eq* comprises all data types which can be compared, the class *Ord* all which possess an intrinsic order, the class *Num* all numeric data types, etc. Thus, in a HASKELL program we can state explicitly whether or not, for example, a given enumerative type supports additions.

- **Pointer**
  Here we leave the hygienic area. Pointers are nothing but the addresses of other values – where an address is simply the point of memory at which a variable is located[8] – and they can be used for building structures like lists and trees. In fact, lists are generally implemented using pointers (as shown before), but working with explicit pointers is a painful matter. On the lower level of programming, pointers are the only way of dealing with dynamic data: a system function generally

---

[7] Not to be confused with the "classes" of OO programming!
[8] That is to say, where it begins.

known as *malloc* returns a pointer to a chunk of hitherto unused memory and marks this chunk now as used, and another function releases such memory again. Let it suffice to say that most crashes of C and C++ programs may be attributed to erroneous pointers. We are not delving further into details here, as PYTHON does all pointer handling for us in a very clandestine fashion. However, two important C-type pointer operations should be mentioned: `p = &v` yields a pointer to the variable v, and `w = *p` gets the thing which p points to.

& and *
Reference

- **Reference**
Something akin to a pointer.

- **Handle**
A pointer to a pointer. In theory, using handles instead of pointers has the advantage that the actual location of the the data object may be shifted when this is necessary. Imagine you allocate memory for 1000 objects, each 100 kilobytes in size; then you process them, and all objects with odd numbers are discarded. That will free a total of 50 megabytes, yet within the entire memory area no object of more than 100 kilobytes may be allocated because of the fragmentation. When using handles instead of pointers, the system may redistribute the surviving objects so that they form a coherent block again, creating a monolithic free space of 50 megabytes.

Handle

- **Functions**
We have discussed this in some length before. Let it suffice here to say once more that in PYTHON, as in HASKELL or O'CAML, functions may be treated as normal variables.

Functions

♠ *Start your* PYTHON *and enter the following lines, one by one:*

```
def square( n ):   return n * n
square
square()
square(5)
cube = lambda x:   x * x * x
cube
cube()
cube(5)
```

Explain the output.

Of these, PYTHON possesses only the following: Boolean, Character, Integer, Double, Reference, Function. Most of these will be automatically converted whenever necessary; i.e., when you multiply an integer and a float, the result will automatically be a float[9].

---

[9] This seems so logical that it is often forgotten that many languages, even modern ones, will reject this with a *type error*.

To virtually every variable type there are corresponding *constants*, of course. Be aware that the term "constant" is ambiguous. Normally it means just an invariant expression, such as

- for Boolean: `TRUE`
- for Character: `'X'` (mind the apostrophes, they are not decoration but are necessary to discriminate the letter 'X' from the variable X!)
- for Double: `3.1415927`

Some languages, such as PASCAL, allow the definition of *symbolic* constants in order to increase legibility: `const absoluteZero = -273`, for example. Symbolic constants are pretty similar to variables, except that they may not be changed (in JAVA, they *are* variables which are "frozen" by the keyword *final*). PYTHON does not comprise symbolic constants.

PYTHON is very logically structured in this respect:

- Where a constant is allowed, a variable is also allowed.
- Where a variable is allowed, a constant is also allowed *as long as this is not on the left-hand side of a '=' sign*.

> Σ A simple or scalar piece of data is an atomic piece of information which cannot be dissected further into its components.

5.3 Intermission: 0x07 – the hexcode never dies

Relevance: The decimal system is a human mode of thinking, whereas computers are based on binary algebra, which is more easily rendered using an octal or hexadecimal notation. Thus, an understanding of these systems is indispensable, even with high-level languages.
Keywords: Decimal, binary, octal and hexadecimal notation.

Basically, numbers can be represented in two different ways: by a system that is based on digits, or by a system that merely adds number signs. Of the latter, the Roman mode, in which I stood for 1, V for 5, X for 10, L for 50, C for 100, D for 500 and M for 1000, is probably the best known, though it is also the most complicated, with the relative position of the number signs determining whether they are to be added or subtracted (thus $IX = 9$ and $XI = 11$. The Greeks had a system which used all letters of their alphabet, plus diacritic marks ($\alpha' = 1$, $,\alpha = 1000$, $\beta' = 2$, $,\beta = 2000, \ldots$, $\psi' = 700$, $,\psi = 700\,000$, $\omega' = 800$, $,\omega = 800\,000$), and so had the Hebrews who prophesied that *"those who have understanding may reckon* The Number of the Beast, *for it is a human number; its number is six hundred and sixty six"* (meaning that the letters of the name, when read as numeric signs, can be added to yield 666).

The Number of the Beast

The alternative system based on digits was used only twice throughout history; once in the first millenium BC by the Mesoamericans who devised a vingesimal system, and once in the first millenium AD in India, where a decimal system was designed, which was then adopted by the Arabs about the year 1000, in memory of whom it is still known as "Arabian numerals", and then brought to Europe by the Italian mathematician Fibonacci in the 12th century.

We all know the decimal system. It basically means that every number is represented as a sum of powers of 10, the rightmost digit – the "least significant digit" – meaning $n \cdot 10^0 = n$, the one on its left $n \cdot 10^1 = 10n$, the third $n \cdot 10^2 = 100n$ and so on. Accordingly, in a Mesoamerican text the "least significant digit" means $n \cdot 20^0 = n$, the next one $n \cdot 20^1 = 20n$, the third $n \cdot 20^2 = 400n$, etc. In these systems, 10 and 20 are the *bases*. However, we can use any basis to represent arbitrary numbers, not just 10 or 20.

Internally, every computer uses a **binary** system, which comprises just the "glyphs" for 0 and 1, with the least significant digit amounting to $n \cdot 2^0 = n$, the second to $n \cdot 2^1 = 2n$, the third to $n \cdot 2^2 = 3n$...

Systems which have as basis a power of 2 are related to the binary system and therefore widely used in informatics. For example, there is the **octal** system, which uses the numbers from 0 to 7 (inclusive), with the least significant digit amounting to $n \cdot 8^0 = n$, the second to $n \cdot 8^1 = 8n$, the third $n \cdot 8^2 = 64n$... A widely-observed convention is that **numbers beginning with a 0 are in octal format**.

More important is the **hexadecimal** system ("hexcode"), which uses the numbers from 0 to 9 plus the letters A to F (A = 10, B = 11, ..., F = 15). It is more important because in hexadecimal writing two digits are sufficient for a byte, four for a word and so on. There are two representations of hexadecimal numbers; the one which is in the tradition of the ALGOL languages uses the dollar sign ($) as a prefix, whereas the tradition of the C languages (extendding to PYTHON) uses 0x as a mark for hexadecimal numbers.

So B1100100 = 0144 = 100 = 0x64:

	B1100100 (binary):								**100** (decimal):		
digit	2^6	2^5	2^4	2^3	2^2	2^1	2^0	digit	10^2	10^1	10^0
value	64	32	16	8	4	2	1	value	100	10	1
	1	1	0	0	1	0	0		1	0	0
cumulated	64	32	0	0	4	0	0	cumulated	100	0	0

	0144 (octal):				**0x64** (hexadecimal):		
digit	8^3	8^2	8^1	8^0	digit	16^1	16^0
value	512	64	8	1	value	16	1
	0	1	4	4		6	4
cumulated	0	64	32	4	cumulated	96	4

5.4 Homogeneous groupings

Relevance: The value of computers in science is mostly due to their ability to process large series of data without tiring. The data structures discussed in this chapter are the basis for this ability.
Keywords: Arrays, lists, strings and pointers.

The terms "homogeneous groupings" and "heterogeneous groupings" are not official but are used here to explain what these data types are about. These aggregates are always the plural form of one data type; this is expressed very clearly by some languages: e.g., in PASCAL we find `array of integer` and in HASKELL `[Integer]` (pronounced "list of integer").

- **Array**
  Mystically defined by Peyton–Jones et al. as "a function whose domains are isomorphic to contiguous subsets of integers"[10]. With less hype, an array – occasionally referred to as *vector* – may be defined as *an ordered sequence of values, each of which has a number*.
  It can be thought of as a street with a single name but separate houses for different values, which can be addressed by their number: e.g., when x is an array with 10 members we may refer to these as x_1, x_2, ..., x_9, x_{10} – in almost all programming languages[11] square brackets are used for indexing: `x[i]` means x_i. It is tempting to explain an array as a "list" of items, but also misleading, as a list is something different. Arrays were first introduced by FORTRAN, and they are now to be found in virtually any language.
  For example, when measuring gene expression on the single-cell level in a cytometer (p. 177), we determine the fluorescence of 10 000 cells one by one and store the values in fl_1 ... $fl_{10\,000}$.
  Arrays may have more than one dimension. As a two-dimensional array is represented internally as an array of arrays, there is no effective limit to this. However, uses for arrays with more than three dimensions are extremely rare[12].

  – *One-dimensional array:*

---

[10] For further information please phone your local supermarket.
[11] With the exception of CAML, where it is `x.(i)`, and HASKELL, where it is `x!!i`.
[12] As a general rule, there is nothing that cannot be used for something by someone. In the most extreme case, some lawyer may be found.

– *Two-dimensional array:*

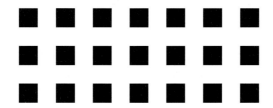

– *Three-dimensional array:*

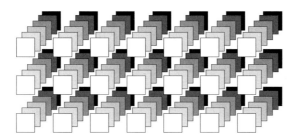

We have already mentioned array-oriented languages (p. 31). These are languages which put special emphasis on the processing of arrays, especially multidimensional arrays.

arrays as pointers

There is close kinship between arrays and pointers (p. 80) in that an array is generally handled via a pointer to its first element. This pointer (p) and the element size (s) are registered in the symbol table, so that the address of element n is simply determined as $p + s \cdot n$.

- **List**

List

By contrast, lists, first introduced by LISP (whose name is in fact an abbreviation for *List Processor*), have made their way into very few languages. The main difference from arrays is that *arrays are of fixed size, lists are of variable size.* For lists, there are functions like "append", "insert", "delete" or "concatenate" – for arrays, there are not. Lists are in fact concatenations of elements, each endowed with a pointer (p. 80) to its successor:

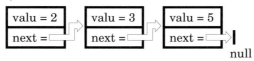

This makes lists rather difficult to implement and consequently slow to handle, whereas arrays are simply contiguous blocks of memory which can be accessed by single processor operations. Luckily, this is of little concern to the user of a language, who will find lists very handy when it comes to working with data of previously unknown size. You will perceive that insertion and deletion of elements, as well as extending and rearranging the list, are very easy to do, since there is no need for the elements to be placed vicinally or in any specific order.

In most programming languages lists are textually represented as runs of values, separated by commas and enclosed in square brackets; o'CAML alone clings to the traditional mathematical notion that a list is just a run of things separated by semicolons.

Usually, for lists there are also functions like "length" and "nth-element", but the classic – and most efficient – way of treating them requires only two functions: the *head* function yielding the first element of a list, and the *tail* function yielding the list without its first element, as in the following interactive HASKELL session:

```
[ophis@zacalbalam:~] ghci
GHC Interactive, version 5.04.2, for Haskell 98.
Prelude> let l = [1,2,3,4,5]
Prelude> l
[1,2,3,4,5]
Prelude> head l
1
Prelude> tail l
[2,3,4,5]
```

Recursive programming (see p. 159) makes extensive use of walking down lists using only these two functions. It is also very easy to see how these functions work.

It must be pointed out that in most programming languages lists may only comprise one type of data, whereas PYTHON, ERLANG and LISP allow arbitrary types to be united[13], including other lists (e.g., [23.666, [0, 8, 15], "Nobody"]), which may be used for the easy formation of nonlinear aggregations (e.g., trees). This may also be done with the more rigid languages such as HASKELL, but this needs a bit more background (see structures, p. 93, for details).

Furthermore, PYTHON does not distinguish clearly between arrays and lists – a fact sometimes considered as an advantage, sometimes as a drawback.

Important tools for lists are[14]:

append(x)
- append(x):
  Add item to end of list; in PYTHON equivalent to a[len(a):] = [x] or, better, a += [x].

extend(L)
- extend(L):
  Extend list by appending all items in the given list; in PYTHON equivalent to a[len(a):] = L.

insert(i, x)
- insert(i, x):

---

[13] Which is not astonishing as these languages are all weakly typed.

[14] Here, the PYTHON names are given, but they are similar in most languages which possess imperative list processing.

Insert an item at a given position. The first argument is the index of the element before which to insert, so a.insert(0, x) inserts at the front of the list, and a.insert(len(a), x) is equivalent to a.append(x).

– remove(x):
Remove the first item from the list whose value is x. It is an error if there is no such item.

remove(x)

– pop() *or* pop(i):
Remove the item at the given position in the list, and return it. If no index is specified, a.pop() returns the last item in the list. The item is also removed from the list.

pop() *or* pop(i)

– index(x):
Return index of first item in list whose value is x. It is an error if there is no such item.

index(x)

– count(x):
Return the number of times x appears in the list.

count(x)

– sort():
Sort the items of the list, in place.

sort()

– reverse():
Reverse the elements of the list, in place.

reverse()

– **Slice:**
The "slice" operator, written as a colon within the index expression (:), returns a sublist: x[a:b] will yield x_a to x_b, x[a:] will yield x_a to the end, x[:b] will yield from the first to x_b. Indices can be negative; then they will be counted from the end.

Slice

As said before, most languages allow lists and arrays to contain only one type of data. However, you may always define an array of records or variant records (see below) so that you are able to store alternative types.

- **Circular lists**

Circular lists

A toy of HASKELL where the last element is followed again by the first (its pointer not directed to "zero" or another terminal value but to the first element). This might be smart for the representation of circular genomes but does not exist in PYTHON.

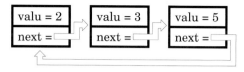

♠ *How could we represent and access circular genomes in* PYTHON, *then?*

- **String**

String

A sequence of characters, usually implemented as an array but sometimes (in ERLANG and partly also in PYTHON) as a list of characters. Languages which put special emphasis on text processing (e.g., PERL) add to the novice's confusion by insisting on treating strings as simple (scalar) values, which they in fact are not. **Please see p. 224 for more on this topic.**

As said before, PYTHON is not very exacting about the difference between arrays and lists; most operators work on both. At any rate, we have the following composition of a string:

$$x = \text{"Anna"}$$
$$x_1 = \text{'A'}$$
$$x_2 = \text{'n'}$$
$$x_3 = \text{'n'}$$
$$x_4 = \text{'a'}$$

For the molecular biologist, strings are invaluable, as they are the natural representation for any kind of linear sequence (DNA or protein). Most of our work will employ strings.

String constants are pieces of text which are included in single or double apostrophes. For example, `"Anna the hannah"` is a valid string constant but `"Anna" the hannah"` is not. Note that there is a special regulation concerning the backslash (\) within string constants (so-called escape processing, p. 337); for the time being, it will be best just to avoid it when experimenting.

There is a special pitfall in PYTHON. Since strings and lists are so similar, the program will not complain when list functions are applied to a string. Now, if we have a function which is supposed to process a list of strings, and for some reason we do not feed a list of strings but a single string into it (which may happen quite easily), the program will not realize that there is in fact only one item to be processed, but instead will digest the string:

```
[ophis@zacalbalam:~] python
Python 2.2.2, bungled by Ophis.  GCC 3.1 on darwin.
>>>def first( list ):  return list[ 0 ]
...
>>>first( [ "alpha", "beta", "gamma" ] )
'alpha'
>>>first( [ "alpha" ] )
'alpha'
>>>first( "alpha" )
'a'
```

Dictionary

- **Dictionary**

Sometimes also named "associative array" or (somewhat bewildering to the German-speaking reader) "hash tab" – a variant of an array where indexes ("keys") are not restricted to numbers. Typically, strings are used as keys, just as in what we generally know as dictionary:

```
vorname = {} # this initializes the variable as an empty
dictionary
```

```
vorname["Holzbock"] = "Hermann"
vorname["Abendroth"] = "Abel"
vorname["Flaig" ] = "Rüdiger"
```

In PYTHON, it is easy to inspect such data types:

```
>>>print vorname
{'Abendroth': 'Abel', 'Flaig': 'Rüdiger',
'Holzbock': 'Hermann'}
>>>vorname[ "Abendroth" ]
'Abel''
>>>vorname["Hussein"]
Traceback (most recent call last):
File "<stdin>", line 1, in ?
KeyError:   'Hussein'
>>>"Hussein" in vorname.keys()
False
>>>"Holzbock" in vorname.keys()
True
```

This is a very high-level structure which may require a hundred to a thousand times more processing time than a genuine array. It is therefore to be found only where speed does not matter overmuch, e.g., in PYTHON or PHP. On the other hand, it is extremely useful for managing complex data. For this reason it is highly advisable to use dictionaries where they are a natural choice.

The first priority must always be stability, together with legibility. Speed is secondary. A program that is not reliable may be worse than no program at all, because wrong results may cause disasters, and an inclination to crash will cost people's time and nerves. A program that is not legible and extensible is *trash*. Do not waste your time devising "workarounds" to avoid the features of programming languages you use, because in some other programming language this feature is not existent or deprecated, because your dog does not like them, or because you have heard rumours that "in America" everything is done differently. Use high-level structures whenever they are available, and don't worry too much about speed. You will find that this will greatly help you to write legible, extensible and *efficient* algorithms. As Donald Knuth, author of LaTeX and a number of fundamental books on information technology, once said: "Early optimization is the root of all evils." In high-level languages such as Python, you may trust in the developers' ability to provide you with well-written solutions. The generally low quality of

trash

contemporary software must be blamed largely on the writers' attention being focused more on nanoseconds than on good writing.

PYTHON offers a further sophistication: so-called "shelf objects". A shelf object (implemented in the module shelve) is basically a "persistent" dictionary – a dictionary whose components are stored in an external file – a complete database! Please see p. 343 for details.

Note here the use of the # sign in the first of these lines. This is a *comment marker* which tells the interpreter to simply ignore the rest of the line. You will read more about this on p. 105.

♠ *Generate an empty dictionary by toplevelling* a = {} *and request information about the associated functionality with* dir(a). *You will see a lot of functions. Request help for those which you deem interesting, but especially for* values *and* keys. *Try to imagine how these could be used for traversing a dictionary.*

- **Set**

Set

A set is an unordered assemblage of a given data type (e.g., set of char or set of integer), very similar to the mathematical notion of sets. As its element are unique (trying to put 2 into a set which comprises 1, 2 and 3 simply will not change the set), it is especially useful to classify a variable value:

```
(* Pascal, Modula & Co.   *)
var legal :  set of char;
legal = [ 'y', 'Y', 'n', 'N' ];
write( 'Please answer Y(es) or N(o)' );
ch = getKey;
if not (ch in legal) then writeln( 'Hey man, Y(es) or
N(o)!!' );
```

In PYTHON, *sets have been added in version 2.4. They can be used pretty much like lists, with the exception that they are denoted by the* set *function rather than brackets:* a = set ([1, 2, 3]) *will assign a set of the three elements* 1, 2 *and* 3 *to* a. *(A syntax using curly brackets { } was proposed but rejected.)*

Bag

- **Bag**
This differs from the set in that elements do not have to be unique... in other words, just an unordered list.

Stack

- **Stack**
Sometimes termed LIFO for "last in, first out". This is a structure which is, on the lower level, used to excess by the processor for dealing with subroutine (function) calls. The idea is that you have two ways of accessing a pile of data that resembles a stack of playing cards: PUSH places an additional item on top of the stack, POP takes the top item

down. Usually there is a third function SIZE which tells us how many items there are on the stack.

– *Initial situation:*

– *Push:*

– *Push:*

– *Pop:*

– *Pop:*

- **Pipeline**

  Sometimes termed FIFO for "first in, last out" – a queue of data items which await processing. You push data at the one end and pop them at the other.

  – *Initial situation:*

  – *Push:*

  – *Push:*

  – *Pop:*

  – *Pop:*

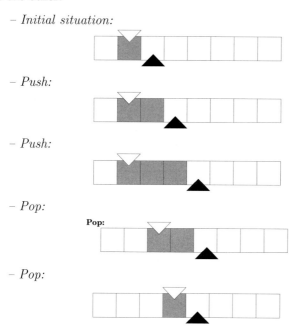

♠ *You have heard about arrays and lists. Which one is more suitable for implementing a pipeline? Why?*

- **Tree**
  An entire group of complex structures which serve for organizing data in a way that is optimized for fast access. We will discuss this in Vol. 2.

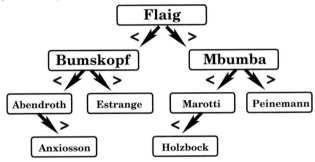

- **File**
  This data type is somewhat obsolescent (COBOL and PASCAL possessing it) but should be mentioned here for the sake of completeness. It is based on the idea that a file is usually a collection of isomorphic "data sets", i.e., a homogeneous aggregation which differs from list and array in that it is actually not kept in memory but outsourced to some persistent (usually magnetic) medium. When turning back to our address file example (p. 32), you will find that there we have defined not only the *addressEntry* data type but also *adressfile* as "file of address". Such aggregate-oriented file handling made perfect sense when memory was limited and persistent storage slow (e.g. magnetic tape). The disadvantage is the lack of flexibility – all data sets must be of identical shape. In the XML age, this data type is rarely used.

PYTHON supports strings and a "one-for-all" aggregate type which is list, array and hashtab at the same time. There is no fixed-size file type[15]; it uses either text files (consisting of lines) or "pickled" files (p. 343), which both offer considerably more flexibility.

 Homogeneous groupings are associations of (simple or compound) data items which convey the notion of "multiplicity". Characteristically, the different members of such an association are anonymous and distinguishable only by an index number.

What in this book is usually referred to as "grouping" is often less concisely termed *abstract data type*: An array, a list, a tree, a stack, etc., is a structure in its own right, independent of the type of data it comprises. Higher programming languages provide abstractions with regard to groupings. The grouping

---

[15] However, PYTHON possesses an interface to low-level file handling, from which such a file type could be constructed if necessary. For the tasks for which PYTHON is eligible, it generally is not.

and its elements can be addressed in identical fashion independent of the actual nature of data, even though, e.g., arrays of integers, arrays of strings and arrays of structures (see below) require very different handling by the processor.

5.5 Intermission: a note on assignment

There is one very important thing to say about '='. From mathematics, we know that this denotes equality. However, in most programming languages including Python it means something completely different. In mathematics, x = x + 1 is obviously absurd; in programming, it means: "Add 1 to the current value of x" – a very, very common thing. The more mathematical-minded language designers have tried to eschew this and used := (Pascal) or <- (o'CAML) instead, reserving the = for comparisons. However, most have followed the example of C and use = for assignment and == for comparison – a choice which is, strictly speaking, not that bad because assignment is a much more frequent thing than comparison.

Finally we have

5.6 Heterogeneous groupings

Relevance: In most cases, a datum is not an individual bit of information but a plurality of those bits which are logically linked, e.g., by referring to different aspects of the same thing, even though they may each be of a different nature. This being logically linked is reflected in that data structures comprise a plurality of such bits.
Keywords: Structures/records, objects, ...

- **Structure**

    Turn back to p. 32 to review what we have said about structures (records). In PYTHON, they are just special cases of objects – objects without code components. One thing must be added here. We have briefly mentioned pointers and references before (p. 80). Any structure may, of course, contain such references to external data. For example, a structure may contain a reference to another structure of the same type, as a page in a book may contain a cross-reference to another.

    This is how lists and trees (Vol. 2) are actually implemented (unlike arrays, which are just blocks of memory): every list item is connected to one or more others. Please turn back to p. 33 where arrays and lists are illustrated. Access proceeds in the direction of the pointers, so for some purposes it is useful to have a *bidirectionally* linked list where every item contains fields "previous" and "next". A traditional LISP/PYTHON style list, however, is linked *unidirectionally*. Every element contains a pointer to its successor (or to zero at the end of the

Structure

list). The *head* function returns the element's contents, the *tail* function the successor pointer[16]. This is why *head* yields a single element, *tail* a list even if there is only one successor (or none at all).

By now we know, or at least are supposed to know, that PYTHON is a weakly-typed language (see p. 53) and does not provide any mechanism for declaring variables. So how do we define the fields of a structure? The simple answer is: We don't – because we don't have to.

A PYTHON structure is in fact an empty object generated by

```
class obj:  pass        # a class of empty objects
structurename = obj() # instantiate (produce object of
                                       that class)
```

Data members can be added at random. There is no need for them to be congruent. Turning back to our address file and that matter about the local representative. Such a representative will be needed only for very few, if any, members. No need therefore to reserve space in each structure for the naming of a representative. Instead, we can plainly write

```
boss.representative = ...
```

wherever and whenever this seems suitable.

Of course, we would get into an awkward situation if trying a field which does not exist in a given structure. To avoid this, PYTHON is equipped with rather sophisticated devices which test for the existence of a field, or return a list of all the fields in a structure.

```
>>> class obj: pass
...
>>> one = obj()
>>> dir(obj)
['__doc__', '__module__']
>>> one.owner
Traceback (most recent call last):
  File "<stdin>", line 1, in ?
AttributeError: obj instance has no attribute 'owner'
>>> one.owner = "American Association of Headbangers"
>>> one.owner
'American Association of Headbangers'
>>> dir(one)
['__doc__', '__module__', 'owner']
>>> "owner" in dir(one)
True
>>> "doener" in dir(one)
False
```

Variant record (union)

- **Variant record (union)**
  This is best illustrated using an example: The PASCAL definition

---

[16] Sometimes head and tail are also referred to as the "car" and "cdr" of a list. This goes back to the very first implementations of LISP, when these operations were performed using different parts of a certain register.

```
type intl_address = record
                fullname, houseandstreet, city: string[ 50 ];
                case nation: ( germany,  uk ) of
                    germany: ( plz: 00000..99999 );
                    uk     : ( bpc: string[ 8 ] );
        end;
```

obviously includes fields for a person's full name, city, house and street and also for the nation – an enumerative type with two values. However, the rest of the record may differ: Depending on nation, there is *either* "plz" as an enumeration (for the ... token, refer to p. 35) *or* "bpc" as a string of up to 8 characters. These two variables do not coexist. In C, a similar instrument is known as "union".

– *A structure or record is a compound comprising a plurality of simple data elements:*

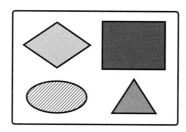

Structure or record

– *An array is a sequence of a defined number of data elements, all of which are of the same type. This type may be simple:*

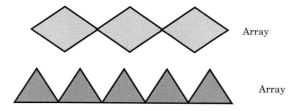

Array

Array

– *An array may also consist of structures or records, all of which must be of the same type:*

Array of structures

– *A list is an association of compound elements, wherein each list member comprises a reference to one other element. The number of list elements is not defined; the end of the list is indicated by the last reference being manifestly invalid. Whereas in an array any element may be accessed simply by its number (index = "house number") within the array, the list consists of "this element" and "all that follows":*

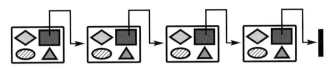

List (of structures)

– A list may be doubly linked to facilitate traversing in both directions (i.e., we can move not only head→tail but also tail→head):

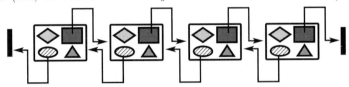

Doubly linked list (of structures)

- **Tuple**

  A tuple is a rather loose affiliation of data, characteristic of a high-level language. It is similar to a record except that its components do not have names (they are just ordered sequentially); or, it is similar to a list except that it is of finite length and may always comprise different types of data. In PYTHON, tuples are sequences separated by commas, and they may or may not be enclosed in parentheses. The interpreter prefers parentheses and will always print them when dumping a tuple.

  Try this:
  ```
  >>>a=1,2,3
  >>>a
  (1, 2, 3)
  >>>x,y,z=a
  >>>y
  2
  ```

  The first line of this is puzzling at first glance, and in fact all lower-level programming languages would plainly refuse this. In PYTHON, this can be read as: "Assign to *a* a packed value which consists of the sequence 1, 2, 3." How this can be passed around and unpacked is seen in the following lines.

  Tuples are handy for returning more than one value at a time, and for simplified handling of multiple variables:
  ```
  >>>import math
  >>>def trigon(x): return math.sin(x), math.cos(x), math.tan(x)
  >>>y=trigon(math.pi)
  >>>y
  (1.2246467991473532e-16, -1.0, -1.2246467991473532e-16)
  >>>hick, hack, schnack = 0, 8, 15
  >>>hack
  8
  >>>foo, bar, blah=trigon(math.pi)
  >>>bar
  -1.0
  ```

```
>>>a = 3
>>>b = 5
>>>a, b = b, a
>>>a
5
>>>b
3
```

The textual representation of tuples generally varies; in O'CAML, they are similar to the PYTHON notation, whereas in ERLANG they are enclosed in braces: {"Anna", "Naja regalis", "University of Heidelberg"}. You notice the similarity to records.

- **Object**   Object
  ♠ *Define this in your own words.*
- **Class**   Class
  The "type" of an object.

All these may be combined at the programmer's pleasure, so it is not uncommon to have an array of structures or a list of tuples.

♠ *Think of a catalogue of restriction enzymes (with name, recognition sequence, and price for each), a telephone book, a GenBank entry, and a diary (in the interest of simplicity, let each entry be a single string). Which kinds of data structure would you use?*

 Heterogeneous groupings are associations of (simple or compound) data items which convey the notion of "forming a larger whole". Characteristically, the different members of such an association are accessible by a name.

5.7 Strings

Relevance: Whereas computers essentialy work on numbers, the human mind has evolved to deal with language – sequences of phonetic or visual tokens capable of conveying meaning. Thus, the ability to deal with such sequences is pivotal to computers being useful in supporting the tasks with which the human mind is faced. In bioinformatics in particular, sequences of nucleic or amino acids are data of utmost importance.
Keywords: Strings.

Strings – sequences of characters forming words in the human sense – have been mentioned before (p. 87). They are extraordinary in that they have features of both simple and aggregate data types. On the one hand, they are clearly compounds of several, sometimes very many, elements of type character. On the other, they are normally handled *en bloc*, unlike arrays of numeric values or structures.

It depends on the language whether the handling of strings bears more resemblance to that of simple or of aggregate data types. In PERL, for example, strings are scalars – natural for a language which is concerned mostly with higher-level string processing. Likewise, BASIC does not even have the character as a distinct data type: a BASIC character is just a string of length 1. On the contrary, a C string is just a pointer (type `char *`); string processing functions work their way through memory until they reach a zero character. It is up to the programmer to make sure that the memory used by the string functions is actually available, and also to release the memory after usage. This is what makes it especially difficult to work with variable-length strings in C. In HASKELL, the expressions `[char]` (a list of characters) and `string` are equivalent; here the interpreter takes care of memory handling, but still a HASKELL string is basically an aggregate data type, which does actually not mean much, given HASKELL's excellent features for handling lists.

JAVA and the scripting languages combine the two approaches, by using objects to represent strings. Please see p. 224 for more.

Strings are what we know as "words" – sequences of letters belonging together. They are problematic because they are simple from the human point of view, but homogeneous aggregates from the computer's point of view.

5.8 Assignments and unary operators

Once more back to the central notion of variables. There are two concepts:

Binding

- **Binding**
  This is used by the purely functional languages (e.g., HASKELL). An identifier[17] is defined by "binding" it to an expression. That is to say, when the source reads `let c = sqrt( a*a + b*b )`, this is a definition of c which must not be changed. The expression `let c = c + 1` is obvious nonsense, as is the attempt to define c without binding some value to it: it would mean "c is one more than c". ERLANG allows re-binding of variables to a different expression, but in neither of these is there any means of "returning" to a previous point in the program.

Assigning

- **Assigning**
  By contrast, the vast majority of languages – including PYTHON, JAVA and C++ – are based on the notion that a variable is a "container" for a value. You reserve the container by declaring it[18], you put in a value, you use that value, and if you feel like it, you replace this value by a new one. That is what is meant by "states". Thus, `c = c + 1` just means: Increase the value of c by one. The functional mujaheddin have coined the pejorative term "destructive assignment" for this.

---

[17] The term "variable" is misleading in this context, as the thing is not variable.

[18] In this case, the content is unpredictable – sometimes it is initialized to zero and sometimes not! JAVA compilers are smart enough to issue a warning if the value of a variable is used before it has been assigned some content.

Now it is obvious that you may assign two kinds of values to a variable: values that do not depend on the previous value of the variable (e.g. c = math.sqrt(a*a + b*b)), and values that do (e.g., c = c + 1). The latter case is so frequent that there are special operators to simplify this:

Binary form	Unary form	Explanation	Languages
a += n	a = a + n	Add n to a	C, C++, JAVA, PYTHON
a++ *or* ++a	a = a + 1	Add 1 to a	C, C++, JAVA
a -= n	a = a - n	Subtract n from a	C, C++, JAVA, PYTHON
a-- *or* --a	a = a - 1	Subtract 1 from a	C, C++, JAVA
a *= n	a = a * n	Multiply n with a	C, C++, JAVA, PYTHON
a /= n	a = a / n	Divide n by a	C, C++, JAVA, PYTHON
a **= n	a = a ** n	Set a to a^n	PYTHON

...and many more (especially for bitwise operations in C, which we are not going to discuss here).

Note that in PYTHON the + operator is also used for string and list concatenation.

♠ *You may also apply ⋆ to a list or string. Any idea what that does? If not, any idea how to find out?*

5.9 Overview

The following table gives a quick overview about features and data types available at the elementary language level of a number of the better-known languages. Of course, "add-ons" may provide additional functionality (e.g., lists in JAVA, arrays in LISP and HASKELL), but in general, these are clumsy and difficult to handle[19]. For languages as ill-defined as BASIC, there may also be differences between dialects.

"TC" = type classes, "Func" = functional variables. "Hash" = dictionaries or associative arrays (hash tables). "Byte, Word, etc." = low-level data types such as byte, word, long, short, unsigned.

	Bool	Int	Byte Word etc.	Float	Enum	Atom	Sub-type	TC	Pointer	Func	Array	List	String	Hash	Set Bag	Struct	Object
TRAN	-	⊗	-	⊗	-	-	-	-	-	-	⊗	-	⊗	-	-	-	-
IC	-	⊗	-	⊗	-	-	-	-	-	-	⊗	-	⊗	-	-	-	-
TH	-	⊗	⊗	⊗	-	-	-	⊗	-	-	-	-	-	-	-	-	-
L	-	-	-	⊗	-	-	-	-	-	-	⊗	⊗	⊗	-	-	-	⊗
A	⊗	⊗	-	⊗	-	-	-	-	-	-	⊗	-	⊗	-	-	-	⊗
ANG	-	⊗	-	⊗	-	⊗	-	-	⊗	-	⊗	-	-	-	⊗	-	-

---

[19] This also holds true for arrays in PYTHON, which are restricted to a very limited data range and therefore not used in the entire course.

C	–	⊗	⊗	⊗	⊗	–	–	–	⊗	–	⊗	–	–	–	–	⊗	–
Python	⊗	⊗	–	⊗	–	–	–	–	⊗	–	⊗	⊗	⊗	⊗	–	⊗
Lisp	–	⊗	–	⊗	–	⊗	–	–	⊗	–	⊗	⊗	⊗	–	⊗	–
Pascal	⊗	⊗	–	⊗	⊗	–	–	–	⊗	–	⊗	–	⊗	–	⊗	⊗	–
C++	⊗	⊗	⊗	⊗	⊗	–	–	–	⊗	–	⊗	–	–	–	–	⊗	⊗
Haskell	⊗	⊗	–	⊗	⊗	–	–	⊗	–	⊗	–	⊗	⊗	–	–	⊗	–
Modula-2	⊗	⊗	⊗	⊗	⊗	–	–	–	⊗	–	⊗	–	⊗	–	⊗	⊗	–
Ada	⊗	⊗	⊗	⊗	⊗	–	⊗	–	⊗	–	⊗	–	⊗	–	–	⊗	⊗
Oberon	⊗	⊗	⊗	⊗	⊗	–	⊗	–	⊗	–	⊗	–	⊗	–	⊗	⊗	⊗
o'CAML	⊗	⊗	⊗	⊗	⊗	–	–	–	⊗	⊗	⊗	⊗	⊗	⊗	–	⊗	⊗

PERL does not discriminate between integers, floats and strings – they are simply "scalars".

In spite of all its complexity, ADA is lacking some of the most powerful structures known, such as lists and functional variables. It has, however, different types for strings of fixed and of variable length, the former being faster in their handling.

5.10 Digression: orthogonality

The term "orthogonality" will be familiar from elementary geometry. It means that two structures form a right angle (90°), i.e., one that is equal from both possible points of view. In informatics, this term is often used to denote that of a given group of elements, all the members are endowed with the same attributes, such as in the following table.

	.A	.B	.C	.D	.E	.F	.G
X_1.	⊕	⊕	⊕	⊕	⊕	⊕	⊕
X_2.	⊕	⊕	⊕	⊕	⊕	⊕	⊕
X_3.	⊕	⊕	⊕	⊕	⊕	⊕	⊕
X_4.	⊕	⊕	⊕	⊕	⊕	⊕	⊕
X_5.	⊕	⊕	⊕	⊕	⊕	⊕	⊕
X_6.	⊕	⊕	⊕	⊕	⊕	⊕	⊕
X_7.	⊕	⊕	⊕	⊕	⊕	⊕	⊕
X_8.	⊕	⊕	⊕	⊕	⊕	⊕	⊕

This, of course, is applicable only to homogeneous groupings, each member of which is an heterogeneous grouping, i.e., to lists of structures, arrays of structures or files of structures, such as arrangements where a data item can be described as item y of structure x. In the description of heterogeneous groupings given above, orthogonality has been taken for granted, as this makes things both easier to understand and to handle[20]. Still it should not be forgotten that a grouping of groupings may well be deficient in orthogonality, as in the following example.

	.A	.B	.C	.D	.E	.F	.G
$X_1.$	⊕		⊕		⊕	⊕	
$X_2.$		⊕		⊕		⊕	
$X_3.$		⊕	⊕	⊕		⊕	
$X_4.$		⊕		⊕	⊕	⊕	
$X_5.$	⊕	⊕	⊕	⊕		⊕	
$X_6.$	⊕	⊕	⊕		⊕	⊕	⊕
$X_7.$		⊕	⊕	⊕	⊕	⊕	
$X_8.$		⊕		⊕	⊕		⊕

Obviously, programming languages with a rigid typing system, such as the C and ALGOL family languages, do not really make provisions for such a situation but rather enforce orthogonality (a notion which is actually incompatible with the concept of "inheritance" (p. 196), a fact that accounts for some of the shortcomings of C++ and JAVA). In the C and ALGOL languages, one will use the most comprehensive structure design imaginable and simply leave blank (or flag as unused) the absent members.

In the object-oriented among the strongly typed languages, non-orthogonal data may be implemented if they are systematically related.

	X coordinate	Y coordinate	Colour	Radius	Radius 2	Beg. angle	End angle
Point	⊕	⊕	⊕				
Circle	⊕	⊕	⊕	⊕			
Ellipse	⊕	⊕	⊕	⊕	⊕		
Ellpie	⊕	⊕	⊕	⊕	⊕	⊕	⊕

Here, "circle" can be derived by inheritance (p. 196) from "point", "ellipse" from "circle" and so on.

It is just as obvious, on the other hand, that in weakly typed languages non-orthogonal data come naturally. The weakly-typed languages emphasize the "associative" nature of heterogeneous groupings. Of course, *all* heterogeneous

---

[20] Orthogonality of a system such as a programming language is Programmer's Paradise, for it allows one to develop efficient and elegant solutions. It is noteworthy that processor designs which have been recognized as particularly elegant are characterized by a high degree of orthogonality, which is used here to say that basically all operations may be applied to all registers in the same way, and that all modes, such as addressing modes, are likewise equally applicable to all instructions (the expression is well established, but arguably "symmetry" might still be more appropriate). Thus, as early as the 1970s PDP-11 was prized for its high degree of orthogonality, and traditionally RISC processors are almost perfectly orthogonal. By contrast, early x86 processors were completely lacking orthogonality, as were the 8-bit processors such as the Z80 and the 6502, as in these machines each of the (very few) registers was associated with a particular group of instructions. This, however, was not due to sloppy engineering but a deliberate feature introduced to minimize complexity at a time when the art of processor design was less advanced than it is today.

groupings are essentially associative in that in any heterogeneous grouping certain "keys", i.e. the names of the compounds (such as $.A, .B, \ldots$ in the table above) are asociated with particular values. Thus, in the primal example of a structure given on p. 33, the key *firstName* is associated with "Dolores", the key *lastName* with "Peinemann" and so on.

The attentive reader will note that this is quite similar to the thing described as "associative array" or "hash" (p. 88), and indeed PYTHON internally uses the same mechanism for the implementation of structures (objects) and associative arrays. As for LISP, structures were a part of this language from the very beginning, implemented simply as lists of lists wherein each constituent list comprised, as lists usually do, a first element and a rest, in which the first element was considered as the key and the rest as the value:

```
ophis@horcrux:~/dox/lehrbuch$ clisp
[1]> (setq boss '((membershipNumber 23) (firstName "Dolores") (lastName "Peinemann")
     (postalCode 69345) (city "Waldwimmersbach") (membershipType 'regular)))
((MEMBERSHIPNUMBER 23) (FIRSTNAME "Dolores") (LASTNAME "Peinemann")
 (POSTALCODE 69345) (CITY "Waldwimmersbach") (MEMBERSHIPTYPE 'REGULAR))

[2]> (defun item (x y) (if (null x) NIL (if (eq (caar x) y) (cdar x) (item (cdr x) y))))
ITEM

[3]> (item boss 'membershipNumber)
(23)

[4]> (item boss 'city)
("Waldwimmersbach")

[5]> (item boss 'IQ)             ;; this element is currently absent
NIL

[6]> (defun new-item (x y value) (cons `(,y ,value) x))
NEW-ITEM

[7]> (setq boss (new-item boss 'IQ 70))
((IQ 70) (MEMBERSHIPNUMBER 23) (FIRSTNAME "Dolores") (LASTNAME "Peinemann")
 (POSTALCODE 69345) (CITY "Waldwimmersbach") (MEMBERSHIPTYPE 'REGULAR))

[8]> (item boss 'membershipNumber)
(23)

[9]> (item boss 'IQ)
(70)

[10]>
```

Here, (2) means: "Item y of list x is defined as follows: If x is empty, there is no y. Otherwise: If the first member of the first member of x is equivalent to y, the item is the rest of the first member of x; otherwise it is item y of the list minus its first member"; and (6): "A new list is the new element and value plus the old list."

This brief (and recursive) definition is completely sufficient to implement heterogeneous groupings in LISP or any related language. As can be seen from (7) to (9), new elements can be added at will.

Error in operator: add beer...

CHAPTER 6

Flow Control

"Can I help you?" Gordon asked. "Are you looking for anything special?"
"No, thank you vewy much, I only want to bwowse", the nancy answered in his soft R-less nancy voice. "I just flowed in, tee-hee – I love bookshops."
Well, my boy, then just flow out again, Gordon thought.

– GEORGE ORWELL: "Keep the Aspidistra Flying"

6.1 Comments

Relevance: Understanding how and why a program should be made intelligible for other human minds as well.
Keywords: Comments and docstrings.

Actually, comments have nothing to do with flow control, but they are a prominent feature of any well-written source text.

A comment is a portion of the source code that is marked for the compiler as "ignore this". Instead, it contains annotations to make the program more legible. There are two kinds of comments: *one-line comments*, which begin at a sign and extend to the end of the line, and *multi-line comments*, which begin and end at a given sign:

- PYTHON: # one-line comments
- C, C++, JAVA, etc.:
  /* multi-line comments */ and also // one-line comments
- PASCAL, O'CAML: (* multi-line comments *)
- HASKELL: {- multi-line comments -} and also -- one-line comments
- LISP: ;; one-line comments
- ERLANG: % one-line comments
- BASIC: REM one-line comments, also known as "remarks"
- HTML (WWW standard format; p. 45):
  <!-- multi-line comments -->

♡ *Beginners tend to underestimate the importance of comments. Some teachers, however, go as far as claiming that "uncommented code is to be considered as unwritten code". For even with a rational choice of variable and function names (which is not always the case) it is extremely difficult to understand other* people's code, and after a few months, one will find it hard to understand one's own code, unless it is well commented and documented. HASKELL *takes this to the extreme by recommending a compiler mode named "Literate* HASKELL*" where in fact the comment is the standard and code lines must be explicitly marked as such. So please, get accustomed to commenting generously!*

D. Knuth (author of LATEX): "Literate Programming (1984)" in Literate Programming. CSLI, 1992, p. 99:

> Let us change our traditional attitude to the construction of programs: Instead of imagining that our main task is to instruct a computer what to do, let us concentrate rather on explaining to human beings what we want a computer to do.
>
> The practitioner of literate programming can be regarded as an essayist, whose main concern is with exposition and excellence of style. Such an author, with thesaurus in hand, chooses the names of variables carefully and explains what each variable means. He or she strives for a program that is comprehensible because its concepts have been introduced in an order that is best for human understanding, using a mixture of formal and informal methods that reinforce each other.

Robert Dunn. "Software Defect Removal". McGraw-Hill, 1984, pg. 308.

<small>true functional modularity</small>

> Common sense also leads us to the recognition of the characteristics of programs that makes the programs maintainable. Above all, we look for programs that exhibit logical simplicity – failing that, at least clarity. The earmarks of simplicity and clarity include modularity (*true functional modularity*, not arbitrary segmentation) and a hierarchical control structure, restrictions on each module's access to data, structured data forms, the use of structured control forms, and generous and accurate annotation.

David Zokaities. "Writing Understandable Code". Software Development, January 2002, pg. 48–49.

<small>people need to understand the software</small>

> Software must be understandable to two different types of entities for two different purposes. First, compilers or interpreters must be able to translate source code into machine instructions. Second, *people need to understand the software* so they can further develop, maintain and utilize the application. The average developer overemphasizes capability and function while undervaluing the human understanding that effects improved development and continued utilization. There should be a description in clear view within the programming medium.

6.2 Docstrings

Another obvious exception from normal flow is frequently seen in well-written PYTHON code:

```
def foo( self, bar ):
    "Implement the fubar algorithm according to Dr. Mabuse,
    J. Low End Res. 1982. The argument must be an object
    of class 'WhiteRabbit'."
    annoyance1 = self.hands.putInPocket();  # demonstrate self-confidence
    annoyance2 = bar.owner.poodle.kick();   # 2do: make sure it's not
                                            # a rottweiler
    annoyance3 = bar.owner.spouse.kiss();   # only if self.sex
                                            #    == bar.owner.sex
    ...
```

You see that there is a string in a position where it actually does not seem to make much sense – after all, PYTHON is imperative, and values without operations attached to them are verboten[1]. So what is this?

A string immediately following a *def* (always make sure it is indented properly!) is exempt from normal flow and is used as a so-called **docstring** – a string serving for doumentation purposes. The PYTHON interpreter carefully keeps track of every bit of code it loads; functions like `dir` and `help` should by now be familiar to you. The docstring is saved as an attribute named __doc__(); you may access it either directly, as `foo.__doc__`, or by `help( foo )`. Accustom yourself to using docstrings from the very beginning!

♡ *Try this at the toplevel (mind the proper indentations), or write it to a file and run that:*

```
class c:
    "complex numbas"
    def __init__(self,img,real):
        "generate one"
        self.i, self.r = img, rea
co = c(3,7)
print dir(c)
print c.__doc__
print co.__doc__
print co.__init__.__doc__
help(co)
help(c)
```

Have a look at p. 211 to find how this is put to use in a real-life example. In the interest of completeness, it should be noted that docstrings made their first appearance in SMALLTALK.

---

[1] Mark this fundamental structural difference between functional and imperative languages: In all functional languages, beginning with LISP, a value that is not involved in anything is considered as a function's return value; the seemingly similar O'CAML line `let foo self bar = "Implement the fubar algorithm..."` makes *foo* return the string "Implement..." upon invocation!

> Comments are indispensable for the legibility of a program – either by others, or by oneself a few months later.

6.3 Blocks

Relevance: As in any natural language, words form sentences, while instructions or definitions form blocks. Both are non-arbitrary associations with an order determined by the grammar of the language which heavily influences the sense, which is the unifying principle of the block. Blocks are logically coherent in that the instructions or definitions forming them serve a common purpose.
Keywords: Blocks.

block marks
significant whitespace

Instructions or equations, the atomic units of programming, rarely walk alone. Usually, several of them are associated and executed under the same condition, usually sequentially in the order in which they are written. Such aggregates must be marked as such. There are two types, namely *block marks* and *significant whitespace*. (For a long time, significant whitespace has been considered as the hallmark of obsolescent languages: in Fortran and Cobol, which were originally still written on punch cards [!], indentation is an important part of the syntax[2]; "modern" languages, beginning with Algol, were those in which the programmer is free to arrange the source code as he pleases. Recently, the significant whitespace has risen again – obviously because it promotes legibility.)

6.3.1 Block marks. If more than one instruction is to be referred to, the block is surrounded by begin and end signs. The C languages, PERL and RUBY use braces ({ and }) to this purpose and the ALGOL languages use the words BEGIN and END. In CAML, we may use either BEGIN and END or parentheses ((and)).

A minority of languages use nonce-terms corresponding to the opening clause of the block to signify its end. For example, SHELL has a DO...DONE structure, and ORCA (a PASCAL derivative) and the mathematical language MAPLE uses DO...OD. In COMAL, a conditional loop (see below) is enclosed into WHILE...ENDWHILE – a homage to BASIC's WHILE...WEND. In MODULA-2, every IF, WHILE or CASE clause (see below) is expected to implicitly begin a block which then ends unceremoniously with END.

---

[2] Update: In FORTRAN, no more so since FORTRAN-90.

Where variables must be declared, this usually happens at the beginning of any block, restricting the variables to this block[3]. See p. 154 for more on this subject.

6.3.2 Significant white space. A different strategy is used in HASKELL, OCCAM and PYTHON: a block consists of all lines which have the same indentation. Of course, this restricts the freedom of arrangement. One statement per line is a "must".

Blocks are usually subordinate to conditional or looping clauses, the discussion of which will occupy the rest of this chapter. If a block is subordinate to this clause, its execution depends on the clause. The clause governs the execution of the block as a whole. A block which is subordinate to an "if" clause is either executed as a whole or not at all, depending on the clause.

Of course, every block may contain sub-blocks. Therefore a block may be described as a *recursive data structure*, and it is processed most easily by *recursive descent*. We will discuss this later. Thus, blocks, even those subordinate to conditional or looping clauses, may again contain conditional and/or looping clauses to which further sub-blocks are subordinate.

recursive data structure
recursive descent

♠ *Look at "class c" and explain the block structure.*

6.4 The IF... THEN... ELSE... construct

Relevance: Like life forms, computers must be able to react to differing conditions. The primary instrument to this end is the if/then/else statement.
Keywords: If, then, else.

6.4.1 Conditions. All programming languages must have the ability to do different things in reaction to different conditions. Otherwise, there would be no sense in programming at all. To this end, the IF... THEN... ELSE... is used. The ELSE clause is optional; there are often situations when the alternative consists in doing nothing. We may emphasize this by writing "else ; " (C/JAVA) or "else: pass" (PYTHON), but this is only for legibility. Primitive languages (such as old BASIC variants) did not even have an ELSE. Before the advent of ALGOL, the general standard was that an "if" clause could be followed only by a single expression, which was usually a "go to" statement (p. 138).

What is a condition? It is an expression that is evaluated to yield a Boolean value, i.e., either TRUE or FALSE. Usually, this is some comparison (float variable greater than threshold value, string variable in string list, ...). A

---

[3] Where there is no declaration of variables, as in PYTHON or RUBY, the scope of variables is a bit of a tricky matter. Theoretically, any variable should be restricted to the block in which it first appears – which would cause great problems with setting them conditionally. The somewhat lukewarm alternative used in PYTHON and RUBY is that scope is not a matter of blocks but of functions.

Boolean value or "flag" may be used directly, it is implicitly compared to TRUE. Not all languages possess explicit Boolean values; those that don't (e.g., C), use integer values to the same purpose, where the general convention is that *0 means FALSE and any other value means TRUE*[4].

6.4.2 Comparisons.
Strictly speaking, a comparison is a function that takes two arguments of equal type and returns a boolean value indicating their relationship. You can implement all comparisons as functions, as in LISP, but most languages use a more familiar infix notation with the comparison operator being a special character separating the terms to be compared.

Common **comparisons** include:

	PYTHON	HASKELL	C, C++, JAVA	PASCAL	O'CAML	ERLANG	ADA	FORTRAN
equal?	==	==	==	=	=	==	=	.EQ.
greater? following?	>	>	>	>	>	>	>	.GT.
greater or equal?	>=	>=	>=	>=	>=	>=	>=	.GE.
lesser? preceding?	<	<	<	<	<	<	<	.LT.
lesser or equal?	<=	<=	<=	<=	<=	=<	<=	.LE.
not equal?	<>, !=	/=	!=	<>	<>	/=	/=	.NE.
identical?		is				==	=:=	
not identical?		is not					=/=	
member of aggregate?	in	'elem'		in	in	member	in	
logical AND	and	&&	&&	and	&, &&	and	and	.AND.
logical OR	or	||	||	or	or, ||	or	or	.OR.
logical NOT	not	not	!	not	not	not	not	.NOT.

O'CAML uses the keyword **and** for the very special purpose of marking mutually recursive definitions (unlike all other languages, recursion is not implied but must be declared explicitly with `let rec`), thus it is not available as a Boolean operator. Initially, it was replaced by a simple **&** ampersand whereas logical *or* remained **or**, but of all the weirdnesses of O'CAML, this was probably the most ugly one. Therefore more recent versions of O'CAML introduced C'ish notation: **&&** and **||**. In the interest of compatibility, the compiler still accepts **&** and *or*, but they are branded as "deprecated" and should no longer be used.

The scopes of some of these operators vary from language to language (for example, C has no built-in comparison operators for strings but uses the library function `strcmp()`). There are also the bitwise operators which are not named here, and they are not to be confused with the very similar logical operators. Strangely enough, a logical XOR[5] seems to be implemented nowhere at all.

---

[4] This explains a rather odd formula often encountered in C programs working with pointers ((p. 80)). Pointers are initialized to 0, so to check them, one has to make sure that they are not 0. This is achieved most easily by just writing: `if (pointer)` ...

[5] a XOR b == (a AND (NOT b)) OR ((NOT a) AND b).

"Identity" is a rarely used thing. Whereas equality tests only for the same value, identity means that this is the same object (or in ERLANG, tuple), not one with the same values[6]. There may be actual applications for this.

PYTHON also allows multiple comparisons in one operation: a > b == c > d is equivalent to (a > b) and (b == c) and (c > d).

6.4.3 Imperative IF–THEN–ELSE. According to the imperative paradigm, conditions are reacted to by performing different actions or different sets of actions.

```
if condition then
    execute-block-1
else
    execute-block-2
```

You may visualize it like this:

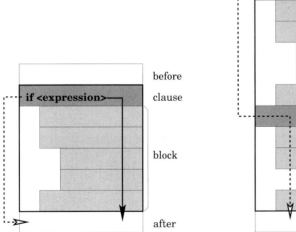
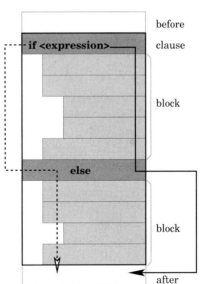

where the solid line indicates the path taken in the case where the condition is fulfilled, and the dotted one when it is not. Note that any block may comprise an arbitrary number of simple or compound statements, including further branchings. If the condition is not true, the subordinate block is skipped; if it is true, the subordinate block is executed, but if an "else" clause follows, the block subordinate to this is skipped.

---

[6] Compare here the German expressions *dasselbe* and *das Gleiche*. In C, one would probably test for equality of the addresses to make sure: if (&a == &b) ... We will discuss this later.

6.4.4 Short circuits. When a condition comprises several logically linked sub-clauses, their evaluation is usually terminated when the result is clear: An *AND*-linked condition need not be evaluated further than to the first "false" result, an *OR*-linked condition not further than to the first "true" result. All modern interpreters and compilers take this into account. This has to be considered if some of the conditions are supposed to perform "side effects" such as input/output.

6.4.5 Functional IF–THEN–ELSE. Naturally, functional languages also require the ability to react differently to different conditions. However, as they reject the notion of actions, this is expressed in terms of selecting a value from several alternative ones:

```
variable = if condition then expression-1 else expression-2
```

This corresponds to the mathematical formalism:

$$f(x) = \begin{cases} e(x) & \text{if condition} \\ e'(x) & \text{otherwise} \end{cases}$$

Of course, this immediately raises the question of what value the variable has if the condition is evaluated to "false" but there is no "else" clause. This depends very much on the language, but a perfect solution has not been found yet. So it is probably best to forswear "else-less" conditional clauses in functional programming altogether: This will at least cause the result to always be defined.

Oddly enough, the otherwise completely imperative C languages have adopted a shorthand for functional IF, which is written like this:

```
m = (a > b) ? a : b;
```

Let m be: if $a > b$ then a, otherwise b.

6.4.6 A closer look. In PYTHON, the IF–THEN–ELSE is basically imperative and looks like this:

```
if condition:
    instruction1
    instruction2
else:
    instruction3
    instruction4
```

e.g.:

6.4 THE IF...THEN...ELSE... CONSTRUCT

```
if a > b:
    max = a
    print "A is greater than B"
else:
    max = b
    print "B is greater than A"
```

If the blocks contain only single statements, these may be written directly following the respective colons:

```
if a > b:  max = a
else:  max = b
```

In RUBY, a C-like shorthand mechanism also exists:

```
m = if a > b then a else b end
```

In PYTHON, it has a somewhat odd appearance:

```
m = a if a > b else b end
```

♠ *What happens if a and b are equal?*

You will have noted the command "print". The use of this word goes back to the times before the general introduction of monitors, when computers were actually conncted to telex machines for output. "Print" sends its arguments **not** to the printer, as one might expect, but to the screen, or more exactly, to the console (which may be a screen in text mode or a text window). (For output redirection, please see the section on Kant & Uexküll [p. 67].) This is known to be confusing to beginners, and a number of languages (e.g., PASCAL) therefore prefer the more neutral "write".

Somewhat bewilderingly, a few languages (LISP, RUBY) also possess an UNLESS feature which is equivalent to "if not".

Now for some real-life examples (assume that *fileaccess* is the internal representation of an open file and *process* the main part of the program which operates on the file).

PASCAL:

```
(* read file, if it exists *)
if exists_file( filename ) then begin
    fileaccess = open( filename );
    if fileaccess <> FILEERROR then begin
        process( fileaccess ); close( file )
    end
```

```
    end else
       errormessage( 'File '+filename+' not found!' );
```

C:

```
   /* read file, if it exists */
   if (exists_file( filename ) ) {
      fileaccess = open( filename );
      if (fileaccess) {
         process( fileaccess ); close( file );
      }
   } else errormessage( "File not found!" );
```

PYTHON:

```
   # read file, if it exists
   if exists_file( filename ):
      try:   fileaccess = open( filename )
      except:  errormessage( "Could not open file %s!" % filename )
      process( fileaccess )
      close( file )
   else:  errormessage( "File %s not found!" % filename )
```

♠ *Try to explain what is happening here! Which instructions are grouped in blocks? What do you think is the purpose of the obscure "try:/except:" clause in* PYTHON *where* PASCAL *and* C *have comparisons? (Hint: this mechanism catches runtime errors, also known as "exceptions". In a real application, we would not just "pass" here but attempt to do something sensible.)*

 The IF–THEN–ELSE structure enables the program to react differently to different conditions. Very often, these conditions consist in comparisons between variables or expressions containing variables.

6.5 Modifiers

In PERL and RUBY, a conditional statement can also follow a one-line expression. Very often this is used in conjunction with the "unless" feature and is thought to improve legibility by gently hinting that the condition is rare.

6.6 Multiple branching: SWITCH *vs* ELIF

Relevance: Conditions may be more complex than simple binary yes–no. Most programming languages comprise features to handle such complex conditions easily.
Keywords: Switch/case, case/of, el(se)if, ...

"Basically, life is different". This applies to programming too. What if there are more than two possibilities? Many languages have the so-called "case mechanism":

PASCAL:

```
n:  integer;
sendMessage( 'Please press 1, 2 or 3 to specify your order.' );
n := getTelephoneKey();
case n of
    1:  order( 'Quattro Stagioni' );
    2:  order( 'Diavolo' );
    3:  order( 'Calzone' );
    else: sendMessage( 'This is not a valid key here.' )
end;
```

C, C++, JAVA:

```
send_message( "Please press 1, 2 or 3 to specify your order." );
int n = get_telephone_key();
switch (n) {
    case 1:  order( "Quattro Stagioni" ); break;
    case 2:  order( "Diavolo" ); break;
    case 3:  order( "Calzone" ); break;
    default: send_message( "This is not a valid key here." );
}
```

C♯ offers the particularly odd feature of optionally combining multiple branching with GOTO, so within a switch/case structure it is possible to directly *goto* another case.

In PYTHON, we do not have a switch/case structure; instead, we can use an extension of the IF–THEN construct known as IF–ELIF–ELSE:

```
send_message( "Please press 1, 2 or 3 to specify your order." )
n = get_telephone_key()
if n == 1:  order( "Quattro Stagioni" )
elif n == 2:  order( "Diavolo" )
elif n == 3:  order( "Calzone" )
else:  send_message( "This is not a valid key here." )
```

This is superior in flexibility, as it allows one to consider more pieces of information at the same time, but it needs a bit more typing:

```
send_message( "Press 1 if you like hot taste, any other key if you
don't." )
likes_it_hot = (get_telephone_key() == 1)
send_message( "Please press 1, 2 or 3 to specify your order." )
n = get_telephone_key()
if n == 1:   order( "Quattro Stagioni" )
elif n == 2 and likes_it_hot:   order( "Diavolo" )
elif n == 2 and not likes_it_hot:   order( "Rucola" )
elif n == 3:   order( "Calzone" )
else:   send_message( "This is not a valid key here." )
```

For the sake of completeness it should be mentioned here that, for some unknown reason, the designers of parallel languages (OCCAM and ERLANG) have such a preference for multiple branchings that their "if" is generally implemented as an If–Elif, every clause containing a condition of its own and the final "'else" being simulated by "TRUE":

OCCAM:

```
IF
    a > b
        jdg := "A greater than B"
    a < b
        jdg := "A lesser than B"
    TRUE
        jdg := "A equals B"
```

♠ *How would you implement the* PYTHON *pizza program in* C *or* PASCAL*?*

> Σ The SWITCH and ELIF constructs enable programs to react to conditions which do not evaluate to simple truth values.

6.7 Applying code to piles of data: FOR, MAP, FILTER and friends

Relevance: We have discussed before (p. 84) that the value of computers in science is mostly due to their ability to process large series of data without tiring, and that homogeneous groupings form the basis for storing such series. However, these series must also be processed.
Keywords: For→next, map(), filter(), reduce(), ...

By far the most common issue in scientific programs is the processing of a large amount of data by some software which operates on single data items. The "large amount of data" can be an array or a list of items, either simple or complex.

Envisage a conveyor belt which is connected to a machine which may process only one piece at each time[7]. A shipload of pieces is processed one by one; for example, when we want to print our address file from the example on p. 32, one entry after another is sent to a printing function. That is all that loops and mappings are about.

More precisely, each item belonging to the data collection becomes in turn the argument of a processing operation. Ideally, this operation is a function which does not change the data item but calculates a result from it; these results may be collected or discarded. However, in imperative programming the operation is often permitted to modify the data item "in place".

Of course, this can lead to insidious bugs. You have to be especially aware of the fact that in most languages there is no way of restricting this modifying access to the one data element which is just "under the pointer", and indeed there are applications where you *must* access other data elements too, for example when sorting[8] a group of data. In this case, take utmost care not to overwrite your data with each other (an example of this will follow).

As images are said to be easier to memorize than words, here is an illustration of the piecewise processing of a row of data:

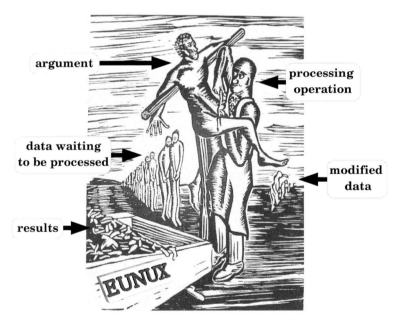

Now let us consider this in more detail.

---

[7] This is a time-honoured concept, known as the "Turing machine". The fact that virtually all programs are based on the one-vice-at-a-time concept is one of the greatest obstacles in the utilization of multiprocessor systems.

[8] Note also that the functional version of the most efficient sorting algorithm does not require this! See p. 264 for details.

6.7.1 Very simple repetition. To the best of my knowledge, the structure `REPEAT n action` has been implemented only in LOGO, a teaching language for children, derived from LISP[9]. As "action" has no way of finding out which pass of the cycle it is currently in, this is of very limited value.

In RUBY, however, there is a comparable structure which is interesting as it is based upon objects[10]. Unlike PYTHON (which is otherwise very similar), RUBY does not comprise any primitive data types; even a simple number constant, such as *1*, is an instance of class integer[11] (see p. 188 if the terms class and instance are giving you the creeps). Thus, there may be more to *1* than its simple value – and there is. Among other things, the integer class comprises a method named *times* which takes a block of code and executes it as many times as the numeric value of the object. Thus,

```
5.times { puts "Repetitio est mater sapientiae!" }
```

will generate the following output:

```
Repetitio est mater sapientiae!
Repetitio est mater sapientiae!
Repetitio est mater sapientiae!
Repetitio est mater sapientiae!
Repetitio est mater sapientiae!
```

(RUBY's *puts* being equivalent to PYTHON's *print*). On p. 213 more will be said about this "spontaneous generation" of objects in the scripting languages.

In REXX, a similar thing is available as a part of the more general *do* structure (see p. 137).

6.7.2 The FOR...TO... loop. A straightforward way of dealing with a pile of equally-structured items is just to count from the first to the last and to access every one in its turn. This is the natural way to deal with arrays, and it is therefore not astonishing to see that the Fortran languages indulge in this. In BASIC, this is known as the FOR–NEXT loop because it ends with the word "next" (meaning, of course, that the next index is now to be used).

In PASCAL, O'CAML and other languages, there are two variants: *FOR n = min TO max* and *FOR n = max DOWNTO min*. BASIC is somewhat more flexible as its FOR–NEXT loop includes an optional "step" clause which indicates by what value to increment the counter variable (1 by default), so it is possible to write not only "FOR i% = 1 TO 100" and "FOR i% = 100 TO 1 STEP -1" but also, for example, "FOR i% = 0 TO 100 STEP 10", thus enumerating 0, 10, 20, ..., 100 if that should be desired.

---

[9] The language, not the children.

[10] A similar structure exists in SMALLTALK, but SMALLTALK's syntax is too confusing to be mentioned here.

[11] Actually, there are two classes, *smallint* and *bigint*, but this is not essential here.

6.7 APPLYING CODE TO PILES OF DATA: FOR, MAP, FILTER AND FRIENDS

```
(* Print an array named 'elem' with 100 elements in PASCAL. *)
var i: integer;
for i = 0 to 99 do writeln( i, ' = ', elem[ i ] );

# Print an array named 'elem' with 100 elements in LUA.
# LUA arrays are hashes, so use 0..99 or 1..100 at your pleasure.
# Could be done more elegantly using an iterator.
for i = 1, 100 do begin
    print( i, ' = ', elem[ i ] )
end

(* Print an array named 'elem' (of any size!) in o'CAML. *)
for i = 0 to (Array.length elem)-1 do
    Printf.printf "%i = %s\n" i elem.( i )
done;;

"Print an array named 'elem' with 100 elements in SMALLTALK."
1 to: 100 do: [ :n | (elem at: n) printNl ]!
```

Note that, unlike human reckoning, which usually begins at 1, array indexing normally starts at 0. Thus, we have to count from 0 to the size of array minus one – something which has been the cause of much grief[12].

You should also keep in mind that in this kind of loop there is also temporal order involved. x_i is guaranteed to be processed before x_{i+1}. This may be of some relevance when you do not restrict your access to x_i. For example, if you want to move a part of the contents of an array to insert a new element (which will of course require removal of other elements, as arrays are of constant size...), this will not do, although it is syntactically correct (the last line represents o'CAML's acceptance of the function in terms similar to those used by HASKELL, see p. 50):

```
[ophis@horcrux ophis]$ ocaml
        Objective Caml version 3.07+2

# let insert field newelement position =
    for i = position to (Array.length field)-2 do
        field.( i+1 ) <- field.( i )
    done;
    field.( position ) <- newelement;
    field;;
val insert : 'a array -> 'a -> int -> 'a array = <fun>
```

When using this function, we will have a nasty surprise:

```
# insert [| 1; 2; 3; 4; 5; 6; 7; 8; 9 |] 666 4;;
```

---

[12] Many of the scripting languages, e.g., LUA, implement arrays as hashes. That is to say, you may use any index number, provided there is an element with that index.

```
- : int array = [|1; 2; 3; 4; 666; 5; 5; 5; 5|]
```

Why? Because first the element x_{n+1} will be set to x_n, then x_{n+2} to x_{n+1} ... which has been set to x_n before! So from *position* onward, all elements will have been replaced by x_n.

Instead, write

```
# let insert field newelement position =
    for i = (Array.length field)-2 downto position do
        field.( i+1 ) <- field.( i )
    done;
    field.( position ) <- newelement;
    field;;

val insert : 'a array -> 'a -> int -> 'a array = <fun>
```

and get

```
# insert [| 1; 2; 3; 4; 5; 6; 7; 8; 9 |] 666 4;;
- : int array = [|1; 2; 3; 4; 666; 5; 6; 7; 8|]
```

 The FOR–NEXT loop performs iteration by using a variable as counter.

6.7.3 The generalized *for (;;)* structure by Kernighan and Ritchie.
In the C superfamily (C, C++, OBJECTIVE C, JAVA, C♯, AWK, SED, PHP) the enumerative loop construct was extended by dividing it into three subclauses: an initializer, a condition and a step action. That is to say, the very common formula

```
for ( int i=min; i<=max; i++ ) { ...; ...; }
```

counts i from *min* to *max*-1 by first setting i to *min*, repeating the following block of instructions while $i \leqslant min$, and performing the operation i++ (i.e. increasing i by 1) every cycle. As the latter may be quite extensive, we may find formulae like

```
for ( int i=0; i<length( source ); sink[ i ] = source[ i++ ] ) ;
```

where the "body" of the loop is reduced to zero.

♠ *What does this C line do? What kind of information must "length" yield? Is this possible without polymorphism?*

In EIFFEL, the loop structure comprises the initializer and condition, but the step action must be named separately:

```
from i := 0 until i >= source.count
loop
    sink.put( source.get( i ), i )
    i := i + 1
end
```

"Two out of three ain't bad."

6.7.4 Implicit loops.

6.7.4.1 *All for one is one for all.* Languages in which arrays are prominent sometimes offer additional support for iterating over arrays without the need for loops. In FORTRAN, such a possibility, approaching PYTHON's use of `for`, is implemented for input and output of arrays:

```
WRITE *,(X(I),I=A,B)
```

will print the array elements $x_a \ldots x_b$. (Intended to resemble the mathematical notation $\sum_{i=a}^{b} f(x_i)$.)

6.7.4.2 *WHERE.* In FORTRAN, the *where–elsewhere* construct mirrors the *if–else* structure but refers to an array, of which every member is treated individually (and optionally in parallel, if the underlying hardware allows this).

6.7.4.3 *FOREACH.* This FORTRAN structure corresponds to *for* as we know it in PYTHON, but does not give any gurantees with regard to the order of processing, which allows for parallelization.

6.7.5 Recursing on lists.
Using functional programming and the *head* and *tail* functions which have already been presented to you (see p. 86), it is possible to *walk down a list* without ever thinking about its size, just by the following processing function:

- If the list is empty, terminate.
- Otherwise do something with the first item (head) of the list, then call the processing function for the list minus its first member (tail).

walk down a list

This avoids much trouble as it reduces the entire thing to only two cases: empty list and non-empty list. For the first, the solution is trivial; for the second, it is easier to implement and to reason about than iteration.

As this is such a frequent thing in functional programming in the widest sense, a special formalism can be used for it. In *pattern matching* (p. 140), the notation [] stands for an empty list, [h:t] (HASKELL), [h::t] (O'CAML) or [H|T] (MIRANDA, ERLANG) for a non-empty list where *h* is the first element and *t* the rest (which may be an empty list, if the whole list contains only one item) and _ for "simply anything". So many recursive definitions are actually defined twice, once for the empty and once for the non-empty list, as in the following piece of ERLANG code:

pattern matching

```
f( _ ,       [ ]    ) -> [ ];
f( Function, [H|T] ) -> [ Function( H ) | f( Function, T ) ].
```

♠ *Describe the effect of* f *in your own words. What values are suitable arguments for* Function*?*

Among the disadvantages, it must be said that this works with considerable overhead[13]. When processing a 10 MB text file character by character, a recursive approach is probably not to be recommended[14]. And, of course, this will process *all* elements of a list. If you want to make exceptions, you will first have to generate a subset. (To this end, genuine functional languages provide "list comprehensions" (p. 128) and the *filter* function which creates a subset of a list, including only those members which satisfy a given criterion.)

Recursing on lists can be used for processing them, and it is done by defining two cases: one for an empty, and one for a non-empty list. The latter calls itself recursively to process what remains of the list after removal of its first element.

6.7.6 The MAP function. Functional programming once more! Imagine a function which takes two arguments – a list of *anything* and a processive function – and returns a new list which is built by applying the processive function to each member of the list.

Thus, map $f\ [x_a, x_{a+1}, x_{a+2}, \ldots, x_n] \longrightarrow [f(x_a), f(x_{a+1}), f(x_{a+2}), \ldots, f(x_n)]$.

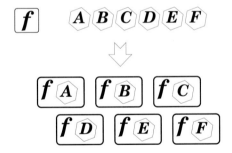

---

[13] Unless the compiler is really good at optimization, which currently holds true for the available ERLANG and CLEAN compilers but definitely not for HASKELL.

[14] More precisely. Storing a string in the form of a linked list is a questionable practice anyway, and certainly not recommendable for large strings. A single character requires 8 bit, the link to the next character at least 32 bit (p. 80), so only 20% of our computer's memory will be available for storing data. For this reason, the authors of ERLANG, which does not possess any aggregate types except for tuples and lists, are eager to point out that processing of large heaps of data is not what ERLANG is built for. In addition, the very paradigm of functional programming prohibits modification of variables and thus also of array elements "in place"; the only way is to return a modified copy. Thus, strictly functional programming is per se not very favourable for processing of large data volumes. Perhaps the most intelligent solution is that of O'CAML, which is functional but permits iterative "in place" modification of arrays, which are distinct from lists, allowing the coder to use the best of both worlds.

6.7 APPLYING CODE TO PILES OF DATA: FOR, MAP, FILTER AND FRIENDS

This is easier to understand by example. Try this in the PYTHON toplevel mode:

```
a = range( 1,10 )
print a
import math
map( math.sqrt, a )
```

An extremely dense and effective way of implementing the "conveyor belt".

Although the *MAP* function is usually described in terms of recursion (map func list = if list == [] then [] else (func (head list))++(map func (tail list))), it is in fact a "black box" whose internal workings do not matter[15]. This may be exploited to great advantage. Unlike a FOR–TO loop, mapping does not require (or guarantee) the individual elements to be processed in any particular order. Thus, this is a good occasion for distributing the workload across multiprocessor systems and doing the processing in parallel.

MAP

In most cases, functions are applied to multiple data not just for the fun of it but because we need some results. Depending on the nature of the operation required, MAP has both advantages and disadvantages when compared to FOR. It is clearly superior if the isomorphic "projection" shown above is what we need. It is not suitable, however, if some interconnection between the different elements has to be made. For example, if we have a list of numbers and want to calculate their sum, we will have to resort to FOR or recursion. Speaking more generally, MAP is not suitable for **collecting** values (p. 129).

♠ *Does this function appear familiar to you?*

We can say that the argument list and the result list are *isomorphic*, that is to say, every element of the one corresponds to exactly one of the other list. Alternatively, the mapping may be described as a *projection*. (This is a fundamental difference from the FILTER function to which it is often compared. This function selects items from a list without modifying them, so the lists can be heteromorphic, with the result list potentially comprising all, some, or no, items of the argument list.)

isomorphic

projection

MAP was developed in the context of functional languages, where lists are the natural form of multiple data. However, there is nothing that argues against applying MAP to arrays as well. In functional languages which support both lists and arrays, e.g., o'CAML, generally both have been implemented.

---

[15] Again, o'CAML is most interesting: it possesses separate MAP functions for mappings of arbitrary functions to lists and mappings of arbitrary functions to arrays, aptly named *Array.map* and *List.map*, respectively. *Array.map* implements an iterative, *List.map* a recursive, mechanism. As o'CAML infers types from their operators, they cannot be united into one universal MAP function.

♠ *Do you now have an idea of how to use the full power of a Sun Fire 15K? (Ace of spades on p. 42).*

 The MAP function applies another function (which is passed as an argument) to all members of a homogeneous data aggregate.

6.7.7 Python's "FOR...:" structure. *MAP* is a thing of beauty, yet somewhat inconvenient in an environment which is not strictly functional, as it requires the "action" to be passed as a parameter. That means that you either have to outsource the action, no matter how minute, into a function of its own (which is not as bad as it might seem, since PYTHON has no objections to defining a function right inside another function), or to squeeze your stuff into a λ function. In both cases, access to local variables is verboten, which is a matter of some concern in a basically imperative language.

For these reasons, PYTHON offers some grammatical sugar. In spite of its name its FOR construct is not to be thought of as an enumerative loop but as a mapping: the code block to which the FOR refers is mapped to the data named in the FOR statement. The following expressions are equivalent:

```
r = []
for i in range(1,10):   r += [math.sqrt(i)]
```

and

```
r = map( lambda( i ):math.sqrt(i), range(1,10) )
```

Which one to use is a matter of personal style – and of details. For example, PYTHON's `print` command is not accredited as function, thus it may not (or only with difficulty) be used in a mapping. On the other hand, when the results are what matters, the mapping is certainly better.

FOR may be applied to anything composite, even to files (provided they have been successfully opened), which are treated as lists of strings. In fact, the only prerequisite for the application of FOR to a structure is that is provides an *iterator* – a function telling the interpreter what is going to be next. Construction of iterators will be discussed in Vol. 2. The following few lines print the complete contents of a file as hex codes:

```
f = open( "thisfile.txt", "r" )
for l in file: # line by line
    for ch in l: print hex(ord(ch)),
f.close()
```

Sometimes you cannot avoid having to use an enumerative loop. In PYTHON, as in the functional languages, this can be emulated by using the RANGE(a,b) function, which returns a list a values from a to b:

6.7 APPLYING CODE TO PILES OF DATA: FOR, MAP, FILTER AND FRIENDS

PYTHON:

```
>>>range( 1, 10 )
[1, 2, 3, 4, 5, 6, 7, 8, 9]
>>>import math
>>>for i in range(1,5):   print math.sqrt(i)
...
1.0
1.41421356237
1.73205080757
2.0
>>>name = "Waldemar"
for i in range( 0, len(name) ):   print i, name[i],
0 W 1 a 2 1 3 d 4 e 5 m 6 a 7 r
```

HASKELL:

```
Prelude>[1..10]
[1,2,3,4,5,6,7,8,9,10]
Prelude>map sqrt [1..5]
[1,1.14121,1.73205,2,2.23607]
```

♠ After much nagging, your eight-year-old nephew has finally succeeded in talking his fashionable parents into buying him a computer. Hearing that you are just taking a course in bioinformatics, they ask you to write something useful for him. As the kid has problems in mathematics, you decide to write a multiplication table for him. Your program is meant to calculate all the products from 1*1 to 20*20 and display them on the 23" screen in the form of 'The product of 2 and 2 is 4. The product of 2 and 3 is 6. The product of 2 and 4 is 8...'. Write it.

♠ The program works, but your nephew is bored. Why should he gaze at a dumb list of numbers when he can bump off nice ugly aliens instead[16]? So you decide to give him a bit of "interaction". Instead of simply printing out the numbers, the program is expected to write "Dear Pubert, please tell me now what the product of 7 and 15 is" and accept a keyboard input.
The program is to display whether the answer is correct or not and to make an internal note so that at the end you can calculate the percentage of correct answers and praise or criticize accordingly – ranging from "Excellent, your parents will be proud of you" at 95 to 100% to "You lazy little brat, better go back to the kindergarten" at 0 to 5%[17].
Hint: There is a pair of functions in PYTHON for obtaining interactive input: *input* and *raw_input*. The difference between these two is that *raw_input* just hands back the user's input whereas *input* tries to evaluate them as PYTHON

---

[16] Not to mention the funny lollipops which some girls on the net use to lick.
[17] Sure it's tough, but who ever said that life was easy?

code – you may even enter complete λ functions here! More about that later. Here, of course, you must use raw_input, otherwise your nephew would just have to answer "7*8" when asked for the product of 7 and 8. Both functions take an optional parameter which is used as the "prompt".
To get familiar with this, start your PYTHON and enter at the toplevel: a = raw_input("Your number?"), then enter any number. Check the value of a. Try this with input. Here enter either a number, an expression such as (3+4)*(5+6) or even the phrase "lambda(x):x*x". In the latter case, what is a, and what is a(6)?

♠ Young Pubert's completely failed yet another math test, and his parents are mad at you. The boy has just memorized the correct sequence of numbers instead of learning the multiplication table properly[18]! So instead of taking Anna to the snake dance downtown tonight, you have to rewrite your program in such a way that the questions are randomized. How would you do this?
Hint: There is a complete module named random for this purpose. You can load this by import random, get a list of its contents with dir(random) and ask questions with the help function (e.g., help(random.random).

6.7.8 Range objects in Ruby. On p. 118, we have described how RUBY employs objects to emulate simple repetition. Interestingly, the FOR structure is also emulated:

```
(1..10).each { |x| puts x }
```

This line has an eerie appearance, and it is not at all clear at first sight why Rubynos consider their favourite language especially beautiful. How does it work?

First of all, the expression (1..10) creates a range object, exactly as PYTHON's range(1,11) would do. The notion (first seen on p. 35) should appear slightly familiar to you by now, as it can also be found in the PASCAL family and in HASKELL. However, in the PASCAL languages this is a definition of a data type, signifying that the integers from 1 to 10 (inclusive) and only these are legal values for the variable, not an actual act of creation; and in HASKELL, generation of data is performed all right (inasmuch as this may be said in a functional language), but the return value of this is a list, not an object, the difference being that HASKELL will evaluate [1..10] to yield [1,2,3,4,5,6,7,8,9,10], whereas in RUBY the object generated by (1..10) will always be printed as 1..10, and list functions won't work with it.[19]

Test this with *hugs* and *irb*:

---

[18] I share Cummings's feelings about so-called "computer-aided learning". Before financial considerations induce us to attempt to replace teachers with machines, there are lots of others things to be cut back first.
[19] RUBY does also possess list functions, and the expression [1..10] is perfectly valid in RUBY, but it is evaluated as [(1..10)], meaning a list with a single item which is a range object.

6.7 APPLYING CODE TO PILES OF DATA: FOR, MAP, FILTER AND FRIENDS 127

```
[ophis@zacalbalam:~] irb
>> 1..4 # range object
1..4
>> [1..4]
[1..4]
>> [1,2,3,4] # list
[1, 2, 3, 4]
>> {1,2,3,4} # hash ("dictionary")
{1=>2, 3=>4}
>> (1..4).each { |x| puts x }
1
2
3
4
=> 1..4
>> 4.times { |x| puts x }
0
1
2
3
=> 4
>> ^D
[ophis@zacalbalam:~] hugs
Prelude> [1..5]
[1,2,3,4,5]
```

On the other hand, there is more to this than just a row of numbers. For example, every range object possesses an *each* method which is similar to the *times* method of integer objects in that it takes an arbitrary piece of code which is executed for each member of the range. Thus, `(1..10).each {...}` is equivalent to PYTHON's `for x in range(1,11): ...` with the notable exception that RUBY's structure does not feature a roving variable (x). However, this is a vital part of the structure, so RUBY's creator has incorporated the `|variable|` notation into the language's grammar, meaning that when a piece of code is put to an iterator, the current iteration value is assigned to the *variable*. This is not restricted to the *each* method; as you see in the example above, it may also be used with the *times* method.

There are two interesting points about this:

- Loops are a prominent structure of procedural languages such as PASCAL and C, and as most so-called OO languages (including C^{++} and JAVA) are actually still basically procedural, they have not vanished; but, in fact, with true object-orientation, they are not really required. The question remains of which approach is easier to understand. Personally, I think that a purely object-oriented language is very difficult to understand unless you know about more conventional structures.

error object
- Object generation does not have to be as conspicuous as Java's `Integer i = new Integer(5);`. The more a language relies on objects, the more clandestine objects there will be. For example, Python's error states are also objects: the `raise` command does nothing but create an *error object*. The efficiency of this is a matter of much debate, as object creation will always involve memory allocation, which is complex and hence slow. As for the raising of error conditions, this is certainly nothing speed-critical, so here the additional flexibility is worth the overhead; and in an interpreted language such as Python or Ruby, it is not much more of a problem to evaluate an object than it is to evaluate an integer variable. Compilers, however, are capable of generating machine code for integers but not for objects, so we need either a bulky runtime system to perform the operations on objects (which was the approach I used for the Python compiler, resulting in a three-megabyte program which ran only twice as fast as the interpreted version) or a *very* sophisticated compiler.

Refer to p. 213 for more about this "spontaneous generation" of objects in the scripting languages.

6.7.9 List comprehension. The function `filter`, implemented in all functional languages – and also in Python –, takes a list and a criterion (described as a function) and produces a list where the members are those of the input list that satisfy the criterion:

```
>>>filter( lambda(x):x>5, range( 1, 10 ))
[6, 7, 8, 9]
>>>import math
>>>a = [1,-4,2,3,-1,0]
>>>map( math.sqrt, filter( lambda(x):x>=0, a ) )
[1.0, 1.4142135623730951, 1.7320508075688772, 0.0]
```

It does not require advanced set theory to understand this. Filtering of a set of objects is such an everyday process that we hardly have to resort to analogies:

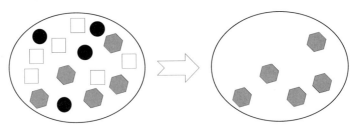

list comprehension
In Python, there is a very functional shorthand for this too. It is known as *list comprehension*:

```
>>>[math.sqrt(x) for x in a if x>=0]
[1.0, 1.4142135623730951, 1.7320508075688772, 0.0]
```

Filtering and list comprehension extract values from a given homogeneous aggregate.

6.7.10 Collecting values. Together, filtering and mapping are sufficient for most of the work to be done on lists, with one very notable exception.

Sometimes the processing of a list is intended to yield a result that is neither a projection (mapping) nor a subset (filtering) of the original list, but represents a "digest" of some kind – a piece of information extracted from, and referring to, the list as a whole, not to just some of its individual components. Length, sum and average are obvious examples. It is often said that values have to be **collected** in such a case.

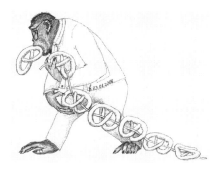

Collection of values can be done either by iteration, in which case a so-called *"accumulator"* is required, or by recursion. The recursive version should appear fairly familiar to you, if you keep in mind our description of the functional paradigm; have a look at the following HASKELL example (for details, please turn to p. 140) which implements the two postulates that the length of an empty list is zero and the length of a non-empty list is one more than the length of this list after removal of its first element:

```
sum [] = 0
sum (x:xs) = x + sum xs
```

We will devote an entire chapter to functions and recursion.

In the iterative version, we need an additional variable for storing intermediate values – the *accumulator*. It is initialized to a value which corresponds to the value defined for the empty list in functional programming (and of course it is the value that will also be returned for an empty list, so this is no coincidence):

```
def sum( list ):
    accum = 0
    for item in list: accum += item
    return accum
```

In PYTHON, this looks quite civilized thanks to poor typing and the expressiveness of PYTHON's FOR structure, but compare this to JAVA:

```
            private static int sum( Vector list ) {
                int accum = 0;
                for ( int i = 0; i < list.size(); i++ ) {
                    accum += ((Integer)list.get( i )).intValue();
                }
                return accum;
            }
```

In the C and ALGOL families, programs have a revolting tendency to be dominated by dreary repetitions of such collecting loops, but there is nothing that can be done about it.

6.8 The REDUCE function

REDUCE

The third, and least well known, function to digest lists is *REDUCE*, also found in all functional languages plus PYTHON. REDUCE is isomorphic to MAP, that is to say, it takes one function f and one list of values l to which this function is applied. However, it works cumulatively, progressing from left to right, so as to reduce the sequence to a single value, and to this purpose f must have two parameters (not one as for MAP). That is to say, REDUCE takes the first two elements from l and applies f to them to create an intermediate resullt: $r = f(l_1, l_2)$; in the next step it proceeds to apply f to this intermediate result and the third element of l: $r' = f(r, l_3)$... until finally $R = f(r^\star, l_n)$. In other words, REDUCE obviates the need for a collecting loop. Thus we can write:

```
            def sum( list ): return reduce( lambda x, y: x+y, list )
```

♠ *Implement the recursive definition of REDUCE.*

6.9 What about the locals?

A common crux about loops is the state of variables after the exit from the loop.

- Variables which have been declared on a higher level or used before (depending on the language) are whatever they were during the last round.
- Variables declared inside the loop (very common in JAVA) simply do not exist outside the loop.

- The problem is with the counting variable. Most languages leave the counter set to its last value, but you should not rely on that.

It also makes things more difficult to read and understand. So the message of this is: when you have a structure like *FOR x = 1 TO 10* or *FOR x IN cool_list*, don't refer to x after the close of the loop.

> *Q: I have a problem with the BugBear program running on my iMac. When I click on the icon in the copyright box while pressing both shift keys, the program will crash. Is there a fix for this bug?*
> *A: There is one: Don't do that.*

6.10 Gimme annather try: WHILE... loops

This is an imperative matter without equivalent in functional programming, because it is about states. Plainly, there is a kind of loop which is basically not enumerative but instead dependent on a certain condition. The body of the loop is repeated over and over either *while* a condition exists or *until* it exists.

6.10.1 Control first. This is sometimes called a "rejecting loop" because it checks for the condition before it executes the loop body for the very first time[20] and thus "rejects" the loop if the condition is not fulfilled. In some languages (PYTHON among them), an additional "else" clause may be defined for the case of rejection.

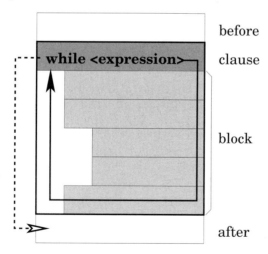

Let us consider the *stack* data type (see p. 90) once more. We have PUSH for placing a "card" on the pile, POP to take off the topmost one and assign it to a variable, and SIZE to see how many cards there are on the pile. Now we want to implement a clearing function which deals with a stack of accumulated data by

---

[20] "Confidence is good, control is better." – Lenin.

printing each item until the stack is empty. However, the stack may be empty from the very beginning, in which case we would like to send a verification that the process has taken place at all.

So that is how we do it:

PYTHON:

```
while size( stack ) > 0:
    item = pop( stack )
    print( item )
else: print "Stack empty"
```

or, if the stack has been implemented as an object (p. 187), as we would normally do it in PYTHON:

```
while stack.size() > 0:
    item = stack.pop()
    print( item )
else: print "Stack empty"
```

Thus we avoid incurring a runtime error by trying to pop from an empty stack.

Actually, this example shows a peculiarity of PYTHON: the ELSE clause for rejecting loops. PYTHON is unique in providing this additional mechanism – a block which is executed once and only once in case the loop is rejected due to its condition yielding a "false" at the very beginning. As in the example here, this will be used mostly for elegant error trapping. In other languages, this will require a few more lines:

PASCAL:

```
if size( stack ) > 0 then begin
    while size( stack ) > 0 do begin
        item := pop( stack ); write_item( item )
    end;
end else writeln( 'Stack empty' );
```

This does not appear really elegant, but there is just no better way!

6.10.2 Control last. This is sometimes referred to as a "non-rejecting loop" because it is executed at least once. Typically, this is a structure which is used (among other things) to collect and process events:

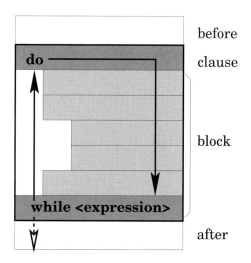

C:

```
main() { /* Quit program with 'Q' */
    char ch;
    init_prg();
    do {
        ch = keyboard_event();
        switch (ch) {
            case 'A':    ...
            case 'Q':    close_files_and_windows(); break;
            case 'R':    ...
            case 'Z':    ...
            default :    ...
        }
    } while (ch != 'Q');
}
```

PASCAL:

```
program fooBar; (* Quit program with 'Q' *)
    var ch: char;
    ...
    initPrg();
    repeat
        ch := keyboardEvent;
        case ch of
            'A':    ...
            'Q': closeFilesAndWindows;
            'R':    ...
            'Z':    ...
            else:   ...
        end
    until ch = 'Q'
```

end.

Unfortunately, such a "control last" loop does not really seem to fit into PYTHON's structure. There is none. You will have to use break instead (see below) – or use a WHILE–ELSE construct.

Historically, it seems that the non-rejecting loop has made its first appearance in COBOL (where it is not immediately recognizable as such, as usual in COBOL) and the rejecting one in FORTRAN:

COBOL: FORTRAN:
    000630 PERFORM UNTIL TOP-VALUE > 0   DO WHILE (A.LT.B)
    000640 ACCEPT TOP-VALUE   READ *,X(A)
    000650 END-PERFORM   A=A+1
                                                END DO

The reasons for these preferences are unknown but may reflect some fundamental difference in the respective mind-set of merchants and scientists.

 Conditional loops repeat the execution of a block depending on a condition, or a group of conditions.

6.11 Take the short cut – to be continued

There are two ways to modify execution within a loop; in modern programming languages, they work with **any** kind of loop, although they are typically used in conjunction with conditional loops.

6.11.1 Break. Leaves the loop immediately and continues right behind it (i.e., after the END, the closing brace or the end of the indentation block)[21].

6.11.2 Continue. This is used somewhat less frequently – it skips the entire rest of the loop and initiates the next cycle[22].

In FORTRAN, the expression CONTINUE is simply the end of an enumerative loop; what the C tradition calls "continue" is here known as CYCLE.

FORTRAN-77 and later also possesses the interesting feature of "named exits", allowing to exit, if desired, several nested loops at once:

---

[21] This is a rather silly feature in the C family: There the switch/case construct also uses the word "break", but to indicate the end of a given case – so when there is a switch/case statement within a loop, it is not possible to break the loop from within the switch/case. Salvation consists in placing the loop within a procedure of its own and to leave that by **return**, which is what most programmers do anyway – or within a do–while loop which tests again for the terminating condition, as in the previous example.

[22] Normally, the test contained in the clause is executed, if it is a control-first loop. But this dilemma is one of reasons why the "continue" statement is less well-liked.

```
outer: DO
inner:     DO i = j, k, l        ! from j to k in steps of l (l is optional)
             IF (...) CYCLE
             IF (...) EXIT outer
           END DO inner
       END DO outer
```

This was necessitated, it seems, by the FORTRAN community's general reluctance to split an algorithm into functions unless absolutely necessary.

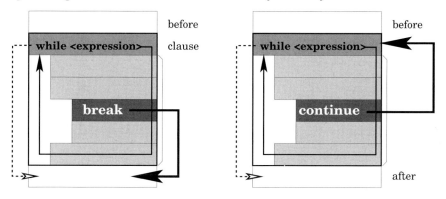

♠ *Try to imagine situations where* break *and* continue *might be useful.*

♠ *Imagine you do not have* continue *but* break*. Would it be possible to do everything?*

♠ *Imagine you do not have* break *but you can* return *from a function. What about that?*

♠ *How might* break *and* continue *be replaced by simple* if...then *constructs?*

♠ *Do a bit of experimenting with the different types of loop. Using the functions from the* time *module, assess how fast they are.*

 BREAK and CONTINUE allow partial execution of blocks subordinate to looping clauses.

6.12 Endless pain

Nitpickers used to point out that there is no such thing as an infinite loop. Every loop will finally come to an end, if only by the demobilization of the computer some day in the 22nd century. Still this is quite a frequent thing; in OBERON and REXX there are even special expressions for this (LOOP...END and do forever...end, respectively). By writing while true: or for (;;) we create a situation where the "head" of the loop does not specify any terminal conditions. The only way to get out of the loop is to break it, to return from the function in which it is contained or to exit from the program as a whole.

Infinite loops are frequently found at the heart of "event-driven" systems. For example, the toplevel of your PYTHON interpreter contains a "for(;;)" loop which is only broken by your pressing Ctrl-D or invoking Sys.exit(). Likewise, graphical programs also used to contain something like

```
main() {
   initialize();
   for (;;) {
      event e = get_mouse_keyboard_or_other_event();
      if (e == menuitem_selected) {
         if (e.detail = filemenu.quitprogram)
            break();
      }
      react_to_event( e );
   }
   finalize();
}
```

Implementing infinite loops in a purely functional language is a touchy subject, which we will not discuss further here. It is sufficient to say that ERLANG was designed explicitly for working with infinitely recursing processes, and that the technique used to this end – "tail recursion optimization"[23] – is also employed by O'CAML, HASKELL and others, but not by PYTHON.

Be warned of recursing into infinity when there is no terminal condition specified. Without a valid termination condition, you will end with something like the age-old German gag:

> Ein Mops lief in die Küche und stahl dem Koch ein Ei,
> Da nahm der Koch den Löffel und schlug den Mops entzwei.
> Da kamen der Möpse viel von überall gerannt,
> Sie setzten ihm 'nen Grabstein, darauf zu lesen stand:
>> "Ein Mops lief in die Küche und stahl dem Koch ein Ei,
>> Da nahm der Koch den Löffel und schlug den Mops entzwei.
>> Da kamen der Möpse viel von überall gerannt,
>> Sie setzten ihm 'nen Grabstein, darauf zu lesen stand:
>>> 'Ein Mops lief in die Küche und stahl dem Koch ein Ei,
>>> Da nahm der Koch den Löffel und schlug den Mops entzwei.
>>> Da kamen der Möpse viel von überall gerannt,
>>> Sie setzten ihm 'nen Grabstein, darauf zu lesen stand:
>>>> *Ein Mops lief in die Küche und stahl dem Koch ein Ei...*

The poor pug's tomb will be more spacious than that of T'chin Shih Huang Di...

---

[23] This is also part of the official language definition for SCHEME. See also p. 136.

6.13 Timer-driven loops

As mentioned on p. 42, there are specialized languages for machine control which possess timer-driven loops like `all 0.5 sec do` . With the exception of PEARL (not to be confused with PERL), none of these languages ever became popular, and even PEARL and the underlying "Real Time Operating System" (RTOS) have been forgotten by now. Still, it is a noteworthy structure.

Of course, here the central problem is what happens if the time allowed for by the loop is, for whatever reason, not sufficient to allow complete execution of the loop body. Will the action be suspended, will it spawn another instance of the action, or will the program simply crash because it is not "reentrant", i.e. incapable of running a piece of code more than once at the same time?

6.14 One Loop To Rule Them All...

As seen above, every loop comprises a terminal condition (the need for this being basically the same as for a terminal condition in recursion). Sometimes, however, there is more than one terminal condition. For example, having asked the pre-defined number of questions or receiving the user's sign to stop might both be sufficient reasons to break the loop. In the *for(;;)* construct of C, C++ and JAVA, this may easily be achieved by *and*ing two conditions:

```
#define strequal( s1, s2 ) (strcmp(s1,s2)==0)
char ans[20];
int n;
for ( n = 1; !strequal( ans, "FOAD" ) && n <= 10; n++ ) {
   scanf( ''%s'', ans );   // C way of getting input
   do_something_with_input( ans );
}
```

A direct combination of several kinds of loops has only been implemented in REXX:

```
do n = 1 to 10 until ans == "FOAD"
   pull ans                 /* Rexx way of getting input */
   call do_something_with_input ans
end
```

The more recent ('85 and later) versions of COBOL have something similar:

```
000880 PERFORM VARYING I FROM 1 BY 1 UNTIL I>10 OR ANS="FOAD"
000890 * Cobol way of getting input:
000900    ACCEPT ANS
000910 * Variables are always global, so we just call a subroutine
000920 * and hope that it will use ANS...
000930    PERFORM SOMETHING-WITH-INPUT
000940 END-PERFORM
```

As described before, the "for" loop of EIFFEL specifies terminal conditions rather than values and no step action, so this is rather close:

```
from i := 1 until i > 10 or ans = "FOAD"
loop
    ans := io.read()         -- Eiffel way of getting input
    do_something_with_input( ans )
    i := i + 1
end
```

In all other languages, the loop has to be broken explicitly (and at the proper point!) if multiple exit conditions are required:

```
for n in range( 1, 11 ):
    ans = raw_input()        # Python way of getting input
    if ans == "FOAD": break
    do_something_with_input( ans )
```

Woe to those who cannot break loops! For example, SMALLTALK's architecture of loops as methods, or LISP's architecture of loops as functions[24], simply do not permit such a thing. Here the only possibility is to use an additional termination flag, or to enforce a termination condition ($i := 10$).

On the other hand, there is a strong tendency in the ALGOL family of languages to reduce the complexity of loops altogether by delegating the exit conditions to explicit breakings, leaving the "infinite loop" as the only remaining structure. This extreme has been implemented only in experimental and didactic derivatives of PASCAL, but the looping construct of ADA has a reductionist flavour (which is grotesque, considering the general complexity of the language).

6.15 "Go to" statements

There is one more elementary control structure to be mentioned for the sake of completeness only. It should have been moved to the ominous position 13 because it is simply *nasty*. It comes from assembly language, plays a great role in BASIC and exists in C (and C♯) as well as PASCAL but has been omitted from JAVA and PYTHON, not to mention the functional languages. This is the "go to" structure, which continues execution somewhere else (at a labeled site or line number). In fact, all loops, conditionals etc. are translated[25] into an assembly language construct full of jumps[26]: The C code `if (flag) action1(); else action2();` is translated into

---

[24] Of course, it is blasphemy to use loops in LISP at all.

[25] Thus, the illustration given for these structures are not mere illustrations – they actually show the inner workings of the program.

[26] A noteworthy exception is the Acorn ARM processor, the instructions of which are executed conditionally, depending on certain processor flags. This obviates the need for most local jumps and thus greatly increases processing speed.

```
            cmpl    $0, 8(%ebp)     ;; compare variable to 0
            je      .L2             ;; if equal, continue at .L2
            call    action1         ;; -> skipped if not equal
            jmp     .L1             ;; continue at L1
    .L2:
            call    action2         ;; -> skipped if equal
    .L1:
```

This has got right into BASIC:

```
1500 if flag% then goto 1520
1510 print "Flag is false"
1520 goto 1540
1530 print "Flag is true"
1540 print "So what?"
```

Obviously, this "spaghetti code" is not a very nice programming style. For the sake of completeness, it should be mentioned that FORTRAN and GONT use labels as integral parts of their looping structures.

Procedural and higher styles do not mix well with "go to" statements. Thus, PASCAL and C allow "go to" statements only locally, i.e., within one procedure, as a "last resort" measure:

```
procedure SalvatoreStupidoStupido( flag: boolean );
    label 1, 2;   (* Labels for jumping to must be declared *)
    begin
        if flag then goto 1;
        writeln( 'Flag is false' );
        goto 2;
    1:
        writeln( 'Flag is true' );
    2:
        writeln( 'So what?' );
    end;
```

However, their use is discouraged, and they have been removed from the successors of these languages (JAVA and OBERON) without being missed. In the ALGOL languages, nested procedures pose a special problem: what happens if we jump out of a local subprocedure?

```
procedure ConfusioSineQuaNon( flag: boolean );
    label 1, 2;
    procedure dumbDump;
        begin
            if flag then goto 1;
            writeln( 'Flag is false' );
            goto 2;
        end;
```

```
begin (* ConfusioSineQuaNon *)
    dumbDump;
1:
    writeln( 'Flag is true' );
2:
    writeln( 'So what?' );
end;
```

If the compiler is not smart enough to prohibit this, it is most likely that this will crash the program. C does not allow nesting of procedures, but most implementations possess the *longjmp()* library function which implements a non-local "go to" – a thing which is simply obscene.

6.16 Pattern matching

pattern matching

At the other end of the spectrum, in functional languages such as ERLANG, HASKELL and O'CAML, we have a mechanism called *pattern matching*. The idea is that a data aggregation is compared to a number of patterns, and actions are taken only when there is a match. For example, the factorial (p. 159) may be written in ERLANG like this:

```
factorial( 1 ) -> 1;
factorial( N ) -> N * factorial( N-1 ).
```

(Note the semicolon, which is used in ERLANG to separate different patterns.)

What does this mean? In this simple case, there are just two patterns: "1" and "N". The latter fits anything. As the comparisons are done sequentially, the argument is first matched against "1"; if it fits (i.e., if the argument value is 1 and nothing else), 1 is returned, and the process ends. If it does not, it is matched against "a variable". This will fit anyway. So the argument is bound to N, and the second definition is used.

Have a look at a more complex example:

```
foo( 1, _, _ ) -> 100;
foo( 2, X, _ ) -> bar( X );
foo( _, _, [] ) -> [];
foo( _, N, [H|T] ) -> [N|T].
```

Here the underscore (_) means "anything – we just don't care", [] an empty list and [H|T] a non-empty list whose head (first element) and tail (rest of list) are then bound to H and T, respectively.

♠ *Explain* foo *in your own words.*

In ERLANG, pattern matching is used extensively in inter-process communication. Messages received by a process are analyzed by matching them to a number of patterns. In spite of ERLANG's weak typing system, type safety may be achieved by "tagging" variables with adequate identifiers, such as the tuple {floatingpoint,3.1415927}. This may be matched like this:

```
case V of
    { floatingpoint, X } -> flopcalc( X );
    { integer, X } -> flopcalc( float(X) );
    { _, _ } -> raise_error( "Numeric value expected" )
end.
```

 Pattern matching can be used either for recursion, or for dealing with heterogeneous aggregates. It relies on the language's ability to discriminate "patterns" of variables, where a pattern may be an arbitrary combination of types and values.

6.17 Choosing a functional variable

Where functional style is supported, there is yet another method of making decisions:

```
(* Dependent on the value of nara, apply either chandra or surya to loka *)
let nara = ... in
   let gupta = if nara then chandra else surya in
       gupta loka;;
```

In PYTHON, this is feasible but less concise:

```
# Dependent on the value of nara, apply either chandra or surya to loka
nara = ...
if nara: gupta = chandra
else: gupta = surya
gupta( loka )
```

A special form of this is the so-called *vector table*. Briefly, this is an array of functional variables. Multiple choices can be made by indexing:

```
def trigfun( x, n ):
    fun = [ math.sin, math.cos, math.tan ]
    return fun[ n ]( x )
```

Of course, both the benefit and purpose of this are hard to see. Actually this will be rarely used in a high-level language like PYTHON. It can be put to very good use, however, when we have to do low-level programming in a language like C. Here this can lead to significant gains in terms of execution time – you will have perceived that, independently of the number of alternatives, the choice will be made in a single move here!

There is one very important application for this, and this is in the field of virtual machines. You remember that a "virtual machine" or VM is a processor that exists only in software, not in hardware, and whose workings will therefore have to be emulated by another processor's hardware. VMs that have been mentioned before comprise the JAVA VM and the ERLANG VM used for running system-independent programs, and the clandestine emulators used by Apple when changing from the Motorola 68k to the PPC processor line and by DEC when changing from the VAX to the Alpha processor. Our final project will be the construction of an own VM which implements a strongly simplified processor for embedded devices (Vol. 2). As you will see, processor instructions are nothing but numbers. In the case of the FOOBAR VM, 0001 fills a value into a register, 0002 fetches the contents of memory location contained in one register into another register, and so on. Thus, the fastest way of "running" VM code uses such an array of vectors, where the first entry contains the function that "loads" a register of the virtual machine with a constant value, the second a function that fetches the contents of memory location contained in one register into another register, etc. In C, which is the most widely used language for such tasks, there are no genuine functional variables but they can be emulated by using pointers to functions.

6.18 Thou Mayst Pass

Occasionally, we want to declare expressly that we are not going to do anything. More often, during program development we need a "stub" or placeholder. In PYTHON, there is the "noperation" pass, equivalent to a lonesome ; in the C languages or a () in O'CAML.

There is also a "nonvariable" with the fanciful name of None, vaguely corresponding to HASKELL's _|_ "bottom"[27]. In both cases the notion is that there is just nothing. Note that None is an *object* (see p. 187 and p. 232 for this topic.)

What is the sound of nothing happening?

---

[27] The token is supposed to represent the mathematical sign $\perp$.

CHAPTER 7

◇ Application: Full Impact

Recently I heard an interview with an elderly London clubman who was fuming that his club had decided to waste money on a professional librarian and said that the books should just be ordered as they used to be. When the reporter asked him what kind of order that was, he roared: «Tallest left, smallest right!»

– RICHARD DAWKINS: "The Blind Watchmaker"

Relevance: This chapter shows how the data types and control structures discussed so far may be used to design a working program.
Keywords: Conditions; loops; homogeneous and heterogeneous aggregates; file access.

We are now going to develop a simple program that is capable of doing some real-life work. You will first see the utilization of the structural concepts learned so far (flow control), then we will continue to show the application of some things which have been mentioned before but not discussed in detail yet. Later, you may return to this chapter and consider how and where advanced structures (objects) might be used.

7.1 The problem

From a complete list of accredited journals, you want to display only those with titles which contain certain keywords. The journals of interest are to be displayed sorting according to their impact factor[1]. The list is a simple ASCII file which looks like this.

```
AARGH BULL                1.419
ABHOR MATH SEM HAMBURG    0.115
ABOMINABLE IMAGING        0.891
AC/DC                     2.904
ACRASIOLOGIA              0.095
ACCOUNTS IRREPROD RES    11.795
```

---

[1] The *impact factor* was created by librarians as a tool for optimizing the shelf arrangement for scientific periodicals. It is defined as the number of quotations referring to this journal divided by the number of articles in this journal, which is thought to be an indicator of the journal's "importance". Nowadays, the quality of scientific work is often judged by the impact factor of the journal in which it is published.

```
  ADVANCED DARK ARTS J      1.748
  (...)
  ZUCKER & EPILEPTIKER      0.003
  ZYZZYX LETTERS            0.815
```

Your program is supposed to take its keywords on the command line (see p. 71, p. 70) and print the output to the console. For example, when looking for all journals which are concerned with molecular and cellular issues, you will probably wish to see something like this.

```
ophis@cirith-ungol:~/natural_killers/lehre/bioinformatik> python imp.py cell mol
              MOL CELL ---> 18.142
         MOL CELL BIOL ---> 9.866
         MOL BIOL CELL ---> 7.527
      MOL CELL NEUROSCI ---> 5.654
      AM J RESP CELL MOL ---> 4.541
       J MOL CELL CARDIOL ---> 2.923
        CELL MOL LIFE SCI ---> 2.891
       BLOOD CELL MOL DIS ---> 2.538
     MOL CELL ENDOCRINOL ---> 2.136
        CELL MOL NEUROBIOL ---> 1.752
      CYTOKINES CELL MOL T ---> 1.709
         MOL CELL BIOCHEM ---> 1.547
            MOL CELL PROBE ---> 1.432
             CELL MOL BIOL ---> 1.172
                 MOL CELLS ---> 0.930
```

7.2 Approach

First of all, let us make up our mind about what is required.

The program will have to be able to read a text file line by line. From each line the name of the journal and the impact factor will have to be derived, and for each journal, the name will have to be stored together with the impact factor. Then the journals of interest will have to be selected, sorted, and printed.

Clearly, this requires both homogeneous and heterogeneous aggregates. In a strongly typed language, we would begin by defining a data type capable of holding journal title and impact factor, e.g., in PASCAL:

```
const maxJournals = 5000;
type jEntry: record
             name: string[ 20 ]; (* this may take up to 20 characters *)
             imfa: real;         (* PASCAL for floating-point *)
             end;
var jx: array [ 1..maxJournals ] of jEntry;
```

and then build our program around this. In PYTHON, there is no need for this. We have to decide on some data structure, of course, but this does not have to be fixed explicitly[2]. Let us simply say that our journal will be recorded as a

---

[2] This means that it is advantageous to describe it in the documentation!

tuple in the form "(name, impact)" and that we will use a list of such tuples for storing all the data.

The advantages are obvious: less red tape and more flexibility – we are not stuck with any self-contracted limitations. The PASCAL program will have to be recompiled to be able either to deal with more than 5000 titles or to run on a machine with less than $(20 + 8) \times 5000$ bytes of free memory (e.g., an "embedded device" such as a palmtops or cellphone) the PYTHON program will use exactly as much memory as the data actually amounts to, plus a little overhead for administrating the lists, typically <5%.

The disadvantages are also obvious: more processing power is required (and more may actually mean several orders of magnitude!), and it is easier to make stupid mistakes, because the PASCAL compiler may detect some kind of nonsense which escapes the PYTHON interpreter's vigilance. However, this is largely balanced by PYTHON's excellent facilities for treating runtime errors.

So first we write a piece of code that is capable of reading our text file, extracting the relevant information and storing it in a form that can be used by the rest of the program. Next, we add a selecting algorithm that discards all entries except for those which we really want to have. Third, we add the code for sorting and printing. Thanks to PYTHON, the latter two will be trivial.

7.3 Rough and ready

This is a clean but coarse solution:

```
import string, sys # load modules for strings and system access
delim = 20 # "magic number" for format

# Get the text file.
fil, jlist, rlist = file( "impact-factor.txt", "r" ), [], []
for line in fil: # scan file line by line
    if line != "": # skip empty lines if there should be any
        journal, fact = string.strip(line[:delim]), string.strip(line[delim:])
        # cut line asunder
        if fact == "": fact = "0.0" # lest shit happens
        jlist += [(journal,float(fact))] # build list of (journal,impact) items

# Screen the jlist for presence of key words:
for j in jlist:
    acceptable = True
    for arg in sys.argv[1:]:
        (a, b) = j # now a will be name and b impact factor
        if string.find( string.lower(a), string.lower(arg) ) < 0:
            acceptable = False # no good
            break # thus no need to test the other args
    if acceptable: rlist += [j] # if found, add this item to rlist

# Now rlist contains a selection of journals matching our key words. Show it:
```

```
list.sort( rlist ) # see documentation
for item in rlist: # the printing function proper
    print "%20s ---> %2.3f" % item # prettyprinting
```

Let us now analyze this line by line.

import
sys
string

(1) `import string, sys`. The *import* statement makes the contents of the modules *sys* and *string* (p. 97) available to the PYTHON interpreter. For details on the contents of these modules, please consult your PYTHON documentation.

(2) `delim = 20`. The next line assigns a simple integer value to the variable *delim* (p. 98). The value we have chosen is determined by the format of the file (the first 20 characters being reserved for the name of the journal), not by "intrinsic" needs of our current program. Of course, we would not really need this variable at all; we could always write "20" instead. However, using a name
- makes the program easier to read and understand
- makes the program easier to maintain, extend and modify.

(3) Do not forget empty lines to separate the parts of your program.

(4) The same holds true for comments (p. 105). *They do matter!*

(5) `fil, jlist, rlist = file( "impact-factor.txt", "r" ), [], []`.

file

Now we initialize a number of variables: *fil* is a *file* (p. 343) which is opened for reading (more exactly: not just "a" file but the file named `impact-factor.txt`), and *jlist* and *rlist* are empty lists (p. 85).

(6) `for line in fil:`. In PYTHON, homogeneous associations may be iterated over using the *for* operator (p. 84, p. 116). A file variable can be used as any homogeneous aggregate would[3], its components being the lines of the text. Thus, the subordinate block processes the file line by line.

for

(7) `if line != "":`. There may be empty lines in the text file. Process only the others.

(8) `journal, fact = string.strip(line[:delim]), string.strip(line[delim:])`. We distil the journal name and the impact factor from the text line. The *slice operator* (p. 87) can be used to access parts of an aggregate (aggregate types including strings): $x[a:b]$ will yield x_a to x_b, $x[a:]$ will yield x_a to the end, $x[:b]$ will yield from the first to x_b. Indices can be negative; then they will be counted from the end.

slice operator

x[a:b]
x[a:]
x[:b]

(9) `if fact == "": fact = "0.0"`. If no impact factor is given in the file, it must be zero.

(10) `jlist += [(journal,float(fact))]`. Convert the impact factor – so far just a number of ciphers – into a floating point number (p. 77), associate it with the journal name to form a tuple (p. 96) and append this tuple to *jlist* (p. 85).

---

[3] This is an example of advanced abstraction and a feature to be found only in the higher-level languages.

(11) We now have a list like this: [("AARGH BULL", 1.419), ("ABHOR MATH SEM HAMBURG", 0.115), ("ABOM IMAGING", 0.891), ..., ("ZUCKER & EPILEPTIKER", 0.003), ("ZYZZYX LETTERS", 0.815)] .

(12) So it has been written, so it shall be done.

(13) `for j in jlist:`. We iterate over all members of *jlist*, assigning in turn each item to *j* (p. 116).

(14) `acceptable = True`. The purpose of the variable *acceptable*, which is set to TRUE (p. 74) at the beginning of each cycle, will become clear soon.

(15) `for arg in sys.argv[1:]:`. We iterate over the *command line arguments* passed to the script (see p. 71 and p. 70), putting each in turn into *arg* (p. 116). sys.argv
command line arguments

(16) `(a, b) = j`. Each *tuple* is investigated, first splitting it into its two parts – the name and the impact factor, which are separately assigned to *a* and *b* (p. 96). tuple

(17) `if string.find( string.lower(a), string.lower(arg) ) < 0`. See the PYTHON documentation for details on *string.find()*. Basically, this tests whether *arg* is not contained within *a*. string.find()

(18) `acceptable = False`. This block is subordinate to the condition that it is not. In this case, the flag is set to FALSE...

(19) `break`: ...and the inner loop – the one iterating over the command line arguments – is broken (p. 134). Why? Because according to the Roman formula regarding legal evidence, "false in one, false in all". If any of the search terms is not contained within the name, this journal is not what we are looking for, so we can save ourselves the trouble of testing the rest too. The program would work fine without this line, but having it here may speed things up considerably.

(20) `if acceptable: rlist += [j]`. When we have arrived here, no matter whether the inner loop has been broken prematurely or not, *acceptable* will be true if and only if ("iff") each of the search terms has been found to be contained within the journal name. Thus, we add the tuple containing name and impact factor to the new *rlist*.

(21) Take a deep breath.

(22) *rlist* now contains the "(name,impact)" tuples of all the journals matching our query.

(23) `list.sort( rlist )`. You may have guessed that this sorts a list. Note that sorting is done "in place" (one of the very few lapses of PYTHON into mathematical impurity). See p. 87 and p. 264.

(24) `for item in rlist:`. We iterate over the list (p. 84, p. 116) containing the results.

(25) `print "%20s ---> %2.3f" % item`. Of course, we might just write "print item", or replace this loop by "print rlist". However, this adds some decent formatting; please see p. 337 for details.

7.4 Refinements

7.4.1 Using a functional parameter for better sorting. So far, so good. However, this program will sort the list only according to journal names.

```
ophis@cirith-ungol:~/natural_killers/lehre/bioinformatik> python imp.py cell mol
      AM J RESP CELL MOL ---> 4.541
      BLOOD CELL MOL DIS ---> 2.538
            CELL MOL BIOL ---> 1.172
        CELL MOL LIFE SCI ---> 2.891
        CELL MOL NEUROBIOL ---> 1.752
      CYTOKINES CELL MOL T ---> 1.709
         J MOL CELL CARDIOL ---> 2.923
             MOL BIOL CELL ---> 7.527
                  MOL CELL ---> 18.142
          MOL CELL BIOCHEM ---> 1.547
             MOL CELL BIOL ---> 9.866
       MOL CELL ENDOCRINOL ---> 2.136
          MOL CELL NEUROSCI ---> 5.654
            MOL CELL PROBE ---> 1.432
                 MOL CELLS ---> 0.930
```

What can we do about this?

- We could implement a sorting algorithm of our own. The "insert sort" might be a sensible compromise between complexity and speed here. Basically, this algorithm inserts one element after another at the proper position by walking along the list until it finds a couple of elements of which one is less-or-equal to the new one and its successor greater-or-equal to the new one, which is then squeezed in between these two.
- We could teach PYTHON's sorting function how to sort our list of tuples properly. This requires not much, simply a function that takes two arguments a and b and returns the following values:
    - -1 if $a > b$
    - ± 0 if $a = b$
    - $+1$ if $a < b$

    Such a function can be passed to PYTHON's sorting function (functional parameter!) and be used as a criterion for sorting.

Our next chapter will treat functions and their applications in more detail, but you have heard enough about them by now to be able to understand the following modified program.

```
import string, sys
delim = 20

def cmpcouple( c1, c2 ):
    "Judgement function for sorting. Takes 2 tuples and compares
the second component of them."
    (a1, b1), (a2, b2) = c1, c2 # split the tuples
    if b1 > b2: return -1
```

```
        elif b1 < b2: return 1
        else: return 0

# Get the text file.
fil, jlist, rlist = file( "impact-factor.txt", "r" ), [], []
for line in fil: # scan file line by line
    if line != "": # skip empty lines if there should be any
        journal, fact = string.strip(line[:delim]), string.strip(line[delim:])
        # cut line asunder
        if fact == "": fact = "0.0" # lest shit happens
        jlist += [(journal,float(fact))] # build list of (journal,impact) items

# Screen the jlist for presence of key words:
for j in jlist:
    acceptable = True
    for arg in sys.argv[1:]:
        (a, b) = j # now a will be name and b impact factor
        if string.find( string.lower(a), string.lower(arg) ) < 0:
            acceptable = False # no good
            break # thus no need to test the other args
    if acceptable: rlist += [j] # if found, add this item to rlist

# Now rlist contains a selection of journals matching our key words. Show it:
list.sort( rlist, cmpcouple ) # see documentation
for item in rlist: # the printing function proper
    print "%20s ---> %2.3f" % item # prettyprinting
```

7.4.2 Legible structure.
A working program is not the same as a good program. To whet your appeptite for the next chapter, where you will learn more about procedures and functions, here is how a "real" program would look.

```
import string, sys

def getfile( fnam, delim ):
    "Read impfac file and convert it to list of tuples"
    fil, jlist = file( fnam, "r" ), []
    for line in fil: # scan file line by line
        if line != "": # skip empty lines if there should be any
            journal, fact = line[:delim].strip(), line[delim:].strip()
            if fact == "": fact = "0.0" # lest shit happens
            jlist += [(journal,float(fact))]
    return jlist

def jn_contains( j, t ):
    "Does journal name contain search term?"
    (a, b) = j
    return string.find( string.lower(a), string.lower(t) ) >= 0

def ldump( l ):
    "Print filtered list"
    def cmpcouple( c1, c2 ):
        "Judgement function for sorting"
        (a1, b1), (a2, b2) = c1, c2
        if b1 > b2: return -1
```

```
        elif b1 < b2: return 1
        else: return 0
    list.sort( l, cmpcouple ) # see documentation
    for item in l: print "%20s ---> %2.3f" % item

# main program begins here:
jl = getfile( "impact-factor.txt", 20 )
for arg in sys.argv[1:]: jl = filter( lambda j:jn_contains( j,arg ), jl )
ldump( jl )
```

This separates blocks of different functions and isolates them quite nicely into discrete functional units (no pun intended). Characteristically, the main program is almost microscopic, with almost all the work being delegated to functions defined previously.

You will have noted the expression `line[:delim].strip()` where previously we had `string.strip( line[:delim].strip )`. This is already an object-oriented style; we will discuss this later. For the moment, just consider these two expressions as equivalent.

7.4.3 Functional biscuits. Instead of the last two lines (p. 24), we might of course write:

```
    def criterion( j ): return jn_contains( j,arg )
    for arg in sys.argv[1:]: jl = filter( criterion, jl )
    ldump(jl)
```

However, use of the λ function saves some typing and makes the purpose clearer. Using a slightly more functional style, we may make this program even more concise.

```
import string, sys
def write( s ): print s # there's no functional "print"... silly

def getfile( fnam, delim ):
    "Read impfac file and convert it to list of tuples"
    fil, jlist = file( fnam, "r" ), []
    for line in fil: # scan file line by line
        if line != "": # skip empty lines if there should be any
            journal, fact = line[:delim].strip(), line[delim:].strip()
            if fact == "": fact = "0.0" # lest shit happens
            jlist += [(journal,float(fact))]
    return jlist

jn_contains=lambda (a,b),t:string.find( string.lower(a), string.lower(t) ) >= 0

def ldump( l ):
    "Print filtered list"
    def cmpcouple( (a1, b1), (a2, b2) ):
        "Judgement function for sorting"
        if b1 > b2: return -1
        elif b1 < b2: return 1
```

```
        else: return 0
    list.sort( 1, cmpcouple ) # see documentation
    map( lambda item:write( "%20s ---> %2.3f" % item ), 1 )

def select( lis, arg ):
    if len(arg) == 0: return lis
    return select( [ li for li in lis if jn_contains( li, arg[0] ) ], arg[1:] )

ldump( select( getfile( "impact-factor.txt", 20 ), sys.argv[1:] ) )
```

This employs list comprehension (p. 128) and mapping (p. 122) in addition to the λ calculus. Naturally, the functional style tends to accumulate function calls, as it does here in the last line. Several purely function languages, e.g., HASKELL, avoid the LISP-like maze of parentheses by means of the "composition operator", usually written as a period: so a(b(c(d(e(f(x,y)))))) is shown more elegantly as a . b . c . d . e . f(x, y). However, this feature is lacking in PYTHON.

7.4.4 Considerations.
Explain the differences between the different versions of the script.

On a 1 GHz processor (both Pentium-III and G4), this program takes approximately 0.2 sec to run. Where would you begin when trying to speed things up?

Hint: Import the profile module and rewrite the last line(s):

```
def main(): ldump( select( getfile( "impact-factor.txt", 20 ), sys.argv[1:] ) )
profile.run( "main()" )
```

or accordingly, and see what happens!

Double-Blind Experiment, n.: An experiment in which the chief researcher believes he is fooling both the subject and the lab assistant. Often accompanied by a belief in the tooth fairy.

CHAPTER 8

Functions and Procedures

Last time you saw me I was walking outta the door,
Got a nasty habit, a coming back for more.
Next time you'll see me I'll be walking right into that door,
All those nasty habits rising up once more.

– OVERKILL: "Shred"

8.1 Subroutines – the foundation of all higher programming

Relevance: We have mentioned before (p. 21) that procedures may be used to "extend the vocabulary" of an imperative language by giving names to blocks, i.e., logically coherent sets of instructions (p. 108). The rationale is that identical groups of actions (e.g., input, output, calculation of intermediate values) may have to be performed in an essentially identical fashion at different points within a program.
Keywords: Subroutines as structural elements within a program.

8.1.1 Purpose. We often encounter a situation where we would like to use a piece of software at different points in our program: e.g., an algorithm for formatted printing of a number is to be used for different variables. (In fact, all calls to the operating system, from the creation of a file to the handling of a window, might be listed here.) The sensible approach to this is outsourcing this piece of software into a *subroutine*.

subroutine

In BASIC, there was the command `GOSUB` which continued execution at a new line number and pushed the old line number onto the *call stack*. It was complemented by `RETURN` which took the topmost line number from the call stack and jumped thither – primitive but effective for small programs. You will find an example of this on p. 166.

call stack

♠ *When you have understood list processing, build a stack system based on lists. Refer to section "The Joy of Stacks" (p. 164) if you want to learn more about the practical applications of stacks.*

As early as the 1950s, it was realized that this was not particularly smart. Instead, the possibility was created of extending the "vocabulary" of a programming language by defining new "words". In 1958 ALGOL, the *Al*gorithmic

Bioinformatics Programming in Python. Rüdiger-Marcus Flaig
Copyright © 2008 WILEY-VCH Verlag GmbH & Co. KGaA, Weinheim
ISBN: 978-3-527-32094-3

Language by Nikolaus Wirth, later developed into the much more famous PASCAL, was the first language to employ this "procedural" structure, after FORTRAN had already paved the way by calling subroutines by names given to them, instead of line numbers.

```
(* In order to avoid rounding errors, money is always dealt with using
integers.  If we need cents, we shall have to do all calculations in
cents -- the computer doesn't care.  But sometimes we want an output in
terms of euros and cents.  *)

procedure write_euro( cent:  integer )
    begin
        write( cent div 100 ); write( '.' ); write( cent mod 100 )
    end;

begin
    sum = gain - loss + prev;
    write( 'previously    :  ' ); write_euro( prev ); writeln;
    write( 'revenues      :  ' ); write_euro( gain ); writeln;
    write( 'expenditures: ' ); write_euro( loss ); writeln;
    write( 'now           :  ' ); write_euro( sum ); writeln
end;
```

♠ *Try to explain this piece of a* PASCAL *program! (WriteLn causes a line feed.)*

♠ *Rewrite this in* PYTHON. *Assume that prev = 10 000, gain = 3000, loss = 2000.*

Functions are closely related to procedures, the difference being that functions always return a result and procedures never do. They are "subroutines", i.e., independent pieces of code which are referred to by their names and may be called from arbitrary points of the program.

8.1.2 Scope of identifiers. Code that is within a procedure may basically access three kinds of data:

Global data
(1) **Global data:**
Consist of variables (and constants) which are defined to be accessible everywhere in the program.

Local data
(2) **Local data:**
Consist of variables (and constants) the scope of which is limited to the procedure. Elsewhere in the program, these local variables are simply nonexistent. When a procedure defines a variable named *counter*, no other procedure may access this counter's values. Another procedure may also define a variable named *counter*, but there is no connection whatsoever between the two *counters*. If the procedure should call

itself recursively, a new variable named *counter* is created with values which are independent of those in the previous instance. When the recursive instance ends, the program returns to the previous level and value of *counter*. (For those who really want to know: this is achieved by putting all local variables on a stack.)

Strongly typed languages identify local variables by the placement of their declaration – if inside a block, they are considered as local to that block. PYTHON is slightly more tolerant in that variables are initialized local to the function in which they appear, unless explicitly declared as GLOBAL. HASKELL and O'CAML are very explicit about the scope of the variables by using the "in" and "where" terms; in both languages, variables are usually local to individual clauses only. The weirdest thing was in COMAL where variables were global by default, but any function or procedure could be tagged as CLOSED, which endowed it with a namespace of its own, into which global variable could be "imported".

(3) **Parameters:** Parameters
are local variables which are initialized with the *arguments* when the arguments
procedure is called. The procedure write_euro has one parameter cent of type Integer; when "write_euro" is called, an integer expression must be supplied as an argument.

- These are legal ways of calling "write_euro":
    - write_euro(750);
    - write_euro(prev);
    - write_euro(prev + gain - loss);
    - write_euro(gain * 2);
- These are not:
    - write_euro(200, 500);
    - write_euro("much money");
    - write_euro();
    - write_euro(3.1415927);

However, things may be more complicated than this. In particular, some languages, notably those of the PASCAL family, allow modification of the arguments. In ADA this has developed into the abominable situation that a parameter may either be read-only, write-only, read-write, or not accessible at all (whatever that may be good for).

♠ *Why is it bad style to use global variables except for where it is inevitable? How do local variables contribute to the ease of programming?*

♠ *Explain **why** the illegal arguments for write_euro are verboten[1].*

---

[1] If you're having trouble pronouncing this, the Oxford dictionary may help you: [fer'bo:t^{e}n].

♠ *Do you think it is good taste to have procedures which are able to modify their arguments (call-by-reference (p. 26)? If not, why was this probably implemented? (Hint: There are no tuples in* PASCAL*.)*

 Functions and procedures may receive a "briefing" in the form of *arguments*; these are values which are assigned to *parameters* – variables valid only within the subroutine (locally). In addition to the parameters, a subroutine may possess additional *local variables* which are nonexistant from a point of view without the subroutine.

8.1.3 Keep the city clean: garbage collection.

We have seen that variables may be restricted to certain areas of the program. Normally, such "local" variables are attached to functions or procedures, but some languages (e.g., C++ and JAVA) permit one to use variables local to individual blocks of code. At any rate, when the program leaves the area for which the variables are defined, they are removed from the namespace, and their space in memory is released.

For "normal", *static* variables, this is easy enough. However, the higher-level languages also possess *dynamic* variables – that is to say, variables of known size which may change during usage. We have discussed lists before; they are the textbook example of dynamic data structures.

Now imagine a local list variable. While the area of its scope is being processed, elements may be added to the list. What happens upon leaving the area? Clearly, all the elements have to be released as soon as the variable is "undeclared" again. When this has been going on for some time, reserving memory for data items and releasing it again is bound to fragmentize the available memory – a very undesirable thing. It is therefore necessary to do some "reform" now and then, with all the released pieces of memory getting collected and unified. This is referred to as a "garbage collection".

Garbage collection is not a trivial issue, and the lower-level languages – such as C and PASCAL – simply avoid it. There are no lists in either. You may build your own lists using pointers (p. 80) and explicit memory allocation (p. 177), but then it's up to you to discard all the elements of a list before exiting a part of the program which uses this list. Failure to do so will result in "memory leaks", meaning that memory will be reserved and reserved and not released properly. Though this will not cause the program to crash or malfunction, this is the kind of bug not to be taken lightly, especially if the program is supposed to work for prolonged periods of time[2].

There are worse things for garbage collectors than lists. A head→tail list is of well-known structure; by moving on from one item to the next, you may reach – and release – each item. By contrast, objects may allocate memory in a completely opaque way. How can obsolete objects be garbage-collected properly,

---

[2] Not only on space probes but on such earthly things as web servers.

then? We will discuss this in detail when talking about objects (p. 190). Let it suffice here to say that an object must know how to remove itself, and upon leaving the area, each of its objects are told to do so.

 When a function/procedure is finished and the program returns to where the subroutine has been called from, the memory occupied by the subroutine's local variables is released.

8.1.4 Returning values. We have described this on p. 23.

8.1.5 Python's keyword "def". Although we have used this several times before and you are probably intuitively familiar with it, I will now give you a comprehensive overview of PYTHON's magic word `def`. The correct syntax is:

   def *name*([parameter[, parameter]]):

e.g.

   def foo(bar, blah):

or

   def critter():

This heading is followed by an indented block of code. Everything that is indented more than the head is considered as belonging to it.

This defines a function. All PYTHON subroutines are functions; there are no explicit procedures. If there is no `return` x clause provided within a function, the function simply returns `None`.

Parameters are those variables defined in the optional list of parameters. All other variables initialized in the body of the function are implicitly local unless explicitly defined as `global`.

People with a background in more conventional languages such as C or PASCAL, where procedures form a kind of "skeleton", sometimes fail to appreciate the dynamic nature of `def`. It can best be understood in terms of the λ calculus, as it actually assigns an otherwise nameless function, the body of which follows the declaration, to the variable *name*. Although it may effectively be put to the same use as a **procedure** in PASCAL, this is markedly different, for PYTHON function definitions may occur not only inside other functions but effectively everywhere. So the following is perfectly legal.

   def gupta(chandra, surya, loka):
       if chandra == surya:
           def nara(x): return guru(x)
       else:

```
        def nara( x ): return raghu( x )
    return nara( loka )
```

Meaning that, depending on the equality or nonequality of *chandra* and *surya*, different code blocks are bound to *nara* – *guru* in case of equality, *raghu* otherwise. In the last line of *gupta*, the code connected to *nara* is executed, regardless of which one it is. In λ notation, this may become clearer.

```
def gupta( chandra, surya, loka ):
    if chandra == surya: nara = lambda x: guru( x )
    else: nara = lambda x: raghu( x )
    return nara( loka )
```

The "variable" or "lambdoid" nature of def is most obvious when entering the following lines at the PYTHON prompt.

```
>>>def foo(): return "bar"
...
>>>foo()
'bar'
>>>foo
<function foo at 0x330b90>
```

The parentheses specify that the variable is to be evaluated by executing the code connected to it. Without these, the plain contents are returned, which means in this case that the function is "marshalled" into this brief description.

8.1.6 Polymorphism. JAVA and C++ have a more useful mechanism than this modifying of arguments as they allow different definitions of procedures and functions for different parameter sets, i.e., we may define *int square(int n)* and *float square(float n)* side by side, or *dump(String x)* and *dump(String x, int line)*. This is known as *polymorphism*[3].

polymorphism

Upon calling the procedure, the compiler decides which one to use: *dump("Out of cheese error")* will lead to using the first definition, *dump("Out of cheese error in line", 513)* the second, and *dump()*, *dump(1.2345)* and *dump("Out of cheese error", 513, "Press F1 to resume")* will all be rejected by the compiler.

In PYTHON, one should correctly speak of "amorphism", since there is no typing of variables. However, for all practical purposes, we may consider all PYTHON functions as implicitly polymorphic.

8.1.7 Default parameters = optional arguments. Some languages, including PYTHON, allow one to provide default values for parameters: def f(a, b=83): means that the function may be called with either two arguments

---

[3] There is also *ad hoc polymorphism*, which is basically that a function is defined for several types – a feature to be found in HASKELL or CAML.

♠ *Suggest some useful applications for polymorphism and for default parameters.*

8.1.8 Variable parameters. In PYTHON, we may also write functions which accept an arbitrary number of arguments by prefixing the last parameter's name with the "one-for-all" parameter with an asterisk. All remaining arguments will be condensed into a tuple.

♠ *Familiarise yourself with this. Then write a function which calculates the area of a rectangle, with the extra feature that squares may be calculated by passing just one parameter.*

8.2 The importance of not getting trapped between two mirrors: recursive programming

Relevance: Recursion has been mentioned before in the context of functional programming languages; the following chapter describes in more detail how recursion can be put to practical use.
Keywords: Recursive programming.

Actually, the comparison between recursion and iteration should have been #1, but it is impossible to understand recursion without knowing about functions...

The textbook example of a recursive function is the factorial (attributed to Gauss as a boy), which is the product of all numbers up to the argument:

$$n! = 1 \cdot 2 \cdot 3 \cdot \ldots \cdot (n-2) \cdot (n-1) \cdot n$$

This line explains what it is, but it takes common sense to understand it, which a computer does not have. Therefore we need a more exact definition:

$$n! = \begin{cases} 1 & (n = 1), \\ n \cdot (n-1)! & (n \geq 2) \end{cases}$$

This definition is called *recursive* (Latin *re-currere* = to run back) because the definition "runs back" to itself. In every recursive definition, there must be a terminal condition where the result of the function is determinable without further recursion[4]! This is sometimes referred to as **well-foundedness**.

In a functional language, one might write this as

```
let fact n -> if n == 1 then 1 else n * (fact n-1)
```

---

[4] Caravan Of Love: the CAML programmer is following a camel which is following a camel which is following a camel which is following a camel which is following Xavier Leroy.

or, still closer to the mathematical formalism,

```
let fact n =
    | when n == 1 -> 1
    | else        -> n * (fact n-1)
```

So now we know what recursion is, but still this does not make it really clear how recursion may be used to process large collections of data.

Let it be our task to find the highest value out of a list of temperature measurements. This is performed most easily, and without any thought about the indexing of the list, by using the following two definitions (which can easily be implemented using any procedural language).

(1) If there are no values in the list, the maximum is absolute zero.[5]
(2) If there are, the maximum is either the first value in the list or the maximum of the rest of the list, depending on which is higher.

♠ *It is not only functions which may be recursive but data structures may be as well. As a matter of fact, any data structure that contains a reference or pointer (p. 80) to a data structure of the same kind can be regarded as recursive. Please explain this.*

There are many instances (no pun intended) of recursively structured data, and processing them is easy to do in a recursive manner. For example, recursive programming and recursively structured head→tail lists (see p. 86) are natural partners.

Let us have another look at the structure of a list. It may be defined like this.

(1) A list may be empty or begin with an item.
(2) An item has a value. Furthermore, it may or may not have a successor, which is a list.

♠ *Analyze this in terms of a recursive function. What are the cases here? Which case is terminating, which one is recursive?*

♠ *Design a function returning the length of a list, using* `head` *and* `tail` *only.*

Thus we may directly transcribe the above definitions into a purely functional language such as HASKELL:

```
findmax [] = -273
findmax list =
    let h = head list in
```

---

[5] In *The Little Schemer*, Friedman and Felleisen devote almost 20 pages to inculcating the absolute necessity of such a "terminal" case into any recursive calculation. If it is missing, the program will run a few hundred thousand recursions and then break down with a "stack overflow".

```
    let tm = findmax ( tail list ) in
        if h > tm then h else tm
```

This is also possible in PYTHON, but it requires a bit more knowledge of the language in order to understand it (e.g., instead of "head list" and "tail list" we have to write list[0] and list[1:]), hence we will postpone this for a while.

♠ *Explain the* HASKELL *program. (Hint: In* HASKELL, *a function may be defined several times for different values; the compiler converts this into an implicit branching. That is to say, one might begin with "findmax list = if list == [] then -273 else..."*)

♠ *When you feel that you are familiar enough with* PYTHON, *try to implement the FindMax function in* PYTHON – *recursively.*

In fact, recursive data types are almost omnipresent, though rarely thought of as such. Consider for one moment that there are two kinds of data types:

- primitive data types: booleans, chars, integers, floats, strings – in a word, all data types which are completely defined in themselves.
- derived data types: lists, arrays, certain kinds of files[6], sets, bags, and most notably structures and objects – all data types which rely on one or more primitive data type(s). In PASCAL, this becomes very clear thanks to the "of" notation:

```
type intarray = array [ 0..99 ] of integer;
type floatfile = file of float;
type charset = set of char;
type coordinate = record x, y:  integer end;
type chsetarray:  array [ 2..8 ] of set of char;
```

The last line of the example already hints at the recursive nature of derived type definitions (which does not imply recursive nature of types, of course):

(1) A *data type* may be either primitive or derived.
(2) Primitive types are Boolean, Char, Integer, Float, String.
(3) Derived types are homogeneous types or heterogeneous types.
   (a) Homogeneous types are List, Array, Set, Bag. Each homogeneous type is based on one *data type*.
   (b) Heterogeneous types are Structure (Record), Tuple and Variant (Union). Each heterogeneous type contains at least one *data type*.

So you may indeed have the box in the box in the box. A tuple comprising a list of sets of chars and an array of strings, or an array of records consisting of two sets of floats each, in PASCAL even a `file of file of boolean` – all of

---

[6] PASCAL's `file of` data type is very similar to an array, except that the information is not stored in memory but on the disk. In other words, an array is related to a file as, in PYTHON, a list is to a shelf object.

this is perfectly legal and may, with relatively little effort, be compiled using recursion.

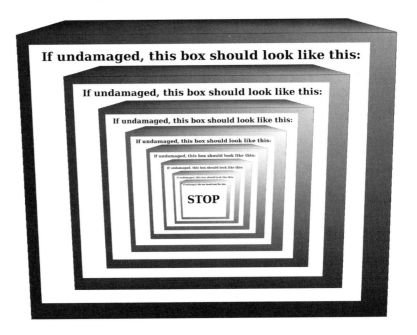

♠ *Draw a diagram that illustrates the definition of data type. Hint: It may remind you of M. C. Escher.*

♠ *Functions and objects have not been mentioned here. How would you classify them?*

♠ *The textbook example for recursion is Fibonacci's function (p. 168), which is 1 for its first two positions and the sum of the two previous ones for all other values: Fibo(n) = 1 for n ⩽ 2, Fibo(n-1)+Fibo(n-2) otherwise. This yields 1, 1, 2, 3, 5, 8, 13, 21, 34... Write a function that calculates Fibo(30).*

The recursive style was introduced with LISP; or rather, LISP was built for recursion (during the development of FORTRAN-I from its predecessors Formac and Comit, dissent arose as to whether recursion was a sensible feature, and the majority of the developers felt it was not good for anything).

Fifty years later, all programming languages support recursion, but its use is still a matter of creed rather than of science. A sensible thing to say is certainly that the power of recursive programming is greatest where we have to deal with data structures which are recursive themselves. All interpreters and compilers use recursion to cope with compound expressions like *sin(cos(tan((a+b)*(c+d))))*. This is rather infrequent with scientific data, which are mostly "flat" repetitions of identically structured pieces of information, and in general, recursion is slower than iteration. Nevertheless, it should be kept in mind as a potential alternative. Once one has got accustomed to it, it is often faster to implement and obtain error-free, but this is a matter of personal taste.

And please, remember the pug from p. 136.

Recursion is an extremely powerful tool but rather alien to the human mind. Therefore, there are no catch-phrases worth remembering here. Dealing with recursion is a skill that requires some training. We will soon have a look at a very recursive program.

8.3 Iterative programming

Relevance: Most program are designed for processing large amounts of data; on a Turing/Neumann machine, this can be done only by processing each item of data in sequential fashion. With the exception of strictly functional and declarative languages, all languages appropriately comprise features to implement such "iterations".
Keywords: Interative programming.

Now let us try this in an *iterative* (non-recursive; Latin *iter* = way) fashion.

(1) Set the maximum to absolute zero.
(2) For all values on the list, do the following:
  (a) Compare to maximum.
  (b) If greater, replace maximum.
(3) Return maximum.

PYTHON's brand of FOR (p. 124) allows us to apply code directly to all members of a sequence:

```
def findmax( list ):
    "Get highest value of a list of temperatures"
    max = -273
    for e in list:
        if e > max:   max = e
    return max
```

This saves us from tinkering with the length of the list, indices and the common "n-1" problem – and it looks quite neat, too. In JAVA, this would make quite a wordy mess, although it is possible (whereas in C or PASCAL it is not: there is no way of finding the size of an array, you have to keep track of that yourself).

```
private static float findmax( float[] list ) {
    float max = -273.0;
    for ( int i=0; i < list.length; i++ ) {
        if ( list[ i ] > max ) max = list[ i ];
    }
```

```
        return max;
}
```

Decide for yourself which of these you prefer.

♠ *What would happen if the line* `return max` *in the* PYTHON *program were indented one position further to the right, i.e., below the "if"?*

♡ *Even if you are sure – test yourself by trying it.*

♠ *Write an iterative version of Fibonacci's function and use it to calculate Fibo(30). Compare its speed*[7] *to that of the recursive one, and discuss the difference.*

♠ *In the 8-bit age, when Anna was a hatchling, a favourite benchmark for computers consisted of calculating square roots, because no floating point hardware was available and floating point operations therefore had to be done with rather complicated software (approx. 5000 instructions had to be executed for the calculation of a single square root). Typical values for a 6502 or Z80 processor, running at 0.9 or 4 MHz respectively, were 10 to 30 minutes for 10 000 square roots (corresponding to a speed of 20 000 to 100 000 instructions per second, where contemporary processors boast of several billions of instructions per second.) Think how you would implement a program to calculate and maybe print 10 000 square roots using*

- *a FOR loop*
- *the MAP function*
- *a WHILE loop*
- *recursion*
- *recursion on a RANGE object.*

 Iterative programming is a style that processes data by the use of loops.

8.4 Digression: the joy of stacks

To obtain a complete understanding of the details of function calls, especially things like tail recursion, it is not absolutely required but may still be helpful to consider the way in which function calls are actually dealt with by the processor. Keep in mind that a functional or declarative programming language is eventually built upon the same processor operations as an imperative one, or even assembly language. So a function call must in fact be emulated internally!

---

[7] On a unixish system, this may be done easily using the "time" command: `time python fibonacci.py`

8.4.1 The program counter. The first thing to keep in mind is that any program code has a one-dimensional structure, or, with less hype, instructions – whether *bona fide* processor instructions or some kind of pseudocode, like that used by PYTHON internally – and are always arranged like beads on a string, one following the next, without a gap, so that every one has a definite position or "address". The processor moves along the program code by means of the program counter – a specialized register containing the address of the next instruction. As the human heart has its diastole and systole, so the heartbeat of the animal inside is a two-phase process, repeated infinitely:

$$\left(\begin{array}{c} \text{Advance program counter to next position} \\ \text{Get instruction at program counter and execute it} \end{array} \right)_\infty$$

8.4.2 Hop, skip, jump. Thus, the simplest way to modify execution of a program is to modify the program counter – resulting in a jump. Jumps may be either **relative** (by adding a positive or negative value to the program counter) or **absolute** (by setting the program counter to a fixed value); this is a question of efficiency which only concerns assembly language programmers. In all higher programming languages, the compiler will know to select the most effective one when translating the program.

A jump is a "go to" statement (p. 138), or rather, a "go to" statement is a jump.

Simply jumping from one point of the program to another usually does not make much sense. Therefore, jumps may be linked to certain conditions. Usually, this is implemented as a two-step process:

Perform comparison operation.
Jump somewhere if result is something.

You have seen a commented example of how conditional jumps are implemented in x86 machine code, and generated from higher-language structures, in the section on "go to" statements (p. 138).

8.4.3 Stack attack. In our chapter on homogeneous data groups, we have encountered stacks (p. 90). A stack is the simplest way of storing variable amounts of data. Think of it as a pile of cards, each holding one piece of information. A new item is added to the stack by placing a new card on top of it (*push*). Access to data happens by taking off and using the topmost card (*pop*). Therefore a stack is also known as a "LIFO" structure for "last in, first out".

Using a register and a pre-allocated piece of memory to the beginning of which the register is set (the *stack counter*), a stack is easily implemented, as long as you do not worry about over-runs or under-runs:

- Push: Write your piece of data to the address denoted by the stack counter; move the stack counter one unit upwards.
- Pop: Move the stack counter one unit downwards; take the piece of data it now points to.

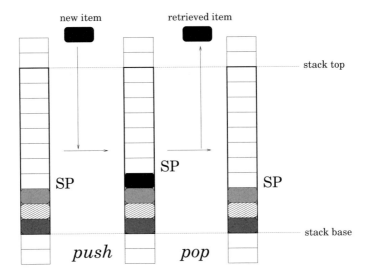

Using C's pointers (p. 80) and unary operators, a complete stack system can be implemented in two lines only:

```
#define PUSH( SP, item ) (*SP++ = item)
#define POP( SP ) (*--SP) // use as function: x = POP(s)
```

8.4.4 The call stack. A stack can be used to keep track of previously visited locations. When jumping from A to B and thence to C and finally D, pushing every location onto a stack when leaving it, we may return from D via C and B to A, just by popping the new location off the stack. In other words, a stack allows one to save and retrieve an arbitrary number of return points.

On the most primitive level, this is used by BASIC's GOSUB and FORTRAN-IV's CALL statements:

```
1210 print "========"
1220 let v% = a%
1230 gosub 1300
1240 let v% = b%
1250 gosub 1300
1260 let v% = c% + d% * e%
1270 gosub 1300
1280 print "========"
1290 end
1300 ' No procedure head or similar. Any line may be entry point for "gosub".
1320 print v%, "  and its square: ", v% * v%
1330 return
```

GOSUB pushes the current line number, RETURN pops it. So at line 1230 the interpreter's "program counter" is set to 1300, after pushing 1230 upon the stack. Execution will continue at 1300, 1310, 1320 and finally 1330, where the new value for the program counter is popped from the stack – 1230. So execution continues at the line following 1230. At 1250, again the line number is pushed, followed by a jump to 1300. This time, the value popped in line 1330 will be 1250, leading to continuation at 1260.

And of course, there is no reason why it shouldn't be legal to perform a second GOSUB from within a subroutine. The program counter will be pushed again, right on top of the first return address, so execution will return first to the original subroutine and only then to the point of the first call.

Now all you have to do is to replace the line numbers by names. This will require only a trifle more work of the interpreter, which will now have to keep a list of the jumping points:

```
1210 print "========"
1220 let v% = a%
1230 gosub OUTPUT
1240 let v% = b%
1250 gosub OUTPUT
1260 let v% = c% + d% * e%
1270 gosub OUTPUT
1280 print "========"
1290 end
1300 OUTPUT:
1310 ' This is the entry point for a gosub statement.
1320 print v%, "  and its square: ", v% * v%
1330 return
```

This is very much what it was like in the QuantumBasic of the Sinclair QL. Now remove the line numbers and let the interpreter keep track of that too (that is to say, have it number the lines only internally):

```
print "========"
v = a
call OUTPUT
v = b
call OUTPUT
v = c + d * e
call OUTPUT
print "========"
exit
OUTPUT:
   print v%, "  and its square: ", v% * v%
   return
```

That's REXX. Now add blocks and let the interpreter keep track of when the subroutine ends:

```
print "========";
$v = $a;           # the $ sign stands for a scalar value
&OUTPUT;
$v = $b;
```

```
    &OUTPUT;
    $v = $c + $d * $e;
    &OUTPUT;
    print "========";

    sub OUTPUT { print $v, "  and its square: ", $v * $v }
```

That's PERL. But why do we have to explicitly call the subroutine by "gosub", "call" or "&"? If the interpreter has to keep track of everything, it may just as well remember that OUTPUT is a subroutine.

```
    PROGRAM OutputDemo;

    var v, a, b, c, d, e: integer;
    /* set variables  to reasonable values here */

    procedure OUTPUT;
      begin writeln( v, "  and its square: ", v * v ) end;

    begin
      writeln( "========" );
      v = a;          OUTPUT;
      v = b;          OUTPUT;
      v = c + d * e;  OUTPUT;
      writeln( "========" );
    end OutputDemo.
```

This is legal in PASCAL. But it is still implemented in the same way, with the program counter being pushed upon invocation of OUTPUT and popped at the close of the subroutine in order to resume work at the next instruction.

Of course, when the subroutine is quit prematurely, the return address is also popped.

What happens at the highest level, when the call stack is empty? Well, actually it usually isn't. This is most obvious in C, where the highest level is the function *main*, which is called by the runtime system, so the call stack *is* never empty – leaving *main* just returns control to the runtime system, which then proceeds to end the program gracefully. But this is also how things are actually handled in other compiled languages too, although in a less transparent way.

8.4.5 The stack for local variables. Using global variables is generally disapproved of. We have seen how a stack can be implemented to deal with subroutines, but how can we achieve proper handling of local variables?

Keep in mind that just hiding a variable from the environment is not enough to make it a local variable. When a function calls itself recursively (see below), as in the Fibonacci function, things have to be kept apart cleanly.

Fibonacci's function is

$$f(n) = \begin{cases} 1 & \text{for } n \leqslant 2, \\ f(n-1) + f(n-2) & \text{for } n > 2. \end{cases}$$

and yields the sequence 1, 1, 2, 3, 5, 8, 13, 21, 34... where every number is the sum of its two predecessors. It is obvious that the function invokes itself twice, and that the values of n must not be mixed up during this process.

Programming languages handle this in a similar fashion, though implicitly. They use a stack for allocating local variables rather than for saving values, but the result is the same. Let us return to the simple example we were discussing before, and use additional local variables to calculate the square and square root of *var*.

```c
#include <stdio.h>  // definitions for printf and friends
#include <math.h>   // we will use mathematics functions too

static int v; // let this be global

void OUTPUT() {
int square = v * v;
float root = sqrt( (float) v ); // convert int to float first
    printf( "%d, its square and its root: %d, %", v, square, root )
    if (v % 2 == 1) { v++; OUTPUT(); }
}

void main() {
int a, b, c, d, e;
    // set variables to reasonable values here
    printf( "========" );
    v = a;          OUTPUT();
    v = b;          OUTPUT();
    v = c + d * e;  OUTPUT();
    printf( "========" );
}
```

This is (poorly written, but working) C ("void" indicates that no value is returned). As we will discuss in more detail on p. 224, a function's local variables are always clustered in one chunk of memory. As the compiler knows which variables are to be used within the function, it may easily calculate the amount of memory needed to accommodate all the local variables – in the above example, one **int**, typically 4 bytes (32 bit), and one **float**, probably 8 bytes (64 bit)[8]. At each invocation of the function, another chunk is placed on top of the previous ones and made the current one, by setting all references to it. After the end of the function, this one will be dropped, and the immediately lower one will become current:

---

[8] The exact sizes are implementation-dependent in C.

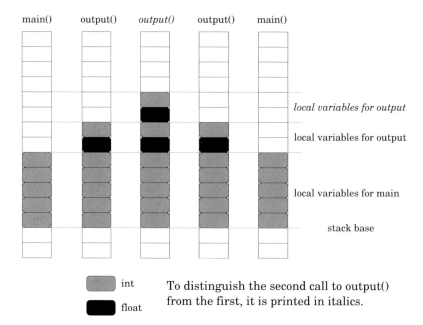

8.4.6 Argument passing. Parameters are largely local variables that must be accessible from the calling function. We will disregard the blasphemous call-by-reference mechanism here and consider call-by-value – the method which agrees with the mathematical concept of functions – only.

There are obviously no special mechanisms required for argument passing, except the initialization, which is usually done using yet another stack. Depending on the implementation, the procedure may vary from tortuous to very straightforward. Most remarkable is the Sun SPARC architecture, which takes measures to ensure that, of the processor's 144 registers, eight are always available for local variables *sensu stricto*, and another eight for parameters. Thus, a function may use up to eight parameters and a total of sixteen local variables and still be ready to go immediately without the need to wait for the tedious stack operations to complete.

```
from math import sqrt
def OUTPUT(v):
    square = v * v
    root = sqrt(v)
    print v, ", its square and its root:", square, root
    if (v % 2 == 1): OUTPUT( v+1 )

# set variables to reasonable values here
printf "========"
OUTPUT(a)
OUTPUT(b)
OUTPUT(c + d * e)
printf "========"
```

Most procedural or higher languages implement a mechanism like the mathematical one for defining the parameters right in the function's head; only PERL and REXX do not, but instead use explicit fetching from a stack or pseudo-variable. This has been described before (p. 25).

8.4.7 The return value. Functions return values, but procedures do not. Actually, that is not much of a difference, especially in highly abstracting languages such as PYTHON where a "no-value" is simply an object named None. The function may either be assigned a value from within, or may use a command like "return" to output its result; the fact remains that there must be some means of returning the result to the calling part of the program.

Luckily, there is no stacking required this time, as the passing of return values happens during the "build-down" phase of a call hierarchy. All we need is a clandestine global variable; in machine-oriented languages such as C, a register is used[9], provided the processor has enough of these.

8.5 Digression: a trick of the tail

As you have learned by now, function call is a subroutine invocation. That is to say, the processor makes a jump to the subroutine, after having pushed the previous address on a return stack (p. 153, p. 166). At the end of the function, the address is taken off the stack, and execution continues just behind the point from which the function was called. So far, so good.

However, if we try to do an infinite recursion, this will fill up the stack pretty soon, as the infinite function never returns! In an imperative language, this can be disregarded, since we have other means of implementing arbitrary repetions there, up to the "infinite loop", so infinite recursion may simply be dismissed as an error. In functional programming, however, there is no other way than infinite recursion to implement endless loopings.

This is where tail recursion joins the fray. Tail recursion optimization means: if nothing is done with the return value, there is no need to push the previous address onto the stack. Thus, the infinite function is effectively reduced to a loop, or, more mystically put, it "operates in constant space". This behind-the-scenes conversion from recursion to looping is a very powerful mechanism even in normal programming and accounts for much of the superior speed of ERLANG.

Compare the following two ERLANG implementations of a function determining the length of a list.

```
length( [] ) -> 0;  % empty list will yield 0
length( [ _ | Tail ] ) -> 1 + length( Tail ).
```

---

[9] This is why in C it is verboten to return arrays, structures or anything else that does not fit into a register. You may, however, return pointers (p. 80) to any of these. The Algol languages have the same ailment, but they solve it in a more gentlemanly way by hiding the pointer mechanism and pretending that actually such things may be passed back.

This is completely correct (and even exemplary), but will need one address to be deposited on the call stack per invocation. Now look at this:

```
length( List ) -> tlength( List, 0 ).
tlength( [], N ) -> N;
tlength( [ _ | Tail ], N ) -> tlength( Tail, N+1 ).
```

Here there is no need to store intermediates, so it is not done. A well-written compiler for SCHEME, ERLANG, O'CAML, MERCURY or CLEAN (where tail optimization is a part of the language standard) will be able to optimize this in such a way that recursive invocation of *tlength* is converted into a mere jump without the need to push the return address on the stack, because after returning from each *tlength*, there are no pending operations to be executed. This is named "tail recursion optimization" because it requires the recursive invocation of the function to be performed last of all the operations in the function.

The two functions are converted into pieces of VM assembly code with a design which could be rendered like this:

```
:length                          ; label for begin
    r1 = POP parameterstack      ; get argument
    TEST-EQUALITY r1, []         ; first clause?
    JUMP-IF-FALSE _nextround     ; skip if list not empty
    SET r0 = 0                   ; r0 holds the return value
    RETURN                       ; end the subroutine
:_nextround:                     ; second clause?
    GETTAIL r1, r1               ; get rest of list...
    PUSH r1, parameterstack      ; ...as argument for...
    GOSUB :length                ; ...recursive invocation
    ADD r0, 1                    ; this is the pending operation!
    RETURN                       ; end the subroutine
```

and

```
:length                          ; label for begin
    SET r2 = 0
    r1 = POP parameterstack      ; get 1st argument (list)
:_linear
    TEST-EQUALITY r1, []         ; first clause?
    JUMP-IF-FALSE _nextround     ; skip if list not empty
    SET r0 = r2                  ; r0 holds the return value
    RETURN                       ; end the subroutine
:_nextround:                     ; second clause?
    GETTAIL r1, r1               ; get rest of list
    ADD r2, 1                    ; not pending any more
    GOTO _linear                 ; no stacking!
```

(The *tlist* label is simply optimized away, unless named in the module's export list.)

Tail recursion optimization allows, under some conditions, legally to recurse into infinity.

8.6 The Travelling Salesman problem

Relevance: Making tracks through possibility space, as in molecule structure prediction, phylogenetic tree construction, etc.
Keywords: Recursion; combinatorics; dynamic programming.

A very nice example for the practical application of recursion is the "travelling salesman" problem. The travelling salesman[10] has to visit a number of cities spread over a certain area. In the interest of simplicity, we assume that connections between all these cities are equivalent, and that the time required to get from one city to another is therefore dependent on the distance only. As time is money, the TS – living in the first city – will do best if he selects the shortest route. So if there are four cities, he may go to either 2, 3 or 4 first. If he opts for 2, he can then do 3 and then 4 or 4 and then 3. If he heads for 3 first...

This is easily implemented using lists and recursions. The way through a number of cities is the way to one of these cities followed by the way to the remaining cities. So our terminal condition is the TS having visited all cities, with no more to be descended upon.

But obviously, with n cities on the map, the number of possible routes is $(n-1)!$ (see p. 159 for an explanation of the factorial). Thus, a complete calculation of a significant number of cities will quickly overcharge any computer. On Zacal Balam, a PYTHON program doing a complete route planning for 10 cities takes about ten minutes to complete, requiring up to 290 MB of memory[11].

What can we do about it? A possible solution lies in *dynamic programming*, meaning that **when we decompose a large problem into its parts and evaluate the subproblems one by one, non-optimal solutions of subproblems will probably not lead to an optimal solution of the entire problem**[12]. This is to say, at every stage we will make a selection of only the more suitable ones.

dynamic programming

---

[10] Political correctness would probably require speaking of a "salesperson" here.

[11] By optimization, this might get *somewhat* better, but considering that addition of one more city would immediately cause processing time and memory usage to rise elevenfold, that would not make much of a difference.

[12] Please do not confuse dynamic programming and dynamic data. The latter are data of variable size, as opposed to static data, which are of fixed size; for example, a list is dynamic, whereas an array is static. Dynamic programming has nothing to do with dynamic data, and in fact it works best with static data. The expression simply denotes a special approach. "Selective programming" might be more appropriate.

The most extreme approach is favoured by our chappie the saleseunuch[13] Thomas "Does Not Compute" W., to whom the nearest city is always the next[14], bringing the total number of routes to be calculated down to n. However, when we take the trouble to do a calculation for a number of randomized cities, we find that the best route is only rarely found. Normally some false decision is made mid-journey, prolonging the route by 5...30%. This is difficult to explain stringently; let us content ourselves by saying that dynamic programming is an approximation, and approximations need not always be correct. A compromise might be to select a number of promising routes, say, to look at the three nearest cities. This would reduce the number of possible routes to approximately 3^{n-1}, or more precisely, $2 \cdot 3^{n-2}$ (less than fourteen thousand for ten cities, or half a percent of the full calculation).

We will return to dynamic programming in Vol. 2. Here is the aforementioned PYTHON script[15]:

```
import random, math, sys, time
def make_citylist( n, mapsize ):
    "Generate random list of n cities spread over a square area
of mapsize * mapsize km. City lists consist of tuples with three
members: x, y and name (number)."
    l = []
    for i in range(n+1):
        x, y = random.randint(1,mapsize), random.randint(1,mapsize)
        l += [(x,y,i+1)]
    return l
def dist( this, next ):
    "Distance between two cities."
    thisX, thisY, dummy = this
    nextX, nextY, dummy = next
    return math.sqrt( (thisX-nextX)**2+(thisY-nextY)**2 )
def findway( citlist, sel=lambda anchor,list:list ):
    class memory:
        "Storing the resulting paths."
        def __init__(self): self.__data = []
        def remember(self,x): self.__data += [x]
        def stored(self): return self.__data
        def sort(self): list.sort(self.__data)
        def show(self,title=""):
            print "\n",title
            if self.__data == []: return
            self.sort()
            for item in self.__data: print item
        def show_extremes(self,title=""):
            print "\n",title
            if self.__data == []: return
```

---

[13] "Business: as cruel as life." OH YEAH!

[14] For non-native speakers, this is also an excellent occasion to appreciate the subtle difference between *nearest* and *next*.

[15] It uses objects (`class memory`) for storing the paths, which have not been discussed in detail so far. Refer to the next chapter (p. 187ff.) for more.

8.6 THE TRAVELLING SALESMAN PROBLEM

```
            self.sort()
            print "min =",self.__data[0]
            print "max =",self.__data[len(self.__data)-1]
    def findways( store, cities, location, citlist, sofar, selector ):
        "The recursive core of findway."
        my_x, my_y, citnum = location
        if citlist == []:
            # If all cities have been visited, calculate
            # the distance and store the path.
            hitlist = []
            for placn in sofar: hitlist += [placn]
            hitlist += [citnum]
            distance = 0
            for i in range(1,len(hitlist)-1):
                loc1, loc2 = hitlist[i-1],hitlist[i]
                a, b = cities[loc1-1], cities[loc2-1]
                distance += dist( a, b )
            store.remember( ("%04d km: " % int(distance)) + str( hitlist ) )
        else:
            # Let's use the selector function to obtain the indices of
            # all the cities satisfying our given criterion:
            eligible = selector(location,citlist)
            for cit in range(len(citlist)):
                if citlist[cit] in eligible:
                    next, remain = citlist[cit], citlist[:cit]+citlist[cit+1:]
                    findways( store, cities, next, remain, sofar+[citnum], selector )
    res = memory()
    findways( res, citlist, citlist[0], citlist[1:], [], sel )
    return res
def best_one( anchor, list ):
    "Select the member of list which is best relative to anchor."
    if list == []: return []
    best, n, mindist = 0, 0, dist( anchor, list[0] )
    for i in list:
        d = dist( anchor, i )
        if d < mindist: d, best = mindist, n
        n += 1
    return [list[ best ]]

li = make_citylist( int(sys.argv[1]), 1000 )
for city in li: print city
t0 = time.clock()
oldway = findway( li )
t1 = time.clock()
newway = findway( li, best_one )
t2 = time.clock()
oldway.show_extremes("Conventional")
print t1-t0,"sec"
newway.show("Best only")
print t2-t1,"sec"
```

♠ *Suggest an implementation for the function* best_three *which returns not a single-item list with the nearest city, but a list of the three nearest cities.*

This is a sample run for ten cities:

```
(283, 648, 1)
(601, 791, 2)
(551, 732, 3)
(871, 798, 4)
(896, 749, 5)
(131, 849, 6)
(983, 70, 7)
(886, 457, 8)
(559, 54, 9)
(752, 381, 10)
(419, 971, 11)

Conventional
min = 2066 km: [1, 6, 11, 3, 2, 4, 5, 8, 10, 9, 7]
max = 6835 km: [1, 5, 9, 4, 10, 6, 8, 11, 7, 2, 3]
682.37 sec

Best only
2479 km: [1, 6, 11, 2, 3, 4, 5, 10, 9, 7, 8]
0.0 sec
```

Now in the age of free enterprise, where scientists strive to emulate shopkeepers and being a pimp makes you almost royal[16], selling is certainly an end in itself, but still one might wonder what this particular exercise is good for.

making tracks through possibility space

The TS problem is the archetype of a problem which can be paraphrased, in Dawkins's words, as *making tracks through possibility space*, and which is of crucial importance in many applications in the sciences, in particular in biology.

Combinatorics is the basis for extreme diversity. Let us put it like this: a small number of a limited set of alternatives may be sufficient to yield a vast number of possible combinations. For the mathematically minded, the number of elements may be seen as the number of dimensions; so a combination of n elements, each of which has m possible "states", will result in m^n possible combinations. For example, for the lowly 5243 base pair genome of the *SV40* virus there is a total of $4^{5243} = 10^{1578}$ possible sequences. By comparison, according to Dirac's estimate, the universe contains "only" about 10^{130} atoms.

SV40

The sequence of a viral genome of this size is easily determined experimentally, but there are tasks that require the identification of one particular alternative among a huge number of possible combinations. Chess is probably the best-known example. There is obviously only a finite number of matches possible, but it is hard even to estimate the number, not to mention what they are.

Two interesting examples from biology are:

---

[16] Among the few more prestigious vocations there are professional boxers, tennis players and racing drivers.

- protein structure prediction;
- generation of phylogenetic trees.

♠ As we all know, saturated non-cyclic hydrocarbons (alkanes) have the formula $C_n H_{2n+2}$. For $n = 1\ldots 3$, only one alkane is possible (methane, ethane, and propane, respectively); for $n = 4$, we have two possibilities (n-butane $= CH_3 CH_2 CH_2 CH_2 CH_3$ and isobutane $= (CH_3)_2 CHCH_2 CH_2 CH_3$); for $n = 5$, 3; and so on. How many forms of "decane" ($n = 10$) and "eicosane" ($n = 20$) are there? And what do they look like? Write a program that generates the structures of all decanes and eicosanes. Hint: A non-cyclic alkane has a recursive structure, wherein each C atom is connected to four residues, each of which may be either an H atom (terminal case) or another C atom (recursive case).

8.7 ◇ Application: cytometry (I)

A widely used method in cell biology is *cytometry* or "facs" (fluorescence-activated cell scanning). Basically, it detects cell surface molecules by using specific antibodies which are chemically coupled to fluorescent groups. The cells are lined up and passed by a detector measuring fluorescence (usually for several colours at the same time) and light scattering (indicative of the size and internal granularity of the cells), so each cell is measured individually. The importance of the technique is due to the fact that almost all cellular actions involve specific cell surface molecules, as the cell surface (rather than the nucleus) is the prime information processing department of the cell.

cytometry

Evidently, rather large amounts of data are generated by this method, which have to be evaluated by a computer in order to classify the cell population as a whole, identify subgroups, shifts and so on. What is the best approach for handling this?

For each and any cell, typically we will have at least the following values, which will be produced ultimately by a call to the cytometer's hardware:

- size – not in nanometers yet but, like the others, as a raw detector result
- granularity
- fluorescence I – usually UV→blue (DAPI)
- fluorescence II – usually blue→green (FITC and GFP)
- fluorescence III – usually green→red (PE and RFP).

For this, a structure or record is the natural form. But how are we going to deal with the plurality? As a matter of fact, we cannot predict how many cells will be passed through the detector; normally the instrument will display a "Press any key to stop"-message and behave accordingly. Thus a list would be most appropriate. However, many widely-used languages do not support lists on the language level. In this case, it would be easiest to use a "homespun" list for collecting the data which is then, after termination of the measurement, converted into an array.

Annasoft, Inc., have just signed a contract with a certain facs supplier whose CEO, having an MBA but not knowing anything about the B which he A-s, originally demanded that everything should be done in COBOL, because that's for business and Business Is Good^{TM}. That could be averted, but now he insists on C, because he wants to sell a *cytometer system* and C was developed for *system programming*[17].

It should not be forgotten that C was co-developed with Unix in the late 1960s. PYTHON incorporates thirty more years of experience, but the concepts developed therein will be the subject of the next chapter. So, before you learn what PYTHON can do for you (p. 240), here is an exemplary solution in C.

First of all, we need the aggregate for a single cell's values:

```
typedef struct {
    // assume the detector yields 16-bit values:
    fc_siz, fc_gran, fc_fluo1, fc_fluo2, fc_fluo3: unsigned short;
} FACS_CELL;
```

We also need an array of cell values. This could be done by writing, e.g., `FACS_CELL data[10000]`. However, we do not know how large the array will be. In C, an array of undefined size can be established using a pointer (p. 85); for practical reasons, this should be accompanied by a number containing the size:

```
typedef struct {
    FACS_CELL *fc_data; // fc_data is pointer to an element of type FACS_CELL
    int fc_array_size;
} FACS_CELL_ARRAY;
```

However, for collecting we should also define a list of cell values:

```
typedef struct {
    FACS_CELL fc_content;     // fc_content is an element of type FACS_CELL
    FACS_CELL_LIST fc_next; // pointer to next element in list
} FACS_CELL_NODE, *FACS_CELL_LIST;
```

At any rate, it must be remembered that C does not permit passing or returning complex values. Instead, pointers to them may be moved around. When we have a pointer variable, a call to the system function *malloc* may be used to allocate memory for it.

malloc

Now imagine that there are the two self-explanatory system functions `read_values_from_cytometer()` and `key_has_not_been_pressed()`:

```
FACS_CELL_ARRAY *get_data() { // create a data block, return pointer to it
   FACS_CELL_NODE anchor;
   FACS_CELL_LIST rover = NULL;
   int listsize = 0;
   /*** get data: ***/
   while (key_has_not_been_pressed()) {
      /* start at list anchor point:                              */
      rover = &anchor;
      /* move to the last item:                                   */
```

---

[17] Many decisions have been made using less sensible reasoning.

8.7 ⋄ APPLICATION: CYTOMETRY (I)

```
      while (rover->fc_next != NULL) rover = rover->fc_next;
      /* write measurement data into the item's components:
       *  (N.B.: C cannot return multiple values nor modify arguments from within
       *  a function, so we pass pointers to the components so that the function
       *  can store the results there)
       */
      read_values_from_cytometer( &(rover->fc_siz), &(rover->fc_gran), &(rover->fc_fluo1),
                                  &(rover->fc_fluo2), &(rover->fc_fluo3) );
      /* append new item to list, or kick the buck if memory full: */
      if ((rover->fc_next = malloc( sizeof(FACS_CELL_NODE))) == NULL) {
         printf( "Out of cheese error" ); exit( 1 );
      } else {
         rover->fc_next->fc_next = NULL; /* no successor to successor yet */
      }
      /* register this by increasing "listsize" by 1:                     */
      ++listsize;
   }
   /*** convert list to array: ***/
   FACS_CELL_ARRAY *result = NULL;
   /* create basic structure */
   if ((result = malloc( sizeof(FACS_CELL_ARRAY))) == NULL) {
      printf( "Out of cheese error" ); exit( 1 );
   }
   /* reserve space for "listsize" elements and register this:   */
   if ((result = malloc( sizeof(FACS_CELL) * listsize )) == NULL) {
      printf( "Out of cheese error" ); exit( 1 );
   }
   result->fc_array_size = listsize;
   /* copy data and degrade list: */
   rover = &anchor;                     /* begin at the beginning */
   int i = 0;
   while (rover->fc_next != NULL) { /* for all dinkum entries       */
      result->fc_data[ i++ ] = rover->fc_content; /* transfer data */
      FACS_CELL_LIST remover = rover; /* remember current item     */
      rover = rover->fc_next;          /* move rover to next one   */
      free( remover );                 /* remove remembered item   */
   }
   /*** that was the show: ***/
   return result;
}
```

The meta-intelligence algorithm identifies intelligence by its attempts to devise algorithms for the identification of intelligence.

CHAPTER 9

Application: Your Most Expensive Pocket Calculator

> If you put garbage in a computer nothing comes out but garbage. But this garbage, having passed through a very expensive machine, is somehow enobled and none dare criticize it.
>
> – ANONYMOUS[1]

Relevance: Conventional arithmetics were designed for the human mind rather than for mechanical processing. It is therefore rewarding to have a look at an algorithm that is suitable for evaluating arithmetic terms.
Keywords: Recursion and other ways of data handling.

You have heard before that iteration is the method of choice for dealing with large, "flat" data collections, whereas recursion is to be preferred for smaller, but structured data accumulations, especially when these accumulations are "nested" and it is therefore not easy to judge how many steps will be required to process them. Programming languages themselves were mentioned as the prime examples for an example of recursive solutions. Remember that the omnipresent block structures (p. 108) were stated to be recursive in nature. A block is a sequence of expressions, each of which may be simple or composite; if it is composite, it comprises at least one "headline" (branching or looping statement, etc.) to which is subordinate at least one block. This is also the deeper rationale for PYTHON's `pass` statement: in the absence of an "end of statement" sign like C's ; this is required to meet the need for a statement.

Later (in the "Lisper" chapters, Vol. 2) we will delve more deeply into this. Here we will apply recursion to solve a more elementary problem which has, however, to be dealt with by any interpreter and compiler (with the exception of those for LISP and FORTH) namely the correct expression of arithmetic expressions. For the time being, we will keep things as simple as possible and not implement things like variable management, but we will write an elementary desktop calculator capable of doing floating-point operations. Of course, this simple calculator will have to be able to cope with operator precedence, and this is the pivotal point where recursion comes into play.

---

[1] One of the most prolific writers on all subjects.

Bioinformatics Programming in Python. Rüdiger-Marcus Flaig
Copyright © 2008 WILEY-VCH Verlag GmbH & Co. KGaA, Weinheim
ISBN: 978-3-527-32094-3

We know from elementary school that $3 + 4 \times 5 + 6 = 3 + (4 \times 5) + 6 \neq ((3+4) \times 5) + 6$. How can we implement this?

Dissecting an input string is easy enough, especially when you have lists at your disposal. The following function does it for you.

```
def scan( string ):
    "Dissect a string into numbers and operators"
    tokens, thistoken = [], ""
    for ch in string:
        if ch in [ "0","1","2","3","4","5","6","7","8","9","." ]:
            thistoken += ch
        elif ch in [ "+", "-", "*", "/", "%" ]:
            if thistoken != "": tokens += [ thistoken, ch ]
            else: tokens += [ ch ]
            thistoken = ""
        else:
            if thistoken != "": tokens += [ thistoken ]
            thistoken = ""
    if thistoken != "": tokens += [ thistoken ]
    return tokens
```

This will transform the string 3+4*5+6 into the list ["3", "+", "4", "*", "5", "+", "6"], and it will handle floating points and interspersed whitespace correctly. Note that this is still a very iterative solution, the proper one for a "flat" problem like this one – transforming a linear sequence of characters into a linear sequence of tokens. But the question remains, how can we process this list of tokens in such a way that operator precedence is respected? If we attempted to do the calculation in a plain iterative way, by performing calculation after calculation, we would inevitably evaluate $3 + 4 \times 5 + 6$ to $((3+4) \times 5) + 6$ like some rural doofus (you know the kind of inbred rednecks who read the *Playboy* at the age of seven but never progress to deciphering the text[2]): $3 + 4 = 7 \longrightarrow 7 \times 5 = 35 \longrightarrow 35 + 6 = 41\ldots$

Operator precedence requires the 4×5 to be evaluated first. But how can this be done in practice?

First of all, let us define the following auxiliary function for processing a list step by step:

```
def HT( list ):
    "Do functional-style [head|tail] separation"
    l = len( list )
    if l == 0: return None, None
    elif l == 1: return list[ 0 ], None
    else: return list[ 0 ], list[ 1: ]
```

---

[2] Luckily for them, these guys may still become famous by boxing, tennis or car racing.

9 APPLICATION: YOUR MOST EXPENSIVE POCKET CALCULATOR

This is a tribute to PYTHON's basically imperative structure – you will remember that purely functional languages do not require such a function but instead use the [] – [H|T] notation for pattern matching (p. 140). The essential thing is that we need something to give us the next in line whenever required. In the olden days, this was usually performed by using a global variable to store all the tokens and a likewise global index pointing at the next token to be processed. More modern compilers are usually based on objects (see next chapter), with the token list being an object that returns the next token whenever the proper method is invoked; this object itself is globally accessible, but the mechanisms for keeping track of the tokens are safely concealed under its hood. Although both of these approaches are somewhat more accessible to human understanding, here we choose the completely functional approach which does not need any global stuff at all but instead relies completely on the passing to and fro of arguments and results. Every function is given the complete list of hitherto unprocessed tokens, and it returns a tuple consisting of the evaluation result for the tokens processed so far and the list of the remaining tokens.

Here the function for evaluating a single term is trivial (in a more complex program, however, this would be the place both for accessing variables and for dealing with additional complexities such as parentheses, which have not been incorporated in the interest of simplicity):

```
def parse_value( expr ):
    "Parse an elementary expression"
    return float(expr)
```

Now for the evaluation proper. **The trick is to have operators with higher and lower precedence evaluated by different functions which are mutually recursive.** If the function which has to evaluate the operators with higher precedence encounters an operator with lower precedence, it will break off and just return the first expression. This is admittedly difficult to understand; let us have a look at it in the self-explanatory graphical form[3]:

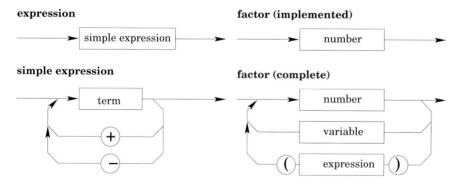

---

[3] The complete definition of the language C comprises sixty such diagrams, to be found in the appendix to the classic "Programming in C" by Kernighan and Ritchie.

term

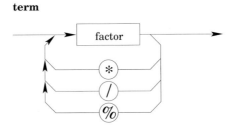

Backus–Naur

Paṇini

This is the visual counterpart of the well-known *Backus–Naur* notation used for describing grammatical structures. It is worth noting here that as early as the 6th century BC (!) the Indian scholar *Paṇini* used a very similar notation to codify Sanskrit grammar.

And this is how it can be implemented:

```
def parse_factor( list ):
    "Parse an arithmetic expression of higher precedence"
    a, rest = HT( list )
    if rest == None: return parse_value(a), None
    else:
        rest_h, rest_t = HT( rest )
        if rest_h in [ "+", "-" ]: return parse_value(a), rest
        else:
            x, y = parse_factor( rest_t )
            if    rest_h == "*": return parse_value(a) * x, y
            elif rest_h == "/": return parse_value(a) / x, y
            elif rest_h == "%": return parse_value(a) % x, y
            else: die_hard( "Unknown Operand '%s'" % rest_h )

def parse_term( list ):
    "Parse an arithmetic expression of any precedence"
    result, list = parse_factor( list )
    if list == None: return parse_value(result), None
    else:
        rest_h, rest_t = HT( list )
        if rest_t == None: return parse_value(rest_h)
        else:
            x, y = parse_term( rest_t )
            if    rest_h == "+": return parse_value(result) + x, y
            elif rest_h == "-": return parse_value(result) - x, y
            else: die_hard( "Unknown operand '%s'" % rest_h )
```

Due to its recursive nature, *parse_term* will always have to return a tuple. When invoked from the top level, the second part of this will always be "None". As we are only interested in the result, which is the first part of this tuple, we may want to write a wrapper function around it:

```
def parse( list ):
```

9 APPLICATION: YOUR MOST EXPENSIVE POCKET CALCULATOR

```
    "Wrapper function for parse_term that returns only the result"
    result, dummy = parse_term( list )
    return result
```

That's it! Now a little more is needed:

```
from sys import argv, exit
def die_hard( msg ):
    "Snuff it (noisily)"
    print msg,"\n"
    exit(1)

# main program
print parse( scan( argv[1] ) )
```

Feel free to extend this by adding the necessary functionality to deal with parentheses. You will see that this will cause the entire program to become highly recursive.

The more advanced, or daring, are cordially invited to have a look at Vol. 2, where the construction of a complete BASIC compiler is described. *Foobasic.py* translates a version of BASIC, very similar to that used on home computers, in the early 1980s, to machine code for the Foobar VM. To this end, it possesses a mechanism for deconstructing, by recursive descent, expressions of arbitrary complexity, including array indices (even nested ones) and comparisons:

```
if 3+alpha*2-(3+beta[7])*gamma==delta[epsilon[4+zeta]*iota]/(8+5*kappa)
then...
```

must be processed properly.

Eggheads unite! You have nothing to lose but your yolks.

CHAPTER 10

The Object-oriented World

> What is this that stands before me,
> A figure in black that points at me?
>
> – BLACK SABBATH: "Black Sabbath"

10.1 What happened so far

Before going on any further, I would like you to review what we have learned about structures and objects so far.

- A structure is an aggregate or compound data type consisting of data of several kinds which belong together by logic (e.g., fields in a form). They may be handled *en bloc* or individually.
- An object extends the notion of structures by incorporating code ("methods") in addition to the data members ("attributes").
- Object members may be visible and accessible to everybody ("public members") or only to members of the same object ("private members").
- Thus, an object may be considered as a "black box" whose internals are hidden (and of no relevance) and which interacts with the surrounding world by receiving and sending "messages".

♠ *Which of the following components must be found in any object that is to make sense? Which should be used sparingly?*

- *public attributes*
- *private attributes*
- *public methods*
- *private methods.*

Explain your answer.

♠ *We have described OO as an extension of imperative programming. Is the purely functional approach also compatible with OO programming; in other words, do you think one day somebody will come up with an* OBJECTIVE HASKELL *compiler?*

It is a bit problematic to talk about OO in general as names for things differ from language to language. So please don't consider any of the following terms as an authoritative *terminus technicus* but focus, as always in this course, on the description of the thing.

For the first time, I will refer almost exclusively to PYTHON. C++ and JAVA have more complex object systems, which are, however, not necessarily more powerful (indeed JAVA itself was born largely out of a desire to reduce the complexity of C++ and to remove features which were found to be more obstructive than constructive). The matter being rather abstract, a comparison between languages would be more confusing than enlightening here.

10.2 Class and instance

It would not make much sense for an object to be the only one of its kind. Objects come in species, as structures (and variables in general) do. Whereas we apply the general expression "type" to structures as well, an object's species is termed its "class". This is partly due to the fact that objects consist mainly of executable code, hence the nomenclature of variables is not too well suited. Moreover, unlike variable types, which are simply abstractions, classes may have a life of their own, but more about that later.

A variable must either be declared (PASCAL, C) or initiated (PYTHON, RUBY) – at any rate, it must be called into existence somehow. Think of a gene which must be expressed. Likewise, a class must normally (exceptions below) be conjured up from the "Platonic" realm of typing possibilities to manifest itself as an object in the real world. With classes, this process is named "*instantiation*". Think of classes as DNA, objects as protein, and instantiation as expression.

instantiation

PYTHON defines and instantiates in the following very straightforward fashion:

```
class ClassName:
    <statement-1>
        .
        .
        .
    <statement-N>

myObject = ClassName()
```

For example,

```
class MyClass:            # this begins the definition
```

```
"A simple example class" # optional documentation string
    i = 12345                # an attribute which is defined by assignment
    def f(self):             # definition of a method
        return 'hello world' # method body

# Instantiate it:
myObject = MyClass()
# Call a method:
myObject.f()
# Access an attribute:
print myObject.i
```

PYTHON instantiates by calling the class as if it were a function that yields an object. C++ and JAVA use the *new* command instead:

```
my_class my_object = new my_class();
```

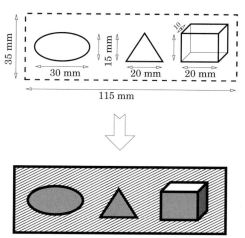

 An object is an instance of its class. *Any* object is an instance of its class. *One* object is *one* instance of its class.

A class is little more than an entry in the compiler's symbol table[1]. Instantiation creates something that is allocated real memory. It is therefore not surprising to see that the opposite is also possible – deletion of objects. Again, this deletion affects only the instance, not the class.

♡ Please make up your mind about the nature of a "type" and its relation to the actual variable. The type is the abstraction (of which there can be only one), whereas the variable is something tangible. One might call the type the

---

[1] For those who really want to know: the code has got to be stored somewhere, of course. What is in the instance is actually implemented as a pointer to the code, i.e., the address where the code is to be found, so multiple instantiation does not produce a litter of identical code fragments.

blueprint and the variable the product. As life scientists, you may compare this to proteins: the type is the amino acid sequence, the actual variable is the individual protein molecule. From all analogies you may conclude that there may be more than one instance of any class. This is indeed so.

10.3 Constructors and destructors

Now you know about instantiation. However, an empty object is rarely of much worth. Usually, it must be filled with values, and quite often, there are default values to be set or other preparations to be made before the object is ready for combat. You may compare this to post-translational modifications. Like these, the code is contained within the object's definition itself.

10.3.1 Baptism. In general, every OO language has a means of defining one method of the object as the standard setup method, also referred to as *initializer* or *constructor*. The syntax of this definition varies greatly. In PYTHON, it is again very easy. The initializer must be called `__init__()`. When a class definition includes a method named `__init__()`, this method will be called (by the interpreter or by the runtime system) for every newly instantiated object. When it does not, no initialization takes place.

initializer
constructor

10.3.2 Extreme unction. As objects can be both instantiated and removed, we can expect the opposite to exist: a function that "executes the testament" of an object before it is sent to Kingdom Come. However, such explicit *destructors* exist only in C^{++}, where the programmer has to take care of the memory management himself – PYTHON and JAVA take care of allocated memory themselves and remove objects automatically once their scope has been left (a process known as "garbage collection"; see p. 156). As de-allocation is the only sensible thing for a destructor to do[2], this completely obviates the need for one[3].

destructors

10.3.3 Godfathers. Sometimes parameters are required for proper initialization. The following is a characteristic piece of code:

```
>>> class Complex:
...     "Library module for complex numbers."
...     def __init__(self, realpart, imagpart):
```

---

[2] Almost. But there may be special objects which represent files, or objects which represent graphical objects. Their destructors will have some additional tasks, of course – making sure that all data are written to the disk, all pending operations finished and the connection to the file system closed properly, or, respectively, causing the part of the screen which they occupied before to be redrawn.

[3] Unless the object reserves explicitly some more memory, of course – then this memory must be released by the destructor `__del__()`. But in PYTHON and JAVA such additional memory also assumes the form of objects which know how to get rid of themselves once their time is over.

```
...            "Generate a complex number"
...            self.r = realpart
...            self.i = imagpart
...
>>> x = Complex(3.0, -4.5)
>>> x.r, x.i
(3.0, -4.5)
```

 Objects are bundles of code and data. Each object is created according to a blueprint named the object's "class". Certain code members are automatically called at setup (constructor) and termination (destructor).

10.4 Self *vs* non-self

In both examples in the preceeding section, you will have noticed a parameter named "self". It will not have escaped your attention, at least in the second example, that this parameter does not correspond to an argument but still seems to be used somehow. What is this?

It is generally necessary for code within an object to specify whether it is accessing variables within the object or somewhere else. PYTHON, where "privateness" of methods and attributes is limited[4], is in special need of this[5]. So a reference to the object itself is passed to each method as the first argument, so the code can access private data via this reference. By convention and in the interest of clarity, the corresponding parameter is usually denoted `self`. (There is no special meaning to this, though. Theoretically, you may use any name, but you are not encouraged to do so[6].)

When you look at the example once more, things will become clearer.

Let us try to see the world from the perspective of the PYTHON interpreter.

(1) I note a definition for a class to be named "Complex"; this definition includes an __init__(a,b,c) method. Will memorize this definition.
(2) Aha, I am supposed to generate an instance of class Complex using 3.0 and 4.5 as parameters and assign that to the variable x.

---

[4] Their name must begin with a double underscore. However, this is not a very powerful mechanism and depends mostly on the politeness of not trying to break it.

[5] And because variables are not declared but just initialized. In JAVA, where they have to be declared, it is usually not necessary to use the "this." self-reference, but in a non-declaring language there is no way of telling whether `a = b` is supposed to initialize *a* as a data member of the object or just a local variable belonging to the method. PYTHON always assumes that this is only a local variable; to initialize it as a data member of the object, use `self.a = b`. RUBY uses a more elegant formalism which will be described later (p. 193).

[6] Editors with syntax support will generally pay special respect to the word "self." when set to PYTHON mode.

(3) First, I allocate enough memory to place the contents of one object there.
(4) Second, I call the __init__(a,b,c) method, using (according to my usual custom) a reference to the newly created object as first, 3.0 as second and 4.5 as third argument. Fits.
(5) Now I am in the __init__(a,b,c) method. The two latter parameters are to be assigned to something, which I can identify as connected to the reference to the object itself. In other words, I am meant to store 3.0 and 4.5 in data members of this very object. The data members are named r and i (which I create at this moment).
(6) My work on the object is completed. It is now lying around in memory, connected to the variable x, until I should be caused to remove it[7].
(7) See, I am requested to dump the members r and i of the objected x. Find them in the symbol table... no problem. There they are: 3.0 and 4.5.

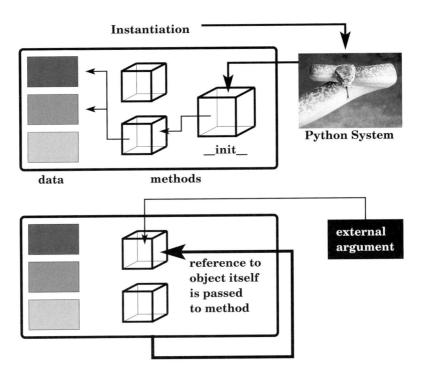

This can be suppressed using the function *staticmethod*, which converts a normal class method to a static method. However, usage of this is not recommended.

---

[7] Which will be at termination time, as this is in the toplevel loop.

The handling of an object's data is more complex than that of a structure's, because code can be "inside" or "outside" the object. In the former case, data belonging to the same object (the "self") have to be discriminated from other data. PYTHON does so by passing a reference to the "self" to each of its methods upon invocation.

10.5 The serpent and the crystal: Python *vs* Ruby

At this point, let us compare PYTHON with another scripting language which is very similar to it, but was designed with the deliberate purpose of avoiding what its author perceived as PYTHON's weaknesses: RUBY. RUBY differs from PYTHON in that it uses lexical conventions to discriminate between scopes of variables: Constants begin with a capital letter, local variables with a small letter, global variables with the prefix $ and instance variables – attributes – with the prefix @. Thus, the need for the awkward "self" parameter is obviated:

```
class Complex
     def initialize( realpart, imagpart )
          @r = realpart
          @i = imagpart
     end

     def value              # return textual representation
          "(#{@r},#{@i})"   # in Python: "(%d,%d)" % (self.r, self.i)
     end
end
=> nil

x = Complex.new( 3.0, -4.5 )
=> <Complex:0x300748f8 @i=-4.5, @r=3.0>

x.value
=> "(3.0,-4.5)"

x.r
=> NameError: undefined method 'r' for #<Complex:0x300748f8>
        from (irb):15
```

At the same time, RUBY's instance variables are *eo ipso* strictly private and can be accessed only via methods (hence the need for the method *value*). A general convention is to define "getter" and "setter" methods for accessing the attributes, the former of the very name of the attribute, the latter suffixed with an equality sign (i.e., Complex.r and Complex.r= for @r, respectively). This can be automatized using the *attr_accessor* class, which will implement these functions of its own accord when incorporated into a class. Further conventions are that methods with a name which ends with a question mark should return boolean information about the object, and methods whose name ends with an exclamation mark should be such as modify the internal state of the object.

Many things in RUBY depend on such somewhat whimsical conventions; for example, $! is the global error object that contains the last error, and $!.message returns its textual representation.

Instantiation is less "functional" than in PYTHON, where a class is considered as a function that returns an object. Instead, classes are objects themselves, as in SMALLTALK, and thus new objects are created by invoking the *new* method. This allows one to define alternative constructors (whoever would need such a thing?).

Inheritance (p. 196) occurs using the < sign:

```
class ImprovedComplex < Complex
    ...
```

In PYTHON, objects can be modified "on the fly" by adding new attributes; in RUBY, this behaviour is restricted to the special class *OpenStruct*. Decide for yourself which approach you like better.

10.6 Primitive types *vs* objects

A "primitive type" is not a redneck but simply a non-compound data type. For example, the type *Integer* is primitive. The coexistence of primitive types and objects is somewhat strained. In SMALLTALK and RUBY, the plainest way has been followed. Here all types are classes, and all variables are objects – so there are no primitive types at all. In JAVA and PYTHON, this is not so. JAVA has the problem of being strongly typed, so it ends up with a class `Integer` corresponding to the primitive type `int` and so on. In PYTHON, this is avoided by weak typing.

♠ *With the class `Integer` corresponding to the primitive type `int`, what would you expect the constructor of `Integer` to look like? What other members are required for this class? (Hint: The entire thing ought to be nominated for the Very Silly Award if it were not absolutely necessary with JAVA's skewed type system.)*

♠ *JAVA also implements strings and lists as objects. What do you think about that?*

 A primitive type is a type that is not an object. Some OO languages (PYTHON, JAVA, EIFFEL) have a coexistence of primitive types and objects, others (RUBY, SMALLTALK) use objects only.

10.7 Modules and statics

Let us look at something superficially different. We have been using PYTHON's module system for quite a while without wasting much thought on it. A module is a collection of software snippets thematically belonging together; e.g., the `math` module contains all the code for higher mathematical functions. And of course there are internal variables, too, because executable code must always include variables.

In other words:

Module = pieces of code + variables

Class = pieces of code + variables

So can we say that modules and classes are the same thing? Not quite: Classes have to be instantiated to objects, whereas modules are immediately available in the real world (with the logical consequence that there can only be one specimen of each module activated, unless the interpreter gets screwed up really badly). We may say that modules are *static*.

The two concepts were developed quite independently too. SMALLTALK originally featured objects but not modules, whereas MODULA-2, as the name implies, has a powerful module system (pretty much like PYTHON) and structures, but no objects. However, in the 1980s the similarity was noted, and C++ and JAVA programs sport a wild mixture of "static" and "non-static" components within a single class, as in these languages the class is the one and only structure[8].

A PYTHON class may contain both static attributes (class variables) and instance attributes. The difference is that the former are initialized outside of methods:

```
>>> class Complex:
...     "Library module for complex numbers."
...
...     precision = 10
...
...     def __init__(self, realpart, imagpart):
...         "Generate a complex number"
...         self.r = realpart
...         self.i = imagpart
...
>>> x = Complex(3.0, -4.5)
>>> x.r, x.i
```

---

[8] Unless you write your C++ stuff in a thoroughly non-OO fashion – an amazingly large number of former C folks actually do this, without being able to explain why they are using C++ at all, considering the infuriating sloth and sometimes insiduous output of C++ compilers (it is not for nothing that Apple for a long time preferred OBJECTIVE C.).

```
(3.0, -4.5)
>>> y = Complex(2.0, 1.0)
>>> y.r, y.i
(2.0, 1.0)
>>> x.r, x.i
(3.0, -4.5)
>>> Complex.precision
10
```

This means that r and i, as instance variables, are parts of the individual manifestations of the class. $x.r$ is a different entity from $y.r$, stored in a different place. When changing $x.r$, $y.r$ will not be touched, and vice versa. Neither of the two exists before instantiation.

precision, however, is a part of the class itself and may be accessed prior to any instantiation, by `Complex.precision` . A real-life example of such a class variable is `math.pi` .

You will not find class variables in the previous illustrations, and they are actually among the features whose use I personally discourage. Especially in PYTHON, where there is no way to protect class variables from being modified[9], they are a likely source of trouble. If somebody should think it necessary to reset `math.pi = 4`, the results will probably not be really amusing.

As for the way of declaring instance variables and class variables, remember what we said about the dynamic or "lambdoid" nature of PYTHON functions. This also holds true for PYTHON classes. A PYTHON class is created by executing all the statements of the indented block subordinate to the class header, thus the "free-floating" initializations are executed immediately, whereas the methods are not run before instantiation.

10.8 Inheritance

The crowning achievement of OO is that classes may be used as a basis on which to build new classes. In other words, classes may be extended. This process is named *inheritance*:

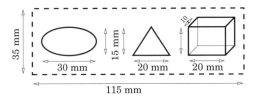

---

[9] Excepting built-in modules – or at least these *should* be write-protected. I have not been able to figure out why at least some of them can still be modified.

10.8 INHERITANCE

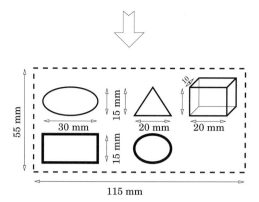

Thus, objects are always arranged in a hierarchy. There is always a *root class* whose objects do not contain anything, and which is not fit for anything, except that it serves as a base class for everything else. In PYTHON, this root class is named `object`. See p. 232 for more details.

♠ *What do you think inheritance is good for? What is to be preferred: single inheritance, where one class may be derived from exactly one other, or multiple inheritance, where one class may incorporate several others? What about recursive inheritance: is it possible, is it sensible?*

In PYTHON, inheritances focuses on methods more than on attributes.

♠ *Why?*

The opposite philosophy is to be found in OBERON, which avoids the very keywords "class" and "object" and insists on treating objects simply as an extension of the old "record" concept. Thus, records may be derived by inheritance, and functions are simply attached to them. Of course, you can use "records" in PYTHON too – a record is simply a class without any code members:

```
class record: pass
```

To this, data members may later be added at will, thanks to PYTHON's weak typing:

```
boss = record()
boss.membershipNumber = 23
boss.firstName = "Dolores"
boss.lastName = "Peinemann"
```

(Thus, individual manifestations of the class may have different data members. To check which ones are available, use the `dir` function, which will yield a list containing the names of the object's members. This is a pecularity of LISP which the scripting languages have inherited.)

♠ *Compare* OBERON, PYTHON *and* JAVA.

An important aspect of inheritance is that the derived class may not only add new functionality but may also modify the old one. Methods from the base class may be *overwritten*. We will see the purpose of this when discussing the use of OO for graphical user interfaces. Finally, PYTHON supports multiple inheritance – meaning, in effect, that a class may contain the components of more than one base class, plus the additions. But this is bound to cause problems due to the potential for ambiguities – what is to be done if we wish to derive a class from two bases which define methods with the same name but different meanings[10]?

 Inheritance is the act of creating a new class by extending an existing one.

10.9 Ram It Down!

We have just learned that a derived class may replace (overwrite; some also say override) its base's methods with its own ones. However, what if we need *both* the new and the old methods?

At first glance, this seems to be a rather academic question, because it is hard to imagine what situation might make this necessary. It is an issue, however, where standardized methods are concerned. For example, __init__(), the constructor. There can be only one per class. The PYTHON interpreter will execute the derived class's __init__() method upon instantiation of the derived class, *y lo basta*. So what can we do if the base class has a constructor that provides vital functionality? Do we have to rewrite it?

Let us have a look at the traditional graphics example:

```
class point:
    def __init__( self, x, y ): self.x, self.y = x, y

class paintstyle:
    def __init__( self, fillcolor, hatch ):
        self.fillcolor, self.hatch = fillcolor, hatch

class circle( point, paintstyle ):
    def __init__( self, x, y, radius, fillcolor, hatch ):
```

Now what is that constructor going to look like? We may, of course, write self.x, self.y, self.radius, self.fillcolor, self.hatch = x, y, radius, fillcolor, hatch and wonder what inheritance is good for. Luckily, there is a way out:

---

[10] O'CAML seems to be the only language which has found a satisfactory, albeit complex, solution to this problem. See http://caml.inria.fr/ocaml/htmlman/manual005.html for details.

```
point.__init__( self, x, y )
paintstyle.__init__( self, fillcolor, hatch )
self.radius = radius
```

It is easy enough to understand what the lines `point.__init__(self,x,y)` and `paintstyle.__init__(self,x,y)` do: they call the respective base class's constructor.

Some programming languages (e.g., JAVA) have the general term `super` to refer to a class's base class. Of course, this does not mix well with multiple inheritance, as we have applied it in this brief example.

It would also be conceivable that the base class's constructor could be invoked automatically. This feature is not desirable for two reasons. First of all, it might slow down the generations unduly if the class's pedigree is long (and generally, there tends to be more inheritance than the programmer is aware of; see below on the `object` class). Second, it would, of course, reduce the control one has. Sometimes one may wish *only* to have the new method. PYTHON actually combines these things naturally – if there is no constructor defined for the derived class (overwriting the base class's constructor), the base class's constructor is still valid, because there is no reason why it should not be, and it is invoked upon instantiation of the derived class.

In general, this is one of the areas where OO languages differ most from each other, so do not take anything for granted here.

As an aside, it will be noted that inheritance potentially destroys orthogonality (p. 100) by introducing new structure members. A hallmark of good OO design is that this undesirable effect is minimized, i.e., that descendant objects are as similar as possible in behaviour to their ancestors.

10.10 Inheritance, abstractness, and privacy

Basically, a method or an attribute will always be visible from within its class but may be visible *or* invisible from without. This leads us to the question what "outside" means here.

JAVA, for example, allows a further differentiation than does PYTHON. Here it can be described neatly as to whether a method may be overwritten, and if so, how. PYTHON's approach relies more on a programmer's politeness not to overwrite methods without good reasons for doing so.

In Python, all methods and attributes are considered to be public unless their names begin with a double underscore (__) – these are private. Private methods are not accessible except from within the object.

JAVA also allows definition of *abstract* classes which may not be instantiated but can only be used as base classes for inheritance, and of *final* classes which can only be instantiated, not inherited. There are no such things in PYTHON.

No contemporary OO language offers the possibility of "disinheriting" features, that is to say, dropping them during inheritance[11]. When you have a look at the visual example presented below, you will agree that this might sometimes be a useful feature. However, this is impossible to reconcile with a central concept of OO programming:

Whatever can be done with or to the objects of a certain class, this can be done with or to the objects of all classes derived from this one.

In a scripting language like PYTHON, there *might* be workarounds, but in any compiled language the possibility of having features vanish would almost certainly result in references to Kingdom Come.

10.11 A different point of view: a look at Smalltalk

The only difficulty with OO programming is that the concept of classes and objects is rather far away from anything we encounter in our daily lives, so it is not easy to find suitable similes, in the same way as you may explain a function in terms of a slot machine taking a value and returning another one. This is also the reason why nomenclature varies strongly between different languages.

So far, we have followed a historical or mechanical approach, by tracing the development of the notion of objects and explaining them mostly in terms of structures which we have encountered previously. This is the most suitable way, as many OO languages, including PYTHON, C^{++} and JAVA, are built on top of a non-OO core. But there are also the mandatory OO languages such as RUBY (strongly influenced by PYTHON), EIFFEL and SMALLTALK, where *everything* has to do with objects. Their authors (especially those of SMALLTALK, which is by far the oldest of the three) decided to deliberately break continuity and impose a new mindset, a different attitude to programming, on people. As the British humorist Terry Pratchett put it: "Ninety-nine percent of witchcraft consists of that 'Z' stuff, whaddaya call it, I think it's named *zychology*". This also holds true for programming. However, people generally object to changing attitudes (no pun intended). This is why EIFFEL and SMALLTALK are still considered as outsider languages, and SMALLTALK has been ousted from its very centre of gravity, the programming of graphical user interfaces.

However, the nomenclature of SMALLTALK may be found enlightening. Keep in mind that SMALLTALK classes and objects are the same things as PYTHON

---

[11] Unless you consider the refusal to bequeath private members unless explicitly marked as inheritable ("protected"), as in C^{++} and JAVA. However, this is restricted to members which are private anyway and do thus not appear on the "surface" of the objects. For the reason given in this paragraph, public members must by all means be inheritable.

classes and objects. Still SMALLTALK refers to "messages" where in PYTHON we have "methods". The rationale is that SMALLTALK's approach is more based on "zychology", trying to inculcate a certain image into the programmer.

According to SMALLTALK, an object is a "black box" – an indivisable entity, an *a-tómos* in the literal sense. You do not call a black box's methods or read its variables. Instead, you send a *message* to the black box, telling it to do something. The black box will react by changing its attributes and returning information.

Imagine the following PYTHON lines:

```
my_wonderful_file = file( mwf_name, "r" )
my_wonderful_data = my_wonderful_file.readlines()
```

As you see, we first create an object (the constructor taking an important parameter, the file's name), then we invoke the `readlines` method to obtain data from this object.

According to the SMALLTALK philosophy, this is interpreted as the object *my_wonderful_file* being sent a message and condescending to respond to this:

(1: object creation is invoked; 2: system provides object; 3: object is sent a message; 4: object yields data.)

This is somewhat reminiscent of the Homunculus in Goethe's *Faust*, which is created by Wagner's alchemy and has for its short life span the power to answer virtually any question.

In fact, SMALLTALK is so extreme in this that there is no other way to do even the simplest of operations. Consider the line:

    'Hello, world' printNl !

You may guess that it prints the "Hello world" message[12], although at first sight it looks somewhat twisted. Actually, it is not, but deals with things in a special way: `'Hello, world'` is a string object which is sent the message `printNl` – in PYTHON's terminology, the string object possesses a method named "printNl" which is invoked. There are no methods, functions or other actions which are not attached to any object, as is PYTHON's `print` command[13].

10.12 The coffee machine: a visual example

Imagine you have to design the control elements for a coffee machine. Basically, there are two kinds of people in contact with this: the normal coffee drinker who can get his poison out of the machine (including a selection of available brands, a display of price and money inserted so far and the possibility of cancelling his order and getting the money back), and the manager who is in charge of the device. The manager will also want to get a cup of coffee now and then like the other folks, but in addition he must be able to check whether the machine needs refilling, and if so, it would be nice if he could alert the guy in charge. The latter functionality ought to be hidden from the rank and file to avoid fooling around with the machine.

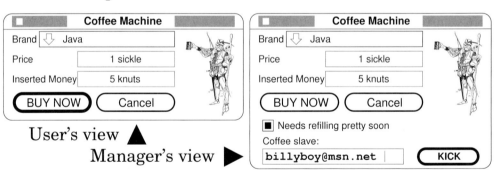

It is obvious that the manager interface can be derived by inheritance from the user interface, which in turn is derived from a base class named "DialogBox" or something similar, by adding graphical objects. This is by no means a pun; members of objects may be objects in themselves. In this case, a suitable class for the user interface box will be derived from "DialogBox" by adding

---

[12] One of the invaluable contributions, apart from C, by Kernighan and Ritchie.

[13] Of course, this immediately raises the question of which object a method involving two or more different objects should be attached. In this case, when writing a string to a file, should the writing method be a part of the string object, as `string.write_myself_to( file )`, or of the file object, as `file.write_into_me( string )`?

elements of type "TextElement", "TextField", "PopUpMenu" and "Button"[14]. In pseudocode, this would look somewhat like the following text (please keep in mind that this is just meant to illustrate the concepts and is not based upon any genuine package!):

```
from WIMP import TextElement, TextField, PopUpMenu, Picture
from HardwareInterface import GetPrice, GetInsertedMoney

class UserInterface inherits DialogBox
   {- DialogBox possesses a title, which is defined by string  Title.
    - It also provides the routine necessary for displaying the whole thing.
    - Furthermore, it comprises to buttons, "OkButton" and "cancelButton",
    - each of which possesses several attributes, e.g. name, "defaultness"
    - (indicated by a fat borderline), additional actions, etc.
    - CancelButton's text defaults to "Cancel". After displaying the box,
    - pressing "OK" returns 1 and "Cancel" 2.
    -}
   Title := "Coffee Machine"
   OkButton.Text := "BUY NOW"
   OkButton.IsDefault := True
   {--- now for the new elements ---}
   T1 := new TextElement                   {- generate it -}
   T1.text, T1.x, T1.y := "Brand", 10, 10 {- set its attributes -}
   T2 := new TextElement                   {- and so on   -}
   T2.text, T1.x, T1.y := "Price", 10, 30
   {- often there are constructors available comprising the most important
    - attributes. This might look as follows: -}
   T3 := new TextElement( "Inserted Money", 10, 50 )
   I  := new Picture( "/bitmaps/robinbooze.png", 150, 10 )
   PM := new PopUpMenu
   PM.Items, PM.x, PM.y := [ "Java", "Peyotl", "Vomissement de la reine" ], 30, 10
   {- So far, all elements were static. Now we want to introduce a text element
    - which is dynamic in that it may be changed while the box is upon the screen.
    - (Changed by software runing in the background, that is to say.)
    - This can be achieved by setting its property "DynamicUpdater" to some function
    - which yields the data of interest; here this is taken from "HardwareInterface"
    -}
   F1 := new TextField
   F1.x, F1.y := 30, 30
   F1.DynamicUpdater := GetPrice
   F2 := new TextField
   F2.x, F1.y := 30, 50
   F2.DynamicUpdater := GetInsertedMoney
end class

from WIMP import Indicator, TextEditField
from HadrwareInterface import GetneedsRefill
from Network import SendShortEmail

class ManagerInterface inherits UserInterface
   {--- change inherited components: ---}
```

---

[14] Very often (especially in JAVA), there is an intermediate layer responsible for the arrangement, so the coordinates do not have to be set for each element.

```
    OkButton.Isdefault := False
    {--- add new ones: ---}
    R := new Indicator
    R.x, R.y, R.DynamicUpdater := 10, 100, GetNeedsRefill
    T4 := new TextElement
    T4.x, T4.y, T4.text :=" Coffee slave:", 10, 120
    Person := new TextEditField
    Person.x, Person.y, Person.text := 50, 120, "<Enter email here>"
    KickButton := new Button( 100, 120, "KICK" )
    KickButton.IsDefault := True
    {- The attribute "Text" of a textEditField can be used not only for setting
     - but also for getting the current text entered therein. This is OK as long
     - as we want to access it in a conventional way. However, if we want to
     - automatize the sending of emails, we have to resort to a method named
     - GetText, which may be defined simply as "return self.text". What's the
     - difference? Well, the line  ... := SendShortEmail( Person.Text, ... )
     - invokes the constructor upon displaying of the dialog box with the value
     - which "text" has *then*, i.e. "<Enter email here>". By writing
     -    ... := SendShortEmail( Person.GetText, ... )
     - we make sure that the *current* value of "text" is transmitted
     - to SendShortEmail.
    -}
    KickButton.whatIfClicked
        := SendShortEmail( Person.gettext, "Refill coffee machine NOW" )
end class
```

Somewhere else in our program this is then invoked:

```
From SystemInterface import CurrentUser, UsersHashtable
From HardwareInterface import MakeCoffee

if CurrentUser == UsersHashtable( "Manager" ) then
    ubox := new ManagerInterface
else
    ubox := new UserInterface
end
PressedButton := ubox.Display
if PressedButton == 1 then
    MakeCoffee( ubox.PM.SelectedItem.Text )
end
```

The expression `ubox.PM.SelectedItem.Text` gets the string of the selected item of the popup menu. Formally, it means accessing a component of a component of a component of *ubox*, which is an instatiation either of *UserInterface* or the derived class *ManagerInterface*.

The *KickButton* demonstrates the great advantage of OO in GUI programming. This object takes care of itself and performs all required actions whenever a *ManagerInterface* is displayed. It is possible to do GUI programming in non-OO style, of course, but in this case the invoking program will have to watch the *KickButton* (if it is the *ManagerInterface* that has been selected...) and send the mail.

The entire program is not particularly lovely. There are many "axioms" about what is (already) contained within certain classes. Well, this is the trouble about GUI programming in general. Every "toolkit" provides its own classes, with their specific attributes and methods, and this may amount to several dozens or maybe even hundreds of individual classes. Well known GUI toolkits are Gtk+, XView, XForms, FLTK, Newt, Motif and Qt. The JAVA VM wraps up the system functions into its own, but even there you have two different toolkits, AWT and Swing. Have a look at Sun's JAVA page (http://sun.java.com) to get an idea of the full complexity of this. The TCL/TK package aims at unifying things across language barriers.

10.13 Tracking and banking

10.13.1 Purpose. The following is a nice little example of combining objects with functional elements to create "building blocks" with maximum flexibility.

Very often a program has to accumulate values non-locally. For example, the identification of prime numbers by means of the "Ophidian Sieve" requires keeping a scroll of all prime numbers found so far, to be used for checking every new candidate. As a more real-life example, you have a piece of software that recursively traverses an entire directory (by using the library function os.path.walk; please consult the official PYTHON documentation for details), scans all files the name of which matches a given type, and looks for a certain pattern. Even if the program is intended to be no more than a simple command line tool or CGI script (see Vol. 2 for more on CGI scripting) that simply dumps its results to the standard output once it has obtained them and can then be allowed to forget them, we might be interested in keeping track of the number of files and directories visited so far, and maybe also of the data volume processed so far.

Let us focus on these three things:

(1) *directories_visited* – os.path.walk does all the directory walking for you by invoking your function for every directory in the tree, passing its name and the lists of its contents to it. Thus, *directories_visited* will just have to be increased by 1 every time the function is called.

(2) *files_visited* – the function is invoked by os.path.walk with a list of all the files in the respective directory, so by looping or mapping, all the files can be processed. Thus it is easy to increment this by 1 for every file that has been touched.

(3) *bytes_read* – the files whose names are in the list can then be processed; in PYTHON, their contents are most easily obtained by x = file(name, "r").read(), which returns an object whose len is the file's size in bytes. This size should be added to *bytes_read* after the file has been processed.

Now the function is called separately[15] for each directory, and after leaving a function, its local values just evaporate (p. 164). So we have to find another way of accumulating our values.

10.13.2 Strategies. Basically, there are six of them:

(1) **Global variables.** This is probably the most straightforward solution: *directories_visited*, *files_visited* and *bytes_read* are globally accessible, and simply add your increments to them. However, it cannot be said often enough that global variables are simply BAD. Imagine a really big piece of software, with hundreds of thousands of lines of code and maybe several hundred mechanisms performing some kind of recursive descent with accumulation of variables: trouble will be inevitable. Actually, in most programming languages variables are not that global after all, but restricted to the module or section where they are defined, unless explicitly declared to be completely global – but this only alleviates the problem without really solving it. Moreover, there will always be the problem of uninitialized or wrongly initialized globals.

(2) **Static variables.** This is one of the features of the C language family which really strikes newbies as neat, but is soon afterwards found to cause more problems than it solves. A static variable has all the characteristics of a normal local variable, with the exception that it will maintain its value during separate invocations of a function. Also, when a function calls itself recursively, the value will be shared, and modifications made during the recursive call will prevail afterwards.

Statics are actually just semantic sugar for globals. They are placed within the same memory segment as globals, and all the problems about uninitialized or wrongly initialized globals also apply to them. Except for the retriction of their visibility to the local namespace, they *are* globals. Just to cause some more confusion, in C even global variables can be declared as "static", in which case their visibility is restricted to the module in which they occur, whereas globals without this epitaph will be visible from all other modules. Now actually C does not really support modules, but a file which is compiled *en bloc* implicitly serves as a module. Things become even more confusing when looking at JAVA: there a "static" variable is a class variable, and a non-static one an instance variable.

(3) **"Writable" parameters:** This is one of the features of the ALGOL language family *sensu stricto* which really strikes newbies as neat, but is soon afterwards discovered to cause more problems than it solves.

---

[15] No order of execution whatsoever is guaranteed. In theory, if you have a multiprocessor system, it may even happen that several directories are processed at the same time.

In the ALGOL family, a function's parameter list may declare parameters in such a way that the function is entitled to modify the arguments which it has been invoked with[16]. In PASCAL, MODULA-2 and OBERON, inclusion of the keyword VAR into the parameter type defines this as a "writable" parameter. ADA takes things even further, by requiring a declaration for each parameter whether this is intended to be read-only, read-write or write-only (whoever would use such a thing).

The notion of a "writable" parameter, known under different names, clearly violates the mathematical concept of functions and fits only into a *very* imperative background. It offers plenty of opportunity for inexplicable bugs and hard-to-find shifts in any variable's values. Actually, it was introduced in ALGOL-58 to provide a mechanism for returning multiple values, long before the concept of tuples was born – or before computers were powerful enough to run compilers and interpreters sophisticated enough to deal with tuples. Now, almost fifty years later, this unclean concept ought to be abandoned once and for all.

A similar thing can be implemented in C and also in FORTH by the use of pointers (p. 80), but this is really nasty. Let it be sufficient here to remember that a pointer is a variable that holds another variable's address in memory. Thus, by passing addresses of variables to functions, rather than passing the actual variables, we can cause all invocations of the function to access the same variable without having to resort to globals or statics[17]. However, pointers are intrinsically ugly and have a tendency to destabilize programs.

(4) Arguments and return values are The Way when using purely functional languages, and surely they are the cleanest and most precise way of handling things. Functional programming keeps the country clean simply by forbidding non-local variables, changes of values and things like that. There are no back doors; everything has to get in and out in an orderly fashion through the front door.

In the next chapter (p. 243) you will see that it is actually feasible to build such a non-locally accumulating program based completely on recursion, lists, parameters and function values. However, if more than one scroll has to be kept, this is going to require some mental discipline beforehand. When disentangling returned tuples, the type-checker of strongly typed languages is your best friend. HASKELL is

---

[16] Of course, such a parameter needs a variable as an argument. If you try to pass a constant, the compiler will consider this as an error – if it is well-written. Poorly written compilers will not notice this, and the outcome will be system-dependent but surely not pleasant.

[17] For the very curious: the "writable" parameters of the ALGOL languages are actually implemented using pointers, as are object references in compiled languages. However, the compiler takes better care of them than humans do.

very dogmatic about this and will stubbornly refuse to compile anything which does not perfectly match – virtually guaranteeing that once your program compiles, it will run. PYTHON and RUBY, however, are as tolerant as the famous atheist church-goer[18], and this will cause a long run of convoluted runtime errors.

A strong argument against this approach is that it requires the value-collecting mechanism to be woven into the program design from the very beginning; it is next to impossible to incorporate it afterwards. It also requires the libraries and all to be built accordingly; it is possible to rewrite `os.path.walk` in such a way that it fits the purely functional approach, but in its current form it does not. Period.

From a more theoretical point of view, *directories_visited*, *files_visited* and *bytes_read* can be considered as states of the program. Functional programming, however, rejects the notion of states, as it rejects the concept of actions. ERLANG is odd in that it is a purely functional language designed mostly for controlling the states of parallel processes. Actually, it does so by emulating states – the "states" of an ERLANG process are never changed, but the process calls itself with new arguments for parameters. This in turn requires infinite recursion, which is essentially verboten and made possible only by "tail recursion" (p. 171). The long and the short of it is that you have to go out of your way to keep an event log in true-bred functional languages.

(5) Interprocess communication is feasible only in languages that support concurrent programming[19], most notably ERLANG and ADA. In both of these, any part of the program may "spawn" so-called *threads*, a "thread" being a sub-process that runs parallel to the original process and is identified by some reference (a number or something opaque). Each process consists of an infinite loop (in ERLANG by tail recursion, p. 136 and p. 171); whenever that loop is exited, the thread evaporates. Now threads are capable of exchanging data. Thus, the parent process may begin by spawning a banking thread, then send the data to be accumulated to this thread, and finally send a request for information to the banking thread (followed by an order to exit the core loop – for obvious reasons, threads should not be left to their own devices[20]). This concept of "sending messages" will remind you of SMALLTALK's "message passing" to objects (p. 200), and in fact the parallels are

---

[18] This particular guy had written several books on the subject of religion and atheism, eloquently voicing his conviction that there cannot be any Deity. Nevertheless, he was seen in church quite regularly. Being asked by the priest about this obviously contradictory behaviour, he explained: "Who am I to be positively sure of the things I just believe?"

[19] Most modern languages, including PYTHON and JAVA, possess library functions that implement at least some degree of concurrency. However, these things are cumbersome if they are not an integral part of the language itself.

[20] Depending on the nature of the implementation and the operating system, threads may even outlive their parent processes unless termination messages are properly sent.

marked. The spawning of a thread is homologous to an object's instantiation, as the sending of a termination message is homologous to calling its destructor.

(6) Objects represent the very opposite of the functional approach. As we have seen before (p. 200), an object's essence can be reduced to a number of states – its data members, which are changed by messages sent to this object (i.e., calls to its functions). Objects are actually very much like threads, with the exception that objects do not do anything except send messages and again nothing after having completed the reaction to a message, whereas threads always have some "basic metabolism" of their own.

An important fact is that when an object is supposed to be passed as an argument, actually only a reference to the object is passed[21]. Thus,

```
Python 2.3.3 (#1, Apr  6 2004, 01:47:39)
[GCC 3.3.3 (SuSE Linux)] on linux2
Type "help", "copyright", "credits" or "license" for more information.
>>> class schtupid: pass
...
>>> def primchange(x,n): x = n
...
>>> def obchange(x,n): x.valu = n
...
>>> t=int()
>>> t
0
>>> primchange(t,53)
>>> t
0
>>> s=schtupid()
>>> s
<__main__.schtupid instance at 0x40373ccc>
>>> s.valu
Traceback (most recent call last):
  File "<stdin>", line 1, in ?
AttributeError: schtupid instance has no attribute 'valu'
>>> obchange(s,53)
>>> s
<__main__.schtupid instance at 0x40373ccc>
>>> s.valu
53
>>> primchange(s,53)
>>> s
<__main__.schtupid instance at 0x40373ccc>
>>> s.valu
53
>>> obchange(t,53)
Traceback (most recent call last):
  File "<stdin>", line 1, in ?
```

---

[21] Of course, this is implemented using pointers – as objects generally are. But don't you worry about that.

```
            File "<stdin>", line 1, in obchange
        AttributeError: 'int' object has no attribute 'valu'
```

In other words: *int()* returns a variable of the primitive integer type, which refuses to respond to messages, have data members, etc., whereas *schtupid()* returns an object of its own class – the simplest class possible, but still a class, with all the characteristics of such a one. As PYTHON is weakly typed, members may later be added. But the essential thing is that **unlike a primitive variable, an object is an entity which is not multiplied implicitly; it is one and only one. Thus, sending a message to that object will always change the one object's state, regardless of how many times it has been passed from one function to another.**

At first glance, it may seem a rather cranky decision to let objects be treated differently from "normal" variables, and one might expect confusion to result from this. However, this is not the case. It is not a matter of objects being subject to different conventions; it is a matter of treating objects as objects. You cannot treat primitives as objects, but you can always treat objects as primitives. Have a look at the effects of *primchange(s,53)*: you will see that this function call does not affect *s* in the least. Internally, all that happens is that a local value is re-assigned. No change is made to the object denoted by *s*, in contrast to what *obchange* does. Thus, in RUBY, where there are no primitives at all, but even integer and boolean values are objects, these two functions will work in just the same way:

```
        class Schtupid
            def initialize
                @value = 0
            end

            def set(x)
                @value = x
            end

            def get
                @value
            end
        end
        => nil

        def primchange(x,n)
            x = n
        end
        => nil

        def obchange(x,n)
            x.set(n)
        end
        => nil

        s = Schtupid.new
```

```
    => #<Schtupid:0x402725a8 @value=0>

    t = 5
    => 5

    obchange(s,77)
    => 77

    obchange(t,77)
    NoMethodError: undefined method 'set' for 5:Fixnum
            from (irb):18:in 'obchange'
            from (irb):23
            from :0

    primchange(s,99)
    => 99

    primchange(t,99)
    => 99

    s
    => #<Schtupid:0x402725a8 @value=77>

    t
    => 5
```

As you see, *primchange* does not affect the arguments; *obchange* changes the states of proper objects and causes errors with improper ones (i.e., objects lacking the correct members).

The gist of this is that objects are very suitable for the task at hand. Define an object which can hold the state you are interested in and possesses a method for changing this state, then pass this object to the functions which are supposed to do the work and make them call the method to change the state. When you are done, look at the state of the object.

10.13.3 Solution. This will probably become clearer by exemplification.

```
class bank:
    "Class for keeping track of multiple values identified by name."

    class account:
        "A single value, with methods for in- and decrementing."
        def __init__( self, typ, start=0 ):
            "Constructor. Takes type function and optional start value."
            self.__t, self.__b = typ, typ(start)
        def into( self, x ):
            "Pay in, i.e. increment value by x."
            self.__b += self.__t(x)
        def outa( self, x ):
            "Take out, i.e. decrement value by x."
            self.__b -= self.__t(x)
```

```
        def balance( self ):
            "Return current balance, i.e. value."
            return self.__b

    # account objects hold single values. We want hashes:
    def __init__( self, namelist, typ=float ):
        "Constructor for bank. Takes a list of names, for which account
          objects are created, and optionally a type function. If the
          latter is omitted, type <float> is assumed."
        self.__ax, self.__typ = {}, typ
        for n in namelist:
            self.__ax[ n ] = self.account( typ )
    def into( self, n, x ):
        "Add x to account n."
        self.__ax[ n ].into( self.__typ(x) )
    def outa( self, n, x ):
        "Remove x from account n."
        self.__ax[ n ].outa( self.__typ(x) )
    def new1( self, name, value=0 ):
        "Create new account within bank. Type will be the one used
          for all the others. Start value may be specified."
        self.__ax[ name ] = self.account( self.__typ, value )
    def balance( self, n ):
        "Value at account n."
        return self.__ax[ n ].balance()
```

This class is instantiated in the normal way, by calling it as a function, passing a list of strings (and, optionally, a function which converts variables into the desired type, thus defining it) as arguments. It can then be invoked as we have learned:

```
>>> b = bank( [ "aleph", "bayt", "gimal", "daleth" ], int )
>>> b
<__main__.bank instance at 0x40373f2c>
>>> help(b)
Help on instance of bank:

<__main__.bank instance>

>>> help(bank)
Help on class bank in module __main__:

class bank
 |  Class for keeping track of multiple values identified by name.
 |
 |  Methods defined here:
 |
 |  __init__(self, namelist, typ=<type 'float'>)
 |      Constructor for bank. Takes a list of names, for which account
 |      objects are created, and optionally a type function. If the
```

```
 |      latter is omitted, type <float> is assumed.
 |
 |  balance(self, n)
 |      Value at account n.
 |
 |  into(self, n, x)
 |      Add x to account n.
 |
 |  new1(self, name, value=0)
 |      Create new account within bank. Type will be the one used for
 |      all the others. Start value may be specified.
 |
 |  outa(self, n, x)
 |      Remove x from account n.
 |
 |  ----------------------------------------------------------------------
 |  Data and other attributes defined here:
 |
 |  account = <class __main__.account>
 |      A single value, with methods for in- and decrementing.
>>> dir(b)
['__doc__', '__init__', '__module__', '_bank__ax', '_bank__typ', 'account',
'balance', 'into', 'new1', 'outa']
>>> b.balance
<bound method bank.balance of <__main__.bank instance at 0x40373f2c>>
>>> dir(b.balance)
['__call__', '__class__', '__cmp__', '__delattr__', '__doc__', '__get__',
'__getattribute__', '__hash__', '__init__', '__new__', '__reduce__',
'__reduce_ex__', '__repr__', '__setattr__', '__str__', 'im_class', 'im_func',
'im_self']
>>> b.balance( "aleph" )
0
>>> b.into( "aleph", 1000 )
>>> b.balance( "aleph" )
1000
>>> b.outa( "aleph", 23 )
>>> b.balance( "aleph" )
977
>>> b.balance( "alpha" )
Traceback (most recent call last):
  File "<stdin>", line 1, in ?
  File "b.py", line 36, in balance
    return self.__ax[ n ].balance()
KeyError: 'alpha'
>>> b.new1( "alpha" )
>>> b.balance( "alpha" )
0
```

10.14 Generatio spontanea

There are three misconceptions about objects which you should beware of at this point.

The first is that objects do not really matter, as they have not been awarded a prominent place in this book. But even though I am personally convinced that OO programming is much overrated and not up to what is possible with functional programming, OO languages – even the semi-OO ones such as C++ and JAVA – are an important improvement over procedural languages. *Objects do matter.*

The second is that objects are trivial. They are not. On the other hand, I think the devout mysticism generally surrounding them is also a senseless exaggeration. Objects are very logical and certainly much easier to get a grip on than, for example, HASKELL's "monads", especially when you trace the history of their development from structured data types. In theory, the progression from a data structure, which is simple enough when describing it in terms of daily life, e.g., as a registry card with a form, to a genuine object, is not very much of a show. Actually, by using pointers to functions, you may emulate objects even in traditional C, a fact of which K&R themselves were not unmindful. As always, the trouble begins with practice. Whereas to code within a procedure or function there are only two scopes of variables: locals including the parameters and globals, things are much more complicated to code when belonging to an object. Here data members of the object, data members inherited from the object's base class(es) and bits of information relating to the object's creation and lifestyle all have to be taken into consideration. There is also the question of constructing and deleting the object itself. And whereas structures are handled in a very similar fashion in almost all languages (there is little more to this than the . operator for accessing members of structures), the handling of objects is as motley as are languages themselves.

The third is that objects, when they are used, always appear with flourish and sennet like the heroes of Shakespeare's tragedies, being conjured up by a grandiloquent construction phrase:

- JAVA:
  ```
  Walk silly = new Walk();
  ```
- C++:
  ```
  Walk* silly = new Walk();
  ```
- O'CAML:
  ```
  let silly = new Walk;;
  ```
- RUBY:
  ```
  silly = Walk.new
  ```
- PYTHON:
  ```
  silly = Walk()
  ```

Often they are, but they do not have to be. The more a language relies on objects (C^{++} < Java < Python < Ruby), the less likely object constructions are to be eye-catchers[22]. With Python's reduction of instantiation to something syntactically identical to a function call (**a Python class actually being nothing but a function returning objects!**), many things which at first glance appear to be simple functions are in fact instantiations. The fact that the higher-level languages take care of the objects themselves and deallocate them when they are no longer needed contributes to obscuring this fact.

For example, the `range` function returns an iterable object, as we all know by now. In other words, *range* is actually a class which is instantiated whenever we call the function. The range object is disposed of as soon as its scope is left.

Similar with `file`, superficially similar to C's notorious *fopen* function but in fact returning not a "file pointer" but a complete file object. The difference is that when the C programmer forgets to call the *fclose* function for his file pointer, data may be lost, the file will remain open, and maybe there will even be damage to the disk's allocation table if the system is shut down with files still being formally open. The file object, on the contrary, performs a proper seppuku. When it receives word from its master that it is no longer needed (which is the first step towards deallocation, taken automatically when a variable's scope is left), it writes all buffered data to the disk and makes sure the file is closed, before it releases all its memory and allows itself to be thrown on the garbage dump. This is one of the reasons for OO being able to promote reliability of complex programs.

The usefulness of this is most obvious where it comes to graphical objects. When a graphical object is removed from the screen, the area it has occupied so far has to be redrawn. On older computers, such as the Atari ST, this was a constant source of trouble: A slight mistake was enough to leave ugly traces of closed dialog boxes or windows on the screen, which were partly overwritten by new output and messed up the entire output.

People who are familiar with C or Pascal, where strings are treated as arrays, tend to be amazed at the idea of strings as objects (p. 224). The great advantage is that this allows strings to be of variable length without complex memory juggling. Actually, the memory juggling does take place, but it is all within the constructors and destructors of the objects so you never see much of this. This is especially tricky as the syntactical sugar hides it.

Guess how many objects are created in the following lines?

```
def prifu():
    a, b = "Hello ", "world"
    print a + b
```

---

[22] Once more with the exception of Objective C, where a unique syntax is used for the handling of objects – dictated by the need to maintain full backward compatibility with C.

```
    prifu()
```

Six of them!

a and *b* are easy enough to identify as objects, but in fact their creation involves a volatile, nameless tuple object with the value ("Hello ", "world") – a tuple consisting of two string objects[23]. In the second line, the concatenation of two strings also gives rise to a new string object, which is then passed to the printing function. And all of these objects are discarded when *prifu* is finished. That is to say, four orders to commit seppuku are sent out and processed: to *a*, to *b*, to the nameless tuple object and to the nameless string object resulting from concatenation. The tuple object, in turn, makes its two component strings perform harakiri before blowing itself up (you remember "garbage collection" as described on p. 156?).

It may be a bit amazing that error conditions in PYTHON are also objects. In fact, `raise` simply creates a new object of one of the error classes, and the `try` and `except` mechanism just watches out for the production of such an object. The error classes are even generated by inheritance, `StandardError` being the root. From this, e.g., `ArithmeticError` is derived, from which in turn `ZeroDivisionError` stems. This is useful in that it allows the definition of new error conditions whenever required:

```
class OutOfCheeseError(StandardError):
    def __init__(self,errmsg): self.err = errmsg
    def __str__(self): print "\nMore %s required.\n" % self.err
```

As you may know by now, __init__() is the constructor, and __str__() is the standard method which is called whenever a textual representation of an object is demanded (this is sometimes named "object serialization")[24]. You will also understand that the initial line defines *OutOfCheeseError* as a class based upon *StandardError*, thus possessing all the hallmarks required for the try/except system.

```
>>> raise OutOfCheeseError( "edamer" )
Traceback (most recent call last):
  File "<stdin>", line 1, in ?
__main__.OutOfCheeseError
More edamer required.
```

---

[23] All through this book, we have been speaking of "multiple return values" of PYTHON, O'CAML or HASKELL functions. Actually, these functions also return single values, but these single values are tuples which are then broken up. This is a detail which is not really important in understanding the functioning of functions.

[24] The difference between __str__() and __repr__() is not at all clear.

10.15 The octopus and the coffee mug: C++ *vs* Java

C++, C♯ and JAVA are not particularly well-designed languages, but are used so extensively[25] that a look at them is almost mandatory, even though SMALLTALK, EIFFEL and OBERON are much more rewarding to study. If you find this chapter difficult to understand and none too interesting, I will agree. In this case, just skip it.

10.15.1 Objects and pointers in C++. As the name implies, C++ has been developed from C ("an octopus obtained by nailing four wooden legs on a dog"). Unlike OBJECTIVE C, which is just an OO extension to standard C, C++ does not strive to maintain full backward compatibility, but nevertheless maintains some of C's most characteristic traits. To begin with, it is a low-level language – lower than the PASCAL family and certainly *much* lower than the SMALLTALK–PYTHON–RUBY assemblage. For example, it preserves C's nefarious use of pointers (p. 80 and Vol. 2). A pointer is a variable that holds the address of another variable.

C makes extensive use of what is known as **dynamic memory allocation**. That is to say, an array with a variable number of elements, a very frequent thing in the absence of lists, can be defined by setting a pointer variable to a crude chunk of memory obtained by calling the function *malloc*:

```
t_ype *varp = (t_ype *) malloc( n * sizeof(t_ype) );
```

will reserve $n \cdot t$ bytes where n is the number of elements and t an element's size in bytes, which can be obtained using C's special expression `sizeof(...)`. The declaration `t_ype *varp` defines *varp* as a pointer to variables of type *t_ype*, and the expression `(t_ype *)` is a "cast" or "type coercion" which will force the compiler to understand the return value of *malloc* as a pointer to variables of type *t_ype* instead of the default, which is a pointer to single bytes. We may then use *varp* as an array:

```
varp[ i ] = result_of_function( ... );
```

where nothing will prevent us from exceeding the boundaries of the primitive array, thereby crashing the program or doing worse[26]. After use, this memory has to be released explicitly (there is no garbage collection in C or C++):

```
free( varp );
```

C++ has replaced *malloc* and *free* with the keywords **new** and **delete**, which differ only in that they are somewhat more conscious of the purpose of the action: they "know" what kinds of objects the reserved memory is going to be, or was, used for, and call these objects' constructors or destructors accordingly. Thus,

---

[25] At least in one case, this can be attributed solely to the owner's huge investments into marketing. By contrast, PYTHON and RUBY have attracted general attention with a marketing budget of zero.

[26] At the moment, a similar weakness is exploited by many viruses to overwrite parts of a running program's code, thereby being enabled to execute their own code with the program's access rights.

```
t_ype *varp = new t_ype[ n ];
```

will create n objects of class *t_ype* and initialize them by calling each object's constructor. Nevertheless, `new` returns not an object itself but a pointer to an object, which lives somewhere on the "**heap**" and can be referred to by address only. Thus, to access its components, we will have to use the $\rightarrow$ operator:

```
varp->transmogrify( argum );
```

Alternatively, we can declare the object locally:

```
t_ype varl;
```

will do the same for just one object of class *t_ype*, which is placed among the other local variables.

Objects can be passed around only as pointers[27]:

```
t_ype *clone_my_object( t_ype *oldp ) {

    t_ype *newp = new t_ype;

    ...

    return newp;

}
```

will work with both

```
t_ype *dollyp = clone_my_object( varp );
```

and

```
t_ype *lollyp = clone_my_object( &varl );
```

But if we rewrite *clone_my_object* like this:

```
    t_ype newl;

    ...

    return &newl;
```

we will have a nasty surprise. The compiler will accept this, but the program will crash immediately. Why? Because *varl* lives not in the Shangri-Là of the heap but among the local variables, which are destroyed upon ending the function *clone_my_object*.

---

[27] Well, actually there *is* the "reference operator" as an alleged alternative. When the definition reads `t_ype *clone_my_object( t_ype& oldp )`, the argument may be an object rather than a pointer to an object. But this reference is converted into a pointer nevertheless, so this is just syntactic sugar – and confusing into the bargain.

Thus, the result of the function call will be a necromantic pointer to a variable which no longer exists!

And on the other hand, if we create the object to be returned with **new**, we will have to remember that this must be deallocated some day. If we fail to do this, we will get a program which runs but quickly jams our memory with lots of useless old objects. This is called a "**memory leak**".

Confused? Welcome to the club. This is neither easy to understand nor easy to handle. And it is one of the reasons why languages like JAVA and C♯ were created. C♯ is mostly an uninspired imitation of JAVA created for political reasons, so let us focus on JAVA here.

10.15.2 Purging the temple. There are no visible pointers in JAVA, and there is no possibility to either create necromantic references or memory leaks, since all objects are created using **new** and taken care of by the system. When the last variable referring to the object has become obsolete, the JAVA system will delete the object in due and ancient form (by calling its destructor and then wiping it from memory), without any need for human interference. Thus,

```
private static t_ype clone_my_object( t_ype oldo ) {

    t_ype newo = new t_ype;

    ...

    return newo;

}
```

will be invoked

```
    t_ype dollyp = clone_my_object( varo );
```

and there's an end to it. When *dollyp* is kicked out of memory, the object it denotes will be destroyed too. But remember what we have said about PYTHON. **An important fact is that when an object is supposed to be passed as an argument, actually only a reference to the object is passed**[28].

Unlike C++, JAVA enforces OO style. Like C++ and unlike PYTHON, it permits **static** members which belong to the class as a whole rather than to one of its instances (thereby effectively turning the class into a module), but apart from that, everything is a class. There is no code which is not incorporated into a class, as in C++ or PYTHON where at least the starting point is "extraclassal". Instead, JAVA relies on the convention that a public method by the name of **main** is executed first if a class is to be run as a whole. Like PYTHON and unlike C++, it has no structured data types except for classes.

---

[28] Of course, this is implemented using pointers – as objects generally are. But don't you worry about that.

10.15.3 Class structure in C++ and Java. A questionable innovation in JAVA is its reunification of class definitions and implementations. In C++, these are carefully kept apart[29]:

```
class t_ype {
public:            // Here comes what is visible and accessible to the
                   // outside world.
    t_ype( void );  // the constructor has the name of the class and
                    // no return type
    ~t_ype( void ); // similar with the destructor
    int m_ethod( int param );
    int a_ttribute;
private:           // Here comes what is not.
    int c_ryptic( int param );
    int *a_rray;   // If this is allocated, the destructor will have to
                   // deallocate it!!
}
```

and

```
t_ype::t_ype( void )  { a_rray = new int[ 100 ]; }
    // "void" just means absence of parameters or return values
t_ype::~t_ype( void ) { delete a_rray;           }
int t_ype::m_ethod( int p_aram ) {
    int x;
        ...
        return x;
}
int t_ype::c_ryptic( int p_aram ) {
    int y;
        ...
        return y;
}

t_ype *obj = new t_ype();
obj->m_ethod( 666 );
```

The rationale behind this is C's method of consistency checking, which C++ has inherited. In C and C++, definitions are usually placed in so-called "header files" which have the suffix .h (whereas normal program code has .c or .cpp, respectively). For example, the definition of class "t_ype" might be stored in a file named "t_ype.h". All source code files which wish to use this class, or objects of this class, are required to import this definition with the directive `#include "t_ype.h"`. This directive actually does nothing but insert the denoted file right into the text (by a part of the compiler known as "preprocessor"), so it can be read by the compiler. The output by the C++ compiler is just a piece of linkable machine code which looks exactly like that of C, and in fact can be linked with C, FORTRAN-77 or other non-OO languages –

---

[29] At least, this is is the recommended way. You *may* also unite definitions and implementations. But then, you *may* also shoot yourself in the foot.

there is no object-related meta-information in it[30]. Without the header file, proper compilation is impossible.

By contrast, the JAVA compiler produces an output (*.class*) which is rather similar to that produced by PYTHON (*.pyc*) in that it contains all the necessary meta-information describing the class. Thus, we can use a class even if we have nothing but its compiled form. This is claimed to be faster, but the real advantage, from the owner's point of view, is that you do not have to disclose anything of your code to enable others to build upon it. Whatever the reason, JAVA has abandoned the preprocessor and returned to a PYTHONish style of combining declarations and implementations:

```
public class t_ype {

    public int a_ttribute;
    private int a_rray[];

    public t_ype( void )   { a_rray = new int[ 100 ]; }
    // no explicit destructor - the garbage collector will do the job!

    public int m_ethod( int p_aram ) {
        int x;
            ...
            return x;
    }
    private int c_ryptic( int p_aram ) {
        int y;
            ...
            return y;
    }
}

t_ype obj = new t_ype();
obj.m_ethod( 666 );
```

The discrimination of self *vs* non-self is a less severe problem in these strongly typed languages than it is in PYTHON or RUBY, since the declaration serves to identify variables. In the rare cases where a method declares a local variable with the same name as a data member of the object, the interpretation defaults to using the local variable unless the keyword `this` is used – a reference to the object identical to PYTHON's `self`, with the only difference that it does not have to be passed as a parameter.

10.15.4 Inheritance – as troublesome as ever. Inheritance exists in both C++ and JAVA, of course. In C++, it can be characterized in one word: overbred. C++ not only allows things of as doubtful value as multiple inheritance and disruption of the "private" limitation by so-called "friends" of the class, but has an inheritance

---

[30] More precisely: there is very little meta-information of any kind in it. If not all references to functions are satisfied, this will result in a very cryptic message; if arguments and parameters do not match, the linker will not know. Neither will you, but you will *feel*.

mechanism so complex that a full description comprises some thirty pages. The elementary mechanism is naturally simple:

```
class derived_class : base_class { ... }
```

But soon it becomes troublesome. Public members are inherited all right, whereas private ones are not. So in case private members are needed which are supposed to be inheritable, they can be placed in a third category: "protected". Protected members are private members which *are* inherited. Now when we look at PYTHON or, maybe, OBJECTIVE C, we find that in these languages all private members are also inherited and should thus be classified, in C^{++} terminology, as "protected" rather than "private". So what is the point of having "private" members in the C^{++} sense at all – members which are not inherited?

This leads us to the field of "binding", that is to say, deciding which method will be selected from a cascade of inheritances. The scripting languages make this decision at runtime (hence the expression "late binding") and will select from the youngest generation, i.e., prefer taking the method from the derived class to taking it from the base class. C^{++}, on the other hand, preferentially uses "early binding" which allows for generation of much shorter and faster code but may have unexpected results. Imagine we have a class C which contains the methods M and N, where M is called from N. From this class, another one C' is derived by inheritance, and M is overwritten by M' whereas N is not modified. If $C'M$ is called directly, it will always be M' which is invoked. So far, so good. If we, however, call N, what happens depends on the binding. In the "late binding" languages the logical thing will happen and $C'N$ will invoke $C'M$, i.e., M'. In the "early binding" languages, on the contrary, $C'N$ will invoke the M belonging to the same generation in which N itself was defined, that is to say, in the base class and not the derived class:

```
#!/usr/bin/env python
class abc_1st:
    def describe( self ):
        print "The first letters of the alphabet are:"
        self.firstch()
    def firstch( self ): print "Latin - 'A', 'B', 'C', 'D'."

class hebrew_abc( abc_1st ):
    def firstch( self ): print "Hebrew - Aleph, Bayt, Gimal, Daleth."

class greek_abc( hebrew_abc ):
    def firstch( self ): print "Greek - Alpha, Beta, Gamma, Delta."

abc_1st().describe()
hebrew_abc().describe()
greek_abc().describe()

The first letters of the alphabet are:
Latin - 'A', 'B', 'C', 'D'.
The first letters of the alphabet are:
Hebrew - Aleph, Bayt, Gimal, Daleth.
The first letters of the alphabet are:
Greek - Alpha, Beta, Gamma, Delta.

// The same in C++
#include <stdio.h>

class abc_1st {
public:
```

```
  void describe( void ) {
    printf( "The first letters of the alphabet are:\n" ); firstch();
  }
  void firstch( void ) { printf( "Latin - 'A', 'B', 'C', 'D'.\n" ); }
};

class hebrew_abc : public abc_1st {
public: void firstch( void ) { printf( "Hebrew - Aleph, Bayt, Gimal, Daleth.\n" ); }
};

class greek_abc : public hebrew_abc {
public: void firstch( void ) { printf( "Greek - Alpha, Beta, Gamma, Delta.\n" );    }
};

int main( void ) {
  (new abc_1st)->describe();
  (new hebrew_abc)->describe();
  (new greek_abc)->describe();
}

The first letters of the alphabet are:
Latin - 'A', 'B', 'C', 'D'.
The first letters of the alphabet are:
Latin - 'A', 'B', 'C', 'D'.
The first letters of the alphabet are:
Latin - 'A', 'B', 'C', 'D'.
```

Welcome to the world of C^{++} – where you, too, can dance with the compiler instead of writing programs.

```
// Javaanse Jongens

public class earlybird {
  class abc_1st {
    void describe() { System.out.println( "The first letters of the alphabet are:" ); firstch(); }
    void firstch()  { System.out.println( "Latin - 'A', 'B', 'C', 'D'." ); }
  };

  class hebrew_abc extends abc_1st {
    void firstch() { System.out.println( "Hebrew - Aleph, Bayt, Gimal, Daleth." );}
  };

  class greek_abc extends hebrew_abc {
    void firstch() { System.out.println( "Greek - Alpha, Beta, Gamma, Delta." );}
  };

  public static void main( String args[] ) {
    (new abc_1st()).describe();
    (new hebrew_abc()).describe();
    (new greek_abc()).describe();
  }
}

The first letters of the alphabet are:
Latin - 'A', 'B', 'C', 'D'.
The first letters of the alphabet are:
Hebrew - Aleph, Bayt, Gimal, Daleth.
The first letters of the alphabet are:
Greek - Alpha, Beta, Gamma, Delta.
```

Now you see why JAVA has been hailed as a productive simplification of C^{++}. Nevertheless, JAVA has kept the old "protected" and friends attributes, including **final** – a final class may only be instantiated, not inherited from – and **abstract** – the opposite, an abstract class may only be inherited from, not instantiated. It is hard to see the benefit of these features. In general, if a class is just a framework designed to build others upon and instantiating this class might cause trouble, there is a perfect solution to this problem: just don't do it.

10.15.5 More of this. There are further mechanisms for facilitating code generation for complex projects ("templates" in C^{++}, "interfaces" in JAVA, etc.) which will not be discussed here. In PYTHON and related languages, this is superseded by a partly functional approach in combination with weak typing.

10.16 Strings as objects

So far, homogeneous aggregate data types, arrays and lists, have been considered from the point of view of their components. HASKELL's prefab data type string is nothing but sugar for [char], a list of characters[31].

However, this does not really do justice to homogeneous aggregates. As usual, the whole is more than a sum of its parts. Every homogeneous aggregate contains more information than all of its elements taken together. For example, the *size* of a homogeneous aggregate is a piece of information which is particular only to the aggregate as a whole, not to any of its components.

At first glance, this may appear trivial, but in fact the proper sizing of an array has always been a major source of trouble. Remember, in C a string is just a sequence of characters terminated by a zero character. The string variable is actually nothing but a pointer (p. 80), that is to say, it indicates the position in memory where the string begins. Except for the terminal zero character, there is nothing about the "string" variable to indicate the length of the string – and certainly nothing contained within the pointer itself. This is dangerous enough when the string is to be read; it can become positively deadly when the string is to be written[32]:

```
void stringcopy( char *from, char *to ) {
  register int i = 0;
  for(;;) {
    to[ i ] = from[ i ];
    if (to[ i ] == '\0')  break; // end after terminal zero has
                                 //                 been copied
    i++;
  }
}
```

---

[31] The Prelude gives this simple definition – the core of the HASKELL interpreter or compiler does not know anything about strings. In general, this is the attitude which all functional languages take, for obvious reasons.

[32] The example does not show the most elegant way of doing things in C, but the most comprehensible. The following solution is both shorter and swifter: `while (*to++ = *from++); ,` but hardly intelligible.

10.16 STRINGS AS OBJECTS

Now imagine this is called from the following piece of code:

```
void stupid() {
  char sink[10], source[22] = "To be or not to be...";
  stringcopy( source, sink );
  printf("\nSource is '%s'.\nSink is '%s'.\n\n", source, sink );
}
```

This will compile properly and run without a crash, yielding the following nasty result[33]:

```
[ophis@zacalbalam:~] $ ./stupid

Source is 'ot to be...'.
Sink is 'To be or not to be...'.

[ophis@zacalbalam:~] $
```

As you see, the "sink" variable looks perfectly healthy but something eerie has occurred to "source", which was not supposed to be modified at all![34]

What has happened? The stupid function has caused 20 bytes to be allocated for *sink* and 60 bytes for *source*. They are adjacent in memory, that is to say, the first byte of *source* follows the last byte of *sink*. In other words, if you write over the boundaries of *sink*, you will get right into *source*:

– Before:

– While (i = 4):

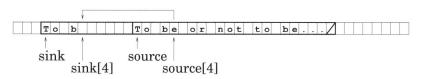

– After:

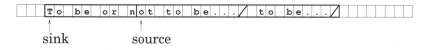

---

[33] For machines with a PPC. If you want to see this on an x86 processor, you will have to declare "source" first, then "sink", because the x86 allocates local variables in a different order. This has something to do with the so-called "endianness"

[34] This also illustrates what range checks causing runtime errors are good for.

This will also happen if you use the C library function *strcpy*.

Things may be even worse than that. If your data organization is more complex, it becomes increasingly difficult to pinpoint the culprit. Debugging tools are not much help, as they normally register only legal access to a variable, not this backdoor stuff.

As usual, there is nothing worse than evil intentions. Many programs, especially those written in lower-level languages like C, read blocks of data from files, or from data sources on the internet, into pre-allocated blocks of memory. For example, when an icon has to be displayed, the graphical representation has to be loaded and stored. Now if you cause the program to read more data than it has space for, this will lead to such a buffer overflow. If you know the binary architecture of a program well enough, you can exploit this clandestinely to modify the program's code. No matter how thorough the system, this modified code will always be executed with the same access rights as the rest of the program, and it may do anything, from sending the email addresses in your electronic address book to *spammaster@pornotheque.biz* to formatting the hard disk. Of course, system designs where all copies of the program are identical on the binary level, as with Windows^{TM} on x86, are extremely vulnerable to this; a system with a comparatively rare processor is much less of a target. Now one will ask why programmers are obviously sloppy about checking the data size, which is generally given in the file to be read. For example, an icon of 100×150 pixels at 24 bit will contain some header that exactly 45 000 bytes of memory are required. How can this be circumvented? Mostly by simply giving negative values. Unless the program has been written carefully, this may lead to the situation in which the entire file is read without any memory allocation! Theoretically, in such a case the address at which the file contents are to be deposited should be expected to be 0x00000000, which is an absolute taboo and should immediately result in the program headbanging (p. 311), which is disconcerting enough but at least will not get you jailed, as may happen if your computer is highjacked and used as a storage place for illegal stuff (you will be held responsible for anything that is on your hard disk). But it seems that such attacks succeed nevertheless; either because some offset is added to the calculated value, thereby always guaranteeing some memory allocation, or due to use of global variables, which will usually have some value from previous operations. I tend to consider the former as much more likely, because the fixed-size "header" generally proceeding the actual data (in the case of our icon file it comprises, among other things, width, height, colour depth, maybe a palette and maybe compression info, sometimes also supplementary information about author and copyright and who knows what else...) will also have to be stored somewhere. Let us assume that our icon format comprises a header of 242 bytes (a number as beautiful as any). Now we give the width as 1 and the height as -1. The system will stupidly try to allocate $1 \times (-1) + 242 = 241$ bytes, which is most likely to succeed. We will thus have a block of 241 bytes allocated at location X. Next, "-1" bytes will be read – a command which causes most data capture functions (i.e., file or network-reading routines) to read all they can. Why? Because -1 (32 bit) is equivalent to 0xFFFFFFFF, the largest unsigned number that exists (p. 76). The capture function will then return the value of bytes which have actually been read, and with a bit of (bad) luck, this will cause the program to "generously overlook" the obviously defective geometry data of the icon and carry on, a thing made even more likely by the fact that most icon designers are in fact very sloppy about their meta-data. Now all the stuff is deposited at X, and from byte 242 on, everything that follows will overwrite something else – in the best case, just empty memory, but often enough either other data or executable code. So when working with a language like C or C^{++}, mind well whether you have signed or unsigned integers, be lavish about explicit type conversions, and do not let anybody in for play. Remember Edward Gibbons (*The Decline and Fall of the Roman Empire*) who admonished that, of all human motives, only the most base one should be considered, and reject data sources which show the slightest irregularity. This matter also illustrates the benefits of higher programming languages, which are generally immune to this kind of abuse, since they take more care of memory allocation at the cost of being a little bit slower. Personally, I think this kind of speed matters only for people who would also jump out of a skyscraper window because the elevator is too slow for them.

The trouble is that there is simply no way of finding out the actual size of the "array". *None*. So obviously it would make sense to tether up the pointer with some additional information:

```
typedef struct AIRBAGGED_STRING {
    char *data; // the characters themselves
    int size;   // indicates how much space there is
} airbagged_string;

void ab_stringcopy( airbagged_string from, airbagged_string to ) {
  register int i = 0;
  for(;;) {
    to.data[ i ] = from.data[ i ];
    if (to[ i ] == '\0' || i > from.size || i > to.size) break;
    i++;
  }
}
```

This, of course, would require some additional code. How are we going to create an airbagged_string, for example? We will need a function that takes a string constant (or a normal "string" variable), allocates the memory (necessarily trusting in the final zero), sets *size* accordingly and initializes *data*. This granted, we will have to keep track of the memory usage of our airbagged_strings and get rid of them when they are no longer used.

You see that this literally calls for implementation as a class, with the two aforesaid tasks being delegated to the constructor and destructor, respectively.

Turn back to the note on SMALLTALK (p. 200). We are now about to encapsulate the string data in a way that is not easily accessible from the outside. Why should we? To avoid tampering with it. If every access to the string data is performed via the class's methods, which can be enforced simply by making all data members *private*, we can make sure that memory is handled properly. No premature releasing of memory that is still needed, no leaving of forgotten garbage, no crashes due to blunders in pointer handling.

In the C version, accessing a character belonging to the string means simply accessing an array element. In the OO version, the string object is a "black box" which can be sent a message: `my_string.elementAt( n )` means that the object is ordered *Please tell me what character stands at position* n, hiding the way in which the object satisfies the request. In JAVA, there is actually a method named `charAt` doing this. PYTHON, on the other hand, has some rather obscure classes (see the previous chapter), and strings belong to them. You write `my_string[ n ]`, whick looks suspiciously similar to C's original array access. But in fact it is not: it is just syntactic sugar for the string class's clandestine *__getslice__()* method. You don't have to worry about *__getslice__()*, the interpreter handles things for you. But you should keep in mind that this *is* a mechanism different from that of C: more flexible, more reliable, and much more time-consuming.

When we implement strings as a class, however, we can add a few refinements; e.g., our copying function might check before whether the target is large enough, and allocate new memory if required:

```
void airbagged_string::copy_to( airbagged_string to ) {
  register int i = 0;
  if (to.size < size) {
      delete to.data; // get rid of the old memory chunk
      to.data = new char[ size ]; // organize more RAM
      to.size = size; // this has to be set, of course
  }
  for(;;) { // the rest as had before
    to.data[ i ] = data[ i ];
    if (to[ i ] == '\0' || i > size || i > to.size) break;
    i++;
  }
}
```

This opens endless (and sometimes senseless) possibilities. Our class airbagged_string could comprise methods for the following purposes.

- atof – the floating point value of an airbagged_string.
- atoi – the integer value of an airbagged_string.
- capitalize – the string with its first letter in uppercase.
- capwords – the string with each word's first letter in uppercase.
- center(length) – the string adjusted to the centre of a given length, with both ends padded with spaces.
- count(pattern) – the number of occurences of a given pattern.
- find(pattern) – the position of the first occurence of a given pattern.
- ljust(length) – the string adjusted to a given length, with the right end padded with spaces.
- lower – the string with all letters in lowercase.
- lstrip – the string with leading whitespaces removed.

10.16 STRINGS AS OBJECTS

- replace(old,new) – the string with all occurences of pattern "old" replaced by "new".
- rjust(length) – the string adjusted to a given length, with the left end padded with spaces.
- rstrip – the string with trailing whitespaces removed.
- strip – the string with leading *and* trailing whitespaces removed.
- swapcase – the string with all uppercase letters converted to lowercase, and vice versa.
- lower – the string with all letters in uppercase.
- levenshtein – the similarity between this string and another according to the Levenshtein algorithm (see p. 230).

Moreover, it could contain a few useful constants:

- whitespace – all whitespace characters (space, carriage return, tab, etc.).
- lowercase – a list of all lowercase letters.
- uppercase – a list of all uppercase letters.
- punctuation – a list of all punctuation letters.
- digits – a list of all numeric letters.

And, oh yes, apart from a constructor which sets up the internals there should also be a destructor releasing the dynamically allocated memory once the airbagged_string is no longer needed (remember "garbage collection", p. 156 and p. 190).

Now have a look at PYTHON's *string* module, by typing `import string` and `help(string)`!

The PYTHON module actually implements a few more functions, such as `split`, which dissects a string into a list of strings, looking for a certain delimiter character.

The *string* module has a somewhat unusual position within the language. On the one hand, you may import it explicitly and access its functions, as with any other library:

```
>>> import string
>>> string.find( "trumpet", "pet" )
4
```

On the other, all PYTHON strings are objects of type *string*, as explained in the previous chapter (p. 213), and the methods described above are thus immediately available:

```
>>> s = "lieber arm dran als arm ab"
>>> s.capitalize()
'LIEBER ARM DRAN ALS ARM AB'
>>> s.split()
```

```
['lieber', 'arm', 'dran', 'als', 'arm', 'ab']
>>> s.upper().split()
['LIEBER', 'ARM', 'DRAN', 'ALS', 'ARM', 'AB']
>>>
>>> "treason".find( "reason" )
1
```

The expression *s.upper().split()* is based on the fact that the output of the string method *upper* is another string object, which possesses all string methods itself. *s.split().upper()* will not work, because *split* produces a list object, not a string object.

As the last line demonstrates, even string constants are string objects.

For more details, please see the official PYTHON documentation.

It should be pointed out that *all* homogeneous aggregates can be implemented as objects. As mentioned above, PYTHON lists should be named more properly *list objects*. Unlike the head→tail lists in truly functional languages, PYTHON lists[35] possess a number of useful methods.

```
>>> a = [1,2,3]
>>> a
[1, 2, 3]
>>> a.reverse()
>>> a
[3, 2, 1]
```

In HASKELL, we would have to apply the function `reverse` to the list, instead of calling a method (note that the HASKELL function *returns* a value, whereas the PYTHON method modifies the list "in place"):

```
Prelude> reverse [1,2,3]
[3,2,1]
Prelude>
```

Where arrays are a distinct feature of object-oriented languages, they may also be implemented as objects. This is realized in JAVA, for example. All JAVA arrays have an attribute named *size*, as all JAVA strings have a method named *length()*.

The **Levenshtein algorithm** has been mentioned before. This straightforward algorithm compares two sequences of elements and returns a value that serves as a measure of their similarity. It can be used as a basis for DNA sequence comparisons, but is more widely used for string analysis. Here is a sample implementation in JAVA (courtesy of DKFZ-IBIOS), which you will be able to understand easily:

---

[35] They *may* be implemented as head→tail lists internally, with some airbagging around. Who knows? Who cares?

10.16 STRINGS AS OBJECTS

```
public static int hamming(String str1, String str2) {
    /*check whether strings are or equal length:*/
    if ( str1.length() != str2.length() )
       throw new ProgrammingError("Error: Hamming distance cannot be calculated!");
    /*compares at each string position:*/
    int hammingDistance = 0;
    for (int i = 0; i < str1.length(); ++i)
       if ( str1.charAt(i) != str2.charAt(i) )  ++hammingDistance;
    /*gives the calculated hamming distance:*/
    return hammingDistance;
}

public static int levensthein(String str1, String str2) {
    /*determine the longest and shortest strings among str1 and str2:*/
    String longest  = null;
    String shortest = null;
    if (str1.length() > str2.length() ) {longest  = str1; shortest = str2;
    } else { longest  = str2; shortest = str1; }

    /*initialization of the levensthein distance*/
    int levenstheinDistance = longest.length() + shortest.length();

    /*complete annealing          -----------------------------        */

    /*determine the first and the last position in the longest string,
      where a complete annealing of the shortest one could occur*/
    int start = 0;
    int end   = longest.length() - shortest.length() ;

    /*anneal the shortest string over the longest one*/
    for (int i = start; i <= end; ++i) {
      String toCompareWith = longest.substring(i,i+shortest.length());
      /*finds the minimum levensthein distance*/
      int temp = hamming(toCompareWith,shortest) + end;
      if (temp < levenstheinDistance) levenstheinDistance = temp;
    }

    /*incomplete left annealing        -----------------------------        */

    /*determine the first and last position of an incomplete left annealing*/
    start = 1;
    end   = shortest.length();

    /*left-anneal the incomplete shortest string over the complete longest one*/
    for (int x = start; x < end; ++x) {
      /*finds the minimum levensthein distance*/
      int restLength = end + longest.length() -2*x;    //rest longest + shortest
      String sub1 = longest.substring(0,x);             //"long"  overlapping end
      String sub2 = shortest.substring(end - x ,end);   //"short" overlapping end
      int temp = hamming(sub1,sub2) + restLength;
      if (temp < levenstheinDistance) levenstheinDistance = temp;
    }

    /*incomplete right annealing       -----------------------------        */
```

```
  /*determine the first and last position of an incomplete left annealing*/
  start = longest.length() - shortest.length() + 1;
  end   = longest.length();

  /*left-anneal the incomplete shortest string over the complete longest one*/
  for (int x = start; x < end; ++x) {
    /*finds the minimum levensthein distance*/
    int restLength = shortest.length() - end + 2*x;   //rest longest + shortest
    String sub1 = longest.substring(x,end);           //"long"  overlapping end
    String sub2 = shortest.substring(0 ,end - x);     //"short" overlapping end
    int temp = hamming(sub1,sub2) + restLength;
    if (temp < levenstheinDistance) levenstheinDistance = temp;
  }

  /*gives the minimum hamming distance over all possible annealings*/
  return levenstheinDistance;
}
```

10.17 None and object in Python

There are two very special objects in PYTHON, the understanding of which requires some mental flexibility.

The first one is plainly named `object`, and it is the root class of all object classes. You remember that we said with respect to inheritance that a class *may* be based upon another, implying that this was a possibility but not a necessity; and for all practical purposes, this is good enough.

However, since the first inception of SMALLTALK it has been felt that lonesome classes do not fit well into social life[36]. It was therefore agreed from the beginning that actually *all* classes must be derived from other classes, making all classes direct or indirect descendants of the same *root class*.

At first, that looks like yet another exercise in weltanschauung. However, the root class may be put to very practical use, by installing some methods (or at least method stubs) in it which will then, by definition, be available for each and any object.

As you may remember from the section on "Generatio spontanea" (p. 213), PYTHON has a tendency to hide some of the more plebeian method calls by causing them to be invoked automatically, and implicitly, when other things happen. Fire up your PYTHON interpreter and type `dir(object)` ; you will get the following list:

```
>>> dir(object)
```

---

[36] Actually, this goes a little bit deeper. As SMALLTALK is "reflective", there is no way of doing anything except by sending messages to objects. Thus, even inheritance is performed by sending a message – inherit!, or in SMALLTALK's vocabulary: **subclass:** – to the base class. Thus, there must be one *primum mobile*.

```
['__class__', '__delattr__', '__doc__', '__getattribute__', '__hash__',
'__init__', '__new__', '__reduce__', '__reduce__', '__repr__',
'__setattr__', '__str__']
```

Apart from its being root (and class x: therefore being implicitly equivalent to class x(object):), object is a class like any other. For example, it may be instantiated:

```
>>> obbo = object()
>>> obbo
<object object at 0xc83c0>
>>> print obbo # this calls __str__
<object object at 0xc83c0>
>>> obbo.__doc__ # get the docstring
'The most base type'
>>> obbo.__class__
<type 'object'>
```

Or derived:

```
>>> class subject(object):
...     "rather pointless class"
...     def __init__(self,x): self.x = x
...     def __str__(self): self.x = x
...
>>> beppe = subject("eh bien")
>>> beppe
<__main__.subject object at 0x132970>
>>> print beppe # this calls __str__
eh bien
>>> beppe.__doc__ # get the docstring
'rather pointless class'
>>> beppe.__class__
<class '__main__.subject'>
>>> obbo.__str__
<slot wrapper '__repr__' of 'object' objects>
```

There is very little to understand about **None**, except for the odd fact that it is a true object possessing all the methods of object, but which may not be put to any normal use. Most notably, it cannot be derived or instantiated. It is just an object indicating the absence of any object. However, it is not a gap – it is an object itself.

> *Nothingness has committed suicide.*
> *Creation was its death.*
> *The world is the grave in which it rots.*
>
> – GEORG BÜCHNER: "Danton's Death"

10.18 Paster of Muppets

So far, we have considered objects mostly from a software builder's perspective. Obviously, the OO approach greatly facilitates reusability of code: you do not have to reinvent the wheel every time. However, the OO approach also opens the intriguing possibility[37] of "remote controlling" software by other pieces of software. This needs some explanation.

There are certain tasks which are both very common and rather complex, and the software to deal with these is easy to procure. Take a word processor. Everybody needs one, and everybody has one. A word processor's job may appear simple, and in fact the first "standard" word processor, WordStar, comprised no more than 16 kilobytes of Z80 machine code, but this program could do little more than edit plain text. A program with the ability to manage different fonts (including proportional spacing), alignments, graphical objects and the like, plus the facilities for storing, reloading and printing documents, is quite a complex piece of software, even if we restrict it to the really essential features[38]. Modern operating systems offer lots of support for such advanced technology, but this support is difficult to understand and use in itself. The same holds true for the "libraries" offered as go-betweens between the operating system and the program. The long and the short of it is: documents of publishing quality *are* a complex matter, and producing them should be left to dedicated software.

At the other end of the spectrum, there are tasks which are almost unique but not too complex, and there will be no prefab "solutions" for them, but they will almost always require custom-made software, ranging from an undergraduate's BASIC hack to a $100 000 software project which is bound to be essentially a BASIC hack with a big bill following. For example, reading data from an external sensor. If our sensor has a modern interface, say USB, we will not even need a "driver": our computer will be able to talk to the sensor without further ado. Let us say our sensor is a microplate reader[39]. The computer

---

[37] And in this case, pure functional programming does not offer a true alternative. However, I maintain a moderately functional approach (with functional parameters, λ, etc., but not with the hardcore-FP restriction to bindings instead of assignments which effectively precludes anything object-like – unless you lean on concurrency and employ threads in lieu of objects, as in ERLANG.)

[38] Up to very recently, the really amazing thing about word processors was how many features can be built in and used as sales arguments without being required. By now, an even more amazing thing is how many features can be built in and used as sales arguments without being ever seen or used.

[39] For those unfamiliar with modern life sciences: a microplate is a standardized piece of plasticware comprising 8×12 "wells" in which reactions may be performed. Very often, such microplates are used for long measurement series, as they provide both ease of handling and a reduction in the volume required. Today, many biochemical measurements are implemented as chromogenic reactions leading to a change in the content's hue, or more technically, an increase or decrease in the "extinction" or absorption of light at a specific wavelength. For example, you may have a "kit" for the determination of soluble proteins in which the presence of proteins leads to the formation of a dye absorbing light of 440 nm. You perform your tests

sends (X,Y,λ) tuples to the sensor, and the device will return the extinction at wavelength λ for the well located at (X, Y). A microplate reader will give you a group of 96 floating-point values per plate. All your program has to do is to send the tuples and receive the data. You may use either high- or low-level programming languages to this purpose; in either case the program core will hardly comprise more than a hundred lines of code, to the tune of:

```
from HardwareInterface import GetUSB
function read_plate( integer wavelength ):   array [ 1..12, 1..8 ] of float
   for x := 1 to 12 do
      for y := 1 to 8 do
         read_plate[ x, y ] := GetUSB( ("GET",x,y,wavelength) )
      next
   next
end function
```

But now imagine you want to tether these two things together: e.g., we may like to print a proper report of our 96 wells, not just a list of numbers. However, any document of more than intermediate complexity will require software which cannot be programmed on the fly:

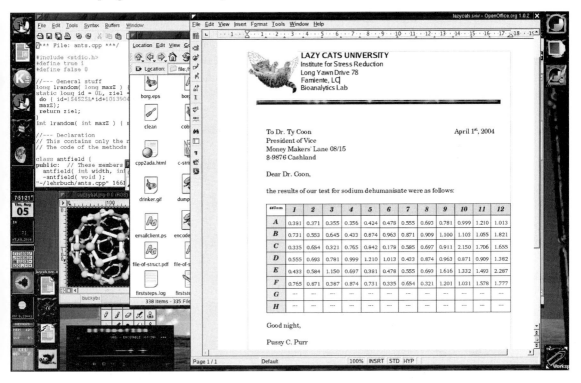

in a microplate and have your microplate reader determine the absorption at 440 nm, for each well. The entire measurement can be performed within a few seconds.

How can we get this job done?

(1) Manually, by providing an electronic form and making your *hiwi* type in all the data. He will curse you for it, and he will wonder why it is not possible to have the d— microplate reader communicate its data directly to the office computer, even though they are both connected to the local network, but that's all right.

(2) By creating a file in the appropriate format which can then be read by the word processor. In general, file formats are much simpler and easier to understand than the complexities of font and image handling (everybody can buy a watch, even though only very few people know how to build one!). If the program uses a markup language (p. 45) for storing its data, this is reasonably simple to do. For example, this is how OPENOFFICE stores its data using XML:

...
```
<text:span text:style-name="T1">LAZY CATS UNIVERSITY</text:span></text:p><text:p
text:style-name="Standard"><text:s/><text:span text:style-name="T2">Institute for
Stress Reduction</text:span></text:p><text:p text:style-name="P1"><text:s/>Long
Yawn Drive 78</text:p><text:p text:style-name="P1"><text:s/>Farniente, LC</text:p>
<text:p text:style-name="P1"><text:s/>Bioanalytics Lab</text:p><text:p
text:style-name="Standard"/><text:p text:style-name="P2">
<draw:image draw:style-name="fr2" draw:name="Graphic2" text:anchor-type="as-char"
svg:y="-1.783cm" svg:width="16.312cm" svg:height="0.309cm" draw:z-index="1"
 xlink:href="../../../usr/OpenOffice.org1.1.0/share/gallery/rulers/blurulr2.gif"
 xlink:type="simple" xlink:show="embed" xlink:actuate="onLoad"/></text:p><text:p
text:style-name="P3"><text:tab-stop/>
To Dr. Ty Coon
```
...

However, this still requires intimate knowledge of the program's data format, and it is poor in terms of flexibility. Any little change requires modification of the program. And with programs that use a proprietary format, the task is really formidable.

(3) By puppetmastering the word processor into writing or modifying an arbitrary file unter the direction of the "custom" software. A text processor – or other application program – is basically a black box that receives messages from the user (keystrokes, menu clicks, etc.) and reacts by changing its internal state. And this is the very description of an object!

Many commercial program packages traditionally come with some built-in "macro language". In general, these so-called macro languages are abominations – crude hacks designed with the sole aim of lowering the "entrance barrier" far enough to allow stenotypists with absolutely no knowledge of or interest in programming to get their everyday tasks partially automatized, at the cost of omitting all structure and order, to the point where the very notion of conscious design is becoming obscured[40]. The invention of the "macro

---

[40] I may be biased in favour of open source software, but it remains a fact that traditional open source applications possess excellent "macro languages" which are, however,

recorder" is what out-herods Herod: a software device that registers the user's interactions and produces some corresponding "macro language" gibberish.

Apart from being amorphous nightmares, the scope and extensibility of "macro languages" are also generally poor. In a "macro language", you just could not write the software for controlling a microplate reader: because (a) there is no interface to the "outside world" and (b) data structures as complex as arrays are no part of a "macro language". So if we decide that we want to do some statistical evaluation of our 96 wells *ex itinere*, there is no way! One of the more praiseworthy deeds of Microsoft was to abandon the original "macro language" concept in favour of VISUAL BASIC which is at least a fully-fledged programming language, though a spongy and unsympathetic one[41].

Now the OO paradigm allows us to use essentially any OO-capable languages for pulling the strings of the program. Why and how? Because the program itself can be handled as an object, as well as the documents it handles – and their components:

```
from Programs import OfficeSuite
from HardwareInterface import GetUSB

nuDoc := OfficeSuite.OpenTextDocument( "LCU_MicroPlateReaderReport.template.sxw" )
resulTable := nuDoc.TableN( 1 )

for x := 1 to 12 do
    for y := 1 to 8 do
        resulTable.setCellContent( x, y, GetUSB( ("GET",x,y,wavelength) ) )
    next
next

nuDoc.sendToPrinter( "ColorLazyJet.isr.lcu.edu" )
```

What does this piece of pseudocode mean? The program "Office Suite" is accessible under the guise of the object *OfficeSuite*, which provides, among other things, the method *OpenTextDocument*. This method is invoked every time a user selects the "Open file..." entry from the "File" menu. It returns a new object of a different type – a text document. This text document again comprises different objects, e.g., tables, which can be accessed by calling the text documents *TableN* method, which returns a reference to the nth table in the document. Such a table object possesses a method *setCellContent(x, y, new_content)*, the meaning of which is obvious. Thus, our twelve-line script will direct the text processor to load the template form and fill in the values

---

not intended for Betty Blonde the three-inch-nailed keyboard killer: e.g., EMACS with its distinctive brand of LISP.

[41] To give the devil his due: considering that it is BASIC, there is little to critizise about VISUAL BASIC. It is not a bad BASIC, but it's bad enough by being BASIC – a language designed to allow a rudimentary kind of software development on 8-bit computers. You don't fit cars with steam engines just because once upon a time this was the accepted standard for transportation, and people got accustomed to it.

obtained from the microplate reader, and then print it automatically. Things can be that simple.

The object model provides an excellent basis for providing interfaces between different languages. We do not need to know which language the objects belonging to *OfficeSuite* are implemented in, and what kinds of private members it may have – all that matters is that they exist and provide public methods which our language can call.

Things are straightforward if we have the source code for Office Suite, but even closed source programs are nowadays designed to accommodate "add-ons" and "plug-ins". Thus, it is not difficult to enrich Office Suite – let us assume it is written in C^{++} – with compiled modules written in C^{++} or EIFFEL, with interpreters that enable Office Suite to "speak" PYTHON or RUBY, or with the virtual machine structure required for executing JAVA or o'CAML bytecode. This is just a matter of a few standardized interface definitions. The very curious are encouraged to consult the official PYTHON documentation which contains another well-written chapter on this subject.

It must be pointed out that programming languages are egocentric. If we fuse Office Suite with a PYTHON interpreter, then from the interpreter's perspective the entire Office Suite package is secondary. It is still PYTHON, and nothing more. To the PYTHON interpreter, the entire functionality of Office Suite is just another module that can be `imported`. The interpreter is the puppet master, and the application program which it has been hybridized with just obeys. We do not get "Office Suite featuring PYTHON", we get "PYTHON featuring Office Suite". The difference is not great, but it facilitates our understanding of the way in which a programming language may be used for controlling a complex application: scripting in the real sense of the word.

We will learn more about this "extending and embedding" in Vol. 2, where we will encounter a simulation program consisting of a very fast core ("engine") written in C and a PYTHON interpreter controlling the engine and used to define scenarios.

10.19 Meyer's Principles: a look at Eiffel

Among the most insightful works on the subject of OO there are those of Bertrand Meyer, the author of EIFFEL. Apart from his famous "design by contract^{TM}" approach, which will be discussed on p. 253, the following concepts should be mentioned:

10.19.1 Command-query separation principle. It goes without saying that data members of any object should be completely encapsulated and not accessible from without, save by means of methods. As access may be either by reading or by writing, two fundamental types of data access methods are imaginable: "getters" or "queries", which inquire into the current value of a data member, and "setters" or "commands" which are messages sent to the object ordering it to change its state (i.e., the value of at least one of its members).

The command–query separation principle states that these two aspects should never be mixed up, i.e., that a "getter" should never change the state of an object and a "setter" should never be used to return the state of an object. The former aspect is by far the more important, as it helps to maintain the stability of objects. Anybody with at least a superficial knowledge of Heisenberg's theorem will appreciate the importance of this.

10.19.2 Uniform access principle.
In Meyer's own words: **"All services offered by a module should be available through a uniform notation, which does not betray whether they are implemented through storage or through computation."**

In practice, this means that (using a terminology which has gained wide acceptance with the advent of JAVA) the *interface* and the *implementation* of any piece of software, that is to say, the specification of the parameters and results on the one and its way of deriving the one from the other, on the other hand, should be kept apart. Why? Because this will allow one to easily design a framework of the desired functionality of a program (or a reusable library of software building blocks), which can then be implemented without having to worry about the overall structure; and because this makes it possible to modify the implementation, if so desired, whether for bug fixing or for optimization.

interface
implementation

An important example of this is in the treatment of lists and arrays. Where the language supports both of these, it is preferable to have the methods of access for both of them as similar as possible. In this respect, the uniform access principle is related to the concept of orthogonality (p. 100). For example, the function or method for determining the length of a list should be the same as the one returning the size of an array, independent of the technique employed for determining it.

10.19.3 Single choice principle.
"Whenever a software system must support a set of alternatives, one and only one module in the system should know their exhaustive list."

This is a most fundamental tenet of OO in general. In a more general wording: *Use OO to avoid repetitions.* If any definition or functionality is repeated in any place, then any change or extension to this will naturally require the corresponding modification of this definition or functionality wherever it occurs, which is bound to become a nightmare even with comparatively small projects.

Use OO to avoid repetitions

The life scientist will understand that dependencies are the ultimate stumbling block to evolution. So-called *conserved structures* are generally conserved because there are other structures which they have to precisely interact with, so

conserved structures

that any change in the one would necessitate a corresponding change in the other component, lest the combination should become useless.

Feel free to turn back to our first example, the address file which is supposed to accommodate postal prefix codes as used in different countries (p. 36). Ideally, here it is only the class `addressEntry` which knows the individual prefix codes (i.e., their representation *and* the way to handle it).

10.19.4 Open-closed principle.

An "open class" is one that may be extended, whereas a "closed" one cannot be modified except by extension (inheritance), which is obviously not mutually excusive.

Originally, the idea was that once completed, the implementation of a class could only be modified to correct errors; new or changed features would require creation of a different class, which could re-use code from the original class through inheritance. The derived subclass might or might not have the same "interface" (p. 239) as the original class.

Recently, the idea has been generally redefined to mean the possibility of alternative implementations of the same "interface".

10.20 Application: cytometry (II)

You remember the task to design a program for obtaining data from a cytometer, as well as the somewhat awkward C solution (p. 177). Now let us consider how to do it in PYTHON.

Most importantly, we do not have to worry about lists and arrays; PYTHON will do all this for us. Likewise, handling of multiple or complex arguments or return values is straightforward in PYTHON. We do not have to declare any structures either:

```
def get_data():
    result = []
    while not key_has_not_been_pressed():
            result += [( read_values_from_cytometer() )]
    return result
```

That's it! No more than this. However, if we want do have more than a shapeless pile of numbers as a result (more precisely: a list of tuples), we ought to include a few more definitions:

```
class fc_data:
    def __init__( self, s, g, fl1, fl2, fl3 ):
        self.size,self.granu,self.fluores1,self.fluores2,
        self.fluores3 = s,g,fl1,fl2,fl3

def get_data():
```

```
    result = []
    while not key_has_not_been_pressed():
        result += [ fc_data( read_values_from_cytometer() ) ]
    return result
```

This makes "result" a list of objects rather than of tuples, so that each cell's different values can be addressed by name, e.g., as `result[ i ].size` .

Within any given paradigm, languages differ mostly with regard to the structures they offer for the storage of variables. Flow control structures are rather uniform; after all, how many things can be done with a branching or a loop?, but variable structures are what provides abstraction (or not).

10.21 And then...

♡ *Have a good meal and a walk in the woods. Then begin this chapter anew. (That's a GOTO statement, by the way[42].)*

♠ *If you feel that you are up to it, try to implement the data types from p. 90 – set, bag and pipeline (and stack, if you have not done this yet) – based on* PYTHON *objects. Hint: The objects should incorporate lists.*

What is the method for creating a serpent object? A Boa Constructor!

---

[42] Iteration would be: REPEAT this chapter UNTIL you have understood it. And recursion. If you have understood it, fine; if not, refer to the explanations given in this chapter.

CHAPTER 11

◇ Exercise: Prime Numbers

> Prime time, nursery rhyme, can't you see the teacher?
> It's all there, black and white; now what's you gonna do?
>
> – Ronnie James Dio: "Strange Highways"

Relevance: Designing a program that handles an unforeseen amount of data.
Keywords: Prime numbers, lists, iteration and recursion.

Now it's time for some exercises. I do not think we are already advanced enough to tackle molecular biology issues, but a bit of sports will do us good.

Your first task is to devise a function named *primetime* with one parameter which returns a list of all prime numbers up to the argument number; i.e., when you call *primetime(20)*, you are expected to get the following output: 2,3,5,7,11,13,17,19.

By the way, this illustrates the benefits of lists. You cannot predict how many elements the result will comprise. Without lists, all you can do is define an array of the size of the argument and know that you are probably going to waste lots of memory.

In this respect, the search for prime numbers may serve as a model for some of the problems to be encountered in bioinformatics. From a large number of raw data, some information has to be extracted the volume of which is largely unknown[1]. Compare the task of identifying potential open reading frames[2] within a nucleic acid sequence. There is little to be assumed before. In any arbitrary stretch, there may be no open reading frames at all; there may be some; or there may be plenty – actually a vast number, given the possibility of alternative splicing and overlapping genes, as employed by certain viruses. So in that case we have the same problem in that we do not know beforehand how much data will be generated.

---

[1] Concerning prime numbers, there are indeed mathematical estimates for the amount of prime numbers within a certain range of integers. However, these are ignored here.

[2] What is sometimes and, less precisely, referred to as "genes" – the portion of a nucleic acid sequence that actually codes for a polypetide.

What is the optimal strategy for finding prime numbers? It is based on a simple definition[3]: A number is a prime number if it cannot be divided by any lower prime number (excepting 1, of course). Obviously, this suggests a recursive solution, but iteration is equally feasible. Try to implement both, if you can – PYTHON is quite up to it.

So we begin with an empty list and the number 2. For every subsequent number, we determine whether it can be divided by any member of the list. If it cannot, we add it to the list.

You are meant to try this on your own so do not skip things if you want to benefit! There is an exemplary solution (recursive) on the next page, but in order not to make things too easy it is written not in PYTHON but in HASKELL.[4]

Let me emphasize that this book is not about learning PYTHON but about learning algorithmic thinking. When you have grasped the concept, you will not have many problems implementing this in any language you like. So first try your own solution and make a hardcopy of it. Then look at the HASKELL program and try to understand the meaning and functioning of the individual clauses, especially the way recursion is used twice for different purposes. Next, re-write this algorithm in PYTHON and test it. Finally, compare the two solutions. Probably you will be astonished at the differences, considering the relative simplicity of the task.

```
-- yield list of prime numbers up to n
primetime n = enum 2 n []

-- count from alpha to omega and collect prime numbers
enum alpha omega primel =
  if alpha > omega then primel else    {- terminal condition: maximum reached
    if not (divisable alpha primel) then enum (alpha+1) omega (primel++[alpha])
    else enum (alpha+1) omega primel

-- test whether a number may be divided by any in a list
divisable _ [] = False        {- terminal condition: list empty
divisable n ilist =           {- "rem x y" yields remainder of x/y
  if (rem n (head ilist))==0 then True else divisable n (tail ilist)
```

---

[3] Of course, this might be refined, and the program greatly accelerated, by making a few other assumptions, but we will try to remain minimalistic here.

[4] What is good enough for Oxford is good enough for you too.

Perhaps this will help you to understand:

	HASKELL	PYTHON
Define function f with parameters x and y	`f x y =`	`def f( x, y ):`
Call function f with arguments x and y	`f x y`	`f( x, y )`
List concatenation	`++`	`+`
Division remainder	`rem x y`	`x % y`
Head of a list	`head list`	`list[0]`
Tail of a list	`tail list`	`list[1:]`
Empty list	`[]`	`[]`
λ function with parameters x and y, adding x and y	`(\ x y -> x+y)`	`lambda(x,y): x+y`
Return value r	`r`	`return r`

Cave: In PYTHON you must make sure that list is not empty and not too short, otherwise you will get an `IndexError: list index out of range`. It is a good idea to define your own `head(l)` and `tail(l)` functions.

Murphy's Law is recursive: washing your car to make it rain doesn't work.

CHAPTER 12

The Cathedral and the Bazaar: Rivalling Strategies

"I can imagine those People's Committees. Probably very few of the people there actually belong to the People."

– TERRY PRATCHETT: "Interesting Times"

By now we have encountered (and hopefully understood) almost all of the major components of programming. We will now have a look at two general approaches to writing a program.

12.1 The TOP-DOWN strategy

You begin with the task as a whole and dissect it into its components, descending layer by layer. Each layer is implemented during this analysis.

Thus, a compiler is generally dissolved into the following phases:

(1) reading the text file
(2) lexical analysis (which letters belong together to form words or symbols)
(3) grammatical analysis (which words or symbols form contiguous phrases)
(4) generation of executable code including variable handling (how information is stored internally)
(5) writing the machine code file.

I mentioned before that OBERON is the second language that is taught at Oxford. Well, here is the actual OBERON compiler they are using, written in O'CAML. It is a genuine except for the addition of a few comments. I direct your attention to the `main()` function:

```
(*****************************************
 * Oxford Oberon-2 compiler
 * main.ml
 * Copyright (C) J. M. Spivey 1995, 1998
 *****************************************)

(* The following lines are equivalent to PYTHON's "import" : *)
open Print
```

Bioinformatics Programming in Python. Rüdiger-Marcus Flaig
Copyright © 2008 WILEY-VCH Verlag GmbH & Co. KGaA, Weinheim
ISBN: 978-3-527-32094-3

```
open Filename
open Tree
open Error
open List

let copyright = "Copyright (C) 1999 J. M. Spivey"
let rcsid = "$Id: main.ml,v 1.2 2002/12/31 13:27:26 mike Exp $"
let print_version () = fprintf stderr "Oxford Oberon-2 compiler version $\n" [fStr Config.version]

let spec = (* this is just a bit of banter *)
  [ "-b", Arg.Clear Igen.boundchk, "Disable runtime checks";
    "-v", Arg.Unit (function () -> print_version (); exit 0), "Print version and exit";
    "-w", Arg.Clear Error.warnings, "Turn off warnings";
    "-x", Arg.Set Error.extensions, "Enable language extensions";
    "-g", Arg.Set Symfile.debug_info, "Output debugging info";
    "-pl", Arg.Set Peepopt.linecount, "Include line numbers for profiling";
    "-I", Arg.String (function s -> Symfile.libpath := !Symfile.libpath @ [s]),
       "Search directory for imported modules";
    "-d", Arg.Int (function n -> Error.debug := n), "Set debugging level" ]
let usage () = Arg.usage spec "Usage:"; exit 2

let error_token lexbuf =
  let tok = Lexing.lexeme lexbuf in if tok.[0] = '\000' then "EOF" else tok

(*** the core of the OBERON compiler ***)
let main () =
  (* preliminaries *)
  Peepopt.show_source := true; Peepopt.optflag := true; Error.warnings := true;

  (* get the command line arguments *)
  let fns = ref [] in
  Arg.parse spec (function s -> fns := !fns @ [s]) "Usage:";
  if length !fns <> 1 then usage ();
  let in_file = hd !fns in
  if not (check_suffix in_file ".m") then usage ();
  let base_name = chop_suffix (basename in_file) ".m" in
  Symtab.current := Symtab.intern base_name;

  (* read file *)
  let chan = try open_in_bin in_file
    with Sys_error s -> fprintf stderr "$\n" [fStr s]; exit 1 in

  (* lexical analysis *)
  let lexbuf = Lexing.from_channel chan in init_errors in_file chan lexbuf;

  (* grammatical analysis *)
  let prog =
    try Parser.program Lexer.token lexbuf with
      Parsing.Parse_error -> exit 1 in
  if !err_count > 0 then exit 1;
  Check.annotate prog;
  if !err_count > 0 then exit 1;
  if !Error.debug > 0 then print_tree stderr prog;

  (* machine code generation and writing *)
  let stamp = Symfile.export prog in
    Igen.translate stamp prog

(*** That's all! ***)

let crash fmt args = (* uh-oh! *)
  fprintf stderr "*** Internal compiler error: $\n" [fMeta (fmt, args)];
  fprintf stderr "***   Please save your program as evidence and report\n" [];
  fprintf stderr "***   the error to '$'.)\n" [fStr Config.bugaddr];
  exit 3

let catch_failure f x = try f x with (* display error, if one occured *)
  | Failure s -> crash "$" [fStr s]
```

```
  | Expr_failure (s, e) -> print_expr stderr e; crash "$" [fStr s]
  | Invalid_argument s -> crash "invalid argument to '$'" [fStr s]
  | x -> crash "exception '$'" [fStr (Printexc.to_string x)]

(*** and now JUST DO IT: ***)
let obc = catch_failure main (); exit 0; ()
```

I do not expect you to be able to fully understand this (o'CAML is very powerful but also rather quirky), but the take-home message is: this is top-down programming. A program of the complexity of a functional compiler is begun by writing a central core, usually surrounded by a few minor auxiliary functions for startup and shutdown, including argument processing, copyright message, etc. The core is minute in size, because it just comprises the steps of the topmost layer. When this is done, we proceed to the next layer. Implementation of the overall functions for lexing, parsing and code generation. And so on.

Top-down is strongly favoured by higher-level programming languages, especially functional and type-inferring ones. It is often considered to be the more powerful approach.

Quoting Chris Pressey of Cat's Eye Technologies (*cpressey@catseye.mb.ca*), in turn quoting *Theories of Programming Languages* by John C. Reynolds (Cambridge University Press, 1998), p.72:

> "At this point, we have an informal account of the first step in a top-down program construction: *we know how to write the program if we can write a while body B with certain properties.*"
>
> In other words: start with the specification, derive the program from it while proving it, and the program will be correct "out of the box".

12.2 The BOTTOM-UP strategy

Actually, bottom-up is largely a contradiction in itself, because you usually begin with a complete project in mind. So in order to know the most elementary buildings blocks of your program, you must first dissect the entire thing mentally. However, there are two exceptions to this:

- Re-utilization of existing code. For example, imagine you have a database access system whose file format is defined in some modules. If the problem at hand is not too complex, you may begin with these modules and climb up to the solution of your task.
- Simulations. This is a far more interesting aspect. Simulations just *cannot* be implemented in top-down fashion, because the very nature of a simulation is that the behaviour of individual units is known but that of whole populations of such units is not. Thus, we begin by defining our individual units – usually, a single class will be sufficient – and then build up a framework for the interactions of these units. You will perceive that this is not programming in the normal sense, because

the behaviour of the complete program is unknown beforehand (and must be, for otherwise we could spare our labour).

Nevertheless, most commonly used programming languages, notably the C and Algol superfamilies, favour bottom-up programming by the need for explicit definition of types and structures beforehand.

12.3 A few notes on style

As in chess, in the world of logic, mathematics and informatics there is no room for surprise presented by empirical data. Thus, everything is obvious at close enough scrutiny. Or to put it differently, practical advice has a tendency to be awfully commonplace, unless based on assumptions which are not commonplace, in which case its general applicability is more than questionable. An example of the first kind is the rule "let variable names be just as long as is required to make them meaningful, but not longer". An example of the second is the notion that classes are always short and many in number, and that a proper IDE must therefore support the parallel handling of dozens of files, but does not need to provide particular help in dealing with long files – an idea which may be correct for most "frontend" programs (which consist mostly of endless definitions of windows, menus and the like) but simply does not hold true for more complex scientific applications, where the algorithms doing the work may be of appreciable size[1]. This will be discussed in more detail below. So basically it is very difficult to make suggestions pertaining to the practical aspects of programming, which are not trite but still generally applicable.

12.4 Splitting a large project

12.4.1 Why split a program?
Every student of martial arts has heard the advice: "the strength is in the centre – move your body, let your hands and feet follow." In a way, this holds true for programming. The secret of writing a good program is in getting a clear understanding of the problem and its solution, everything else will follow. Trying to superimpose artificial demands afterwards will lead nowhere. Thus, a well-structured program will be segmented quite naturally, and the splitting of a large project into a number of small files will reflect the structure and thus increase legibility and intellegibility.

There is also another, more mundane, reason for dividing a program into a number of small files. Let us have a look at some actual compilation times for a very small program, the prime number identificator described in the "Wizards' Sabbath" chapter (p. 359):

---

[1] A mere management issue is that small files are easier to deal with in version control.

Language	Implementation	Source length	Compilation of "primetime"
BASIC**	Foobasic (Vol. 2)	383 bytes	<0.1 sec
PASCAL	GNU	2558 bytes	0.15 sec
OBERON	Oxford	1906 bytes	0.18 sec
FORTRAN-90	GNU	343 bytes	0.28 sec
O'CAML	ocamlopt.opt	494 bytes	0.40 sec
C-99	GNU	1430 bytes	0.42 sec
ADA	GNU (gnat)	1631 bytes	0.55 sec
ERLANG*	erlc	1130 bytes	0.72 sec
C++	GNU	1871 bytes	0.74 sec
CLEAN	cml	814 bytes	0.83 sec
EIFFEL	SmallEiffel	2209 bytes	1.25 sec
JAVA*	javac	745 bytes	3.7 sec
HASKELL-98	Glasgow	776 bytes	5.3 sec
MERCURY	mmc	1276 bytes	32.8 sec

An asterisk (*) indicates that the compiler itself runs on the virtual machine and is thus necessarily slower; two asterisks mean that the compiler is implemented in an interpreted language. All tests were performed on Cirith Ungol (P-III/1 GHz, 256 MB).

Obviously, the intrinsic complexity of a language is of crucial importance for the speed of compilation[2]. It is also noteworthy that these data are by no means correlated with the quality of the generated code. Note that the Wirth languages (PASCAL and OBERON) are by far the fastest in compilation *and* produce good code, in spite of all the differences in implementation[3], indicative of a very consequent and no-nonsense structure of these languages; in fact, early PASCAL compilers could be made to run on 64 kb machines. At the other end, HASKELL and MERCURY are characterized by extreme abstractness (they are both non-strict and lazily evaluating), requiring giant compilers whose output is inflated to grotesqueness (some 650 kb for HASKELL and 1.4 megabytes for MERCURY). As for the time required, processing the entire LaTeX file of this book, together with all graphics, is compiled to form an eleven megabyte PDF file in less time than the MERCURY compiler needs for this tiny program[4]!

At any rate, a program which comprises several megabytes of source code will take quite a while to compile[5]. This is bound to get painful when testing is

---

[2] It is said, "those who do not study LISP are doomed to reinvent it. Poorly." The *g++* C++ compiler uses a LISPish language named KSI internally to represent the program being translated. The fact that the *g++* compiler must do this, and must be itself written in C++, is probably part of what makes *g++* itself so slow.

[3] The GNU PASCAL compiler is written in C, the Oxford OBERON compiler in o'CAML.

[4] The Unix utility **top** was used to monitor CPU and memory usage, and in fact processor power was the bottleneck.

[5] The most extreme case of this which I encountered personally was the "building" of the OPENOFFICE suite on Zacal Balam, which took about thirty (!) hours and yielded more than two hundred megabytes of executable code.

required. A possible remedy lies in the splitting of the program into small building blocks which can be compiled individually. Only those which have actually been changed have to be recompiled; they can then be linked with the others (which is quite fast) or, in the case of VM-based languages such as JAVA and ERLANG, just be run[6].

The third reason for splitting up a program is that there may be fundamentals which are required by all parts of the program, and it facilitates things if these are put into separate files. This is particularly nice when you have a program with an installation script that generates header files tailored to your hardware, so the "building" of the program does not just generate something that runs but, if things are well-written, something that is optimized for your individual computer; a thing which would be impossible if all your source code were packed together into one giant file.

12.4.2 Features of languages that support splitting. In general, such features have evolved over the years. The first versions of FORTRAN and PASCAL did not support any modularization. PASCAL's structure is monolithic, with the file beginning with the word "program" and ending with a period. The only rudimentary kind of modularization allowed was by including other files, satisfying, maybe, the first and the third need but certainly not the second and perhaps most crucial one. TURBOPASCAL was the first dialect of PASCAL to allow references to "external" functions. This was still rather crude, with *all* functions being made public.

C was among the first languages to allow not only inclusion of files (with a special distinction being made between normal source files and *header files* containing only shared definitions, the latter generally bearing the suffix .h instead of the normal .c) separate compilation – when all parts have been compiled, the linker puts them together and arranges them in such a fashion that the program begins at a function named `main`. If there are none or more than one, or if references to functions are not satisfied, linking is cancelled. However, little checking was otherwise provided by K&R; most notably, in the initial definition of C, no comparison between arguments and parameters was made. So when a function was called with a wrong set of arguments, this resulted not in an error message but a program crash. Luckily, around 1990 ANSI-C introduced this kind of control.

Neither PASCAL nor C have ever solved the problem of name conflicts. Imagine a file library and a dynamic string library both providing a function named "new". In C and PASCAL, there is no solution for this. Instead, these languages rely very much on convention and the politeness of using unambiguous names. Thus, in C we will probably have "file_new" and "string_new" as functions.

---

[6] ERLANG was developed with the explicit aim of creating a system where parts of a running program can be exchanged without having to stop the application, not to mention restarting the system.

Of the purely procedural languages, MODULA-2 was the only one ever to feature a module system to cope with this problem. It was also among the first to experiment with the concept of "qualified import", that is to say, it offers the choice whether you want to import a whole module and address its components as *module.function*, e.g., `import file; file.new(...); ,` or whether you explicitly import functions *from* a module: `from file import new; new(...); ,` just as it is available in PYTHON. Otherwise, the procedural languages were, by and large, superseded by object-oriented languages (C itself being the only notable exception), or extended towards object-orientation.

Splitting of projects comes quite naturally with objects and classes, so it is available in all languages which style themselves "object-oriented". Classes are (and are supposed to be) closed systems, so they can easily be separated from other parts of the program. PERL takes the simplest approach in which all components of a class are defined as such by being placed together in a file with the suffix *.pm* (as opposed to the normal *.pl*), the file name being the class name. However, when everything must be an object, it may become tedious to have to open a separate file for each and every class, as in JAVA or EIFFEL, unless you are using one of those fancy IDEs with which you can edit fifty files at once (preferably on a 23" screen).

Some languages, e.g. OBERON, are absolutely consequent in their application of the OO paradigm and allow to use import statements with variables like any other: `IF computer = 'Mac' THEN wilib := AquaBindings ELSE wilib := X11Bindings; FROM wilib IMPORT OpenWindow; ....` PYTHON is more conservative in this respect and actually there are only few uses for this purpose, the principal being (as in this little example) the use of alternative libraries for different systems. As the scripting languages generally tend to provide system-independent libraries, there is really no need for this. So PYTHON considers module names as constants: `import x` will not load a module whose name is in *x* but the module named `x`.

In JAVA and some other OO languages we have rather complex concepts like "interfaces" and "virtual methods" by which it can be enforced that newly created classes are in accordance with formal specifications, that is to say, that they *must* implement certain methods.

EIFFEL is unique in that it provides still further support for splitting by specifying, not only the form of objects, but also value ranges and acceptance criteria by means of its *requires* and *ensure* structures. Eiffelians usually refer to this peculiar feature as *design by contract*, since it allows one to draft a precise outline for objects which can then be implemented independently. That is to say, where other languages only specify that, say, an integer argument is required and a float value returned, an EIFFEL method can be made to demand, for example, an integer argument which is an odd number between 100 and 300 and return a floating point value which must be greater than the argument – preventing lots of logical errors. *Keep in mind that formal correctness alone is*

requires
ensure
design by contract

not enough to guarantee the proper functioning of a program, and that not all functions yield sensible results, or any results at all, for all possible arguments.

In a similar way, ADA's complex subtyping system tries to catch logical errors before they can occur.

Functional languages generally permit easier structuring of programs. All modern functional languages possess module systems with import and export lists, and as there are no variables in the traditional sense, that is all that is conceptionally possible – making available and using immutable functions. Functional programming also avoids all problems with inheritance, which is responsible for the almost baroque long-windedness of declarations in many OO languages like JAVA.

As for the strength of typing, this is perhaps the pivotal drawback of weak typing, apart from its less efficient code generation. Argument–parameter mismatches are pretty likely to occur when a project is split across several files, and as long as there are equal numbers of both, this will result in hard-to-detect logical errors, especially with PYTHON, which strives for maximum compatibility between different data types.

As a general rule, do not rely overmuch on what programs can do for you (most of them are pretty dumb anyway), but begin by working out a clear design and continue by keeping proper protocols on what you have done so far. A carefully written, hardcopied documentation of what a function does and which kinds of arguments it expects is worth more than all of ADA's subtyping, EIFFEL's design-by-contract and JAVA's interfaces put together.

12.4.3 Natural borders. Let me repeat my former admonition: keep clear of mechanical dullness. There are no rules for design which deserve to be engraved into tablets of stone and presented ceremoniously on Babbage's birthday. The strength is in the centre, in the thorough understanding of the problem to be solved. In the second part of this book, the chapters on "The Lisper" will present an exemplary approach on how a programming language itself can be structured. There are: the *scanner* which reads the source file and turns it into a stream of keywords, variables and other "tokens"; the *parser* which makes sense of this stream and identifies errors, and the *back end* which generates machine code or executes the commands, depending on whether your program is a compiler or an interpreter. This tripartite structure was not mystically revealed to some guru in deep meditation but is a natural answer to the requirements of interpreting and compiling. The software of Klingon computers will probably be built along the same lines[7].

---

[7] Those who are familiar with philosophy will in fact find that scanner, parser and back end correspond very neatly to the features of the human mind which Kant termed *understanding* ("Verstand", the power to recognize things), *pure reason* ("reine Vernunft", the power to understand and explain) and *applied reason* ("praktische Vernunft", the power to take reasonable actions).

Other problems will require different structures, and sometimes different design principles may be applicable, but the structure should always reflect the problem.

You are certainly on the safe side if you arrange your project in "diamond structure":

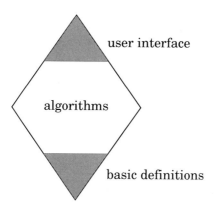

Meaning that the fundamental definitions (including shared constants and essential data types) should be in the most basic section which imports no other, and the user interface in the topmost layer which is imported by no other, with the actual algorithms being in between, as it makes sense.

In this respect, the "Lisper" is far from an optimal design. To improve this, one would place the most elementary class (which one is that?) into a separate file, together with other definitions and declarations, and add a "front end" for selection of the input file and error messages: in the case of LISP, a "read–evaluate–print loop", as we know it from our PYTHON interpreter, would be nice.

12.4.4 "Make".

12.4.4.1 *Module hierarchy.* Any splitting of the souce code introduces the problem of *dependency*. When Module α imports module β, then α is said to be dependent on β. More precisely, when Module α uses definitions given in β, then α is dependent on β. That is to say, changes made to β affect α as well. If module β, in turn, imports module γ and module α also imports δ, then α will depend on β and δ and (indirectly) γ, whereas β depends on γ alone. This is called the "module hierarchy".

dependency

In an interpreted language such as PYTHON, the dependency of modules is of hardly any importance whatsoever, as the source text has to be processed prior to execution anyway.

Things are very different with strictly typed compiled languages, however. Generation of an executable program is a tedious process, which may take several hours for larger projects if everything is rebuilt from scratch. It is thus most expedient to provide a means of reducing this workload.

12.4.4.2 *Definition* vs *implementation* in compiled languages. On p. 16, we have briefly outlined how compiled languages such as FORTRAN, PASCAL, C/C++ or ADA eventually produce executable programs. Individual modules of source code are compiled individually to "object code files", which are then assembled (in the technical sense), together with library files as required.

definitions

It will be appreciated that *definitions* such as data structures (aggregates) or the parameter lists of functions form the "backbone" of the program structure, modifications whereof will require recompilation of everything that makes use of these definitions: If a variable is changed from a 64-bit integer to a 128-bit floating point, all the machine code relating to this variable has to be generated anew. This holds true even for weakly typed languages such as ERLANG. When the *arity* – the number of arguments expected by a function – has been changed, all modules using this function must also be recompiled.

arity

implementation

By contrast, the code proper, which provides the *implementation* of the functionality, may be exchanged quite easily. If the function is supposed to, e.g., sort an array of integers, no calling function will ever notice a difference when the implementation is changed from bubblesort to quicksort (see the chapter on sorting, p. 261ff, for details). Thus, as long as the definition remains unaltered, it suffices to recompile the module comprising the implementation changed, and link the compiled modules together as always, if this is required. In VM languages, such as JAVA, there is no linking at all, as the VM will find on its own the things which are normally sorted out by the linker.

interfaces

Evidently, this also holds true for definitions which are used only internally. When a function is not to be called from outside a module, or a data structure is not to be used by other modules, changes of these definitions do not concern other modules either. The *interfaces* between individual modules, however, are sensitive. And in any great project they will have an "untouchable" status as fruitful cooperation between individual developers largely depends on their inviolateness.

An example for a definition is thus: "a box which fits into a bay as described in section... and is capable of generating a constant output of 6 V (AC)." And for the implementation: "a lead car battery; or a lithium ion battery; or...".

12.4.4.3 *Compiling, linking, and makefiles.* Thus, after any modification of any individual module, let us call it Ω, belonging to a larger project, we have basically one of two different situations.

- The changes have affected (or may have affected) a definition that has consequences outside of module Ω. (Evidently, in this case it will have consequences inside the module too, but this is trivial.) In this case, the following steps will have to be undertaken (in this sequence):

(1) Recompile module Ω.
(2) Recompile all modules using module Ω.
(3) Link all the modules of the project and the libraries as necessary to build a new executable.

- The changes have affected only the implementation of module Ω and/or definitions that have no consequences outside of module Ω. In this case, the following steps will have to be undertaken (in this sequence):
  (1) Recompile module Ω.
  (2) Link all the modules of the project and the libraries as necessary to build a new executable.

Now how can these situations be differentiated? JAVA and C^{++}, for example, are very explicit as to whether any particular definition is to be "visible" from without the module, and changes must therefore be considered as of type (1). In C, the most widely used convention is that all definitions live in so-called *header files* with the extension *.h*, whereas the implementations are found in files with the extension *.c*. The amount to which this information can be utilized differs widely, but it is evident that having to read all the files to resolve the module hierarchy will not be that much faster than recompiling them unconditionally. To this end, it is most expedient to define the module hierarchy in a separate file. This file may be generated automatically, by extracting information from the program files, or manually; at any rate it will facilitate identifying what steps have to be taken in any particular situation.

header files

The best-known tool to this end, from which the generical name for this category is derived[8], is known as *make*. It uses a description of the module hierarchy that can be classified as being declarative (remember pp 39ff!) in nature: For each file, it lists those it depends on, and the steps to be taken if required; thus all files affected can be identified by recursive descent. So, the final executable can be re"made" with the minimum effort required.

make

The importance of "make" can be estimated from the fact that in the GNU project this utility was, together with an editor and a C compiler, one of the very first components to be written by Richard M. Stallman.

12.4.5 Version control. However, this gives rise to another problem. What if, at some level, an error creeps into our program? In the simplest case, with a monolithic source file, we may avoid a sticky situation by making regular (e.g., daily) backups of this file, preferably with the date in a suitable position relative to the name. However, when there are a number of developers each working on a different set of files, all belonging to the same project, this becomes opaque very quickly.

The solution is in using a *version control system*. Suitable software is readily

version control system

---

[8] For example, ant, designed for use with JAVA, describes itself as "a Java based make tool".

available, e.g., CVS. A version control system is basically a database which each and every file is fed into, whenever the developer is of the opinion that this particular version of the file might be valuable on any other day.

♡ *A full description of the capacities of version control systems is beyond the scope of this book, but the reader is strongly encouraged to look at CVS or a related system and get accustomed to using it as early as possible.*

Of course, the repository of the version control system itself has to be backed up regularly!

12.4.6 The trouble with OO and FP. It is quite a problem to get rid of obsolete code as larger projects have a tendency to become more and more like rubbish dumps filled with the leftovers of previous generations. At first glance, this seems simple. However, when you have a look at a program of 10 000 or even 100 000 lines, how are you going to find out which functions have actually been orphaned?

With the relatively strict hierarchy of procedural programming, this is quite easy to test: Begin at the program's entry point, follow every subroutine call and note which functions are called. Actually, this is what the linker does when binding together compiled files and libraries – all functions which are not called at least once are simply omitted from the final product: In the program

```
program optimize;

function a( x, y: integer ): integer;
    begin a := x + y end;

function b( x, y: integer ): integer;
    begin b := x * y end;

var r: integer;
begin
    r := b( 3, 5 )
end.
```

the linker will identify *a* as surplus and discard it.

It is much more difficult for object-oriented or functional languages. Of course, you can make sure that every reference is satisfied. A function or an object that is never referred to can be dropped that is how the linker deals with compiled C++, EIFFEL or OBERON files[9]. On the other hand, the fact that a reference to a function or an object exists does not mean that it is put to any use!

---

[9] There is a tendency to standarize object code format, so the same program – ld for most Unices – can be used for all languages.

```
MODULE optimize;

IMPORT Out;

TYPE dvandva = RECORD x, y, r: INTEGER; END;

PROCEDURE (VAR s: dvandva) a: INTEGER;
    BEGIN s.r := s.x + s.y END;

TYPE troika = RECORD(dvandva) x, y, z: INTEGER; END;

PROCEDURE (VAR t: troika) b: INTEGER;
    BEGIN t.z := t.x * t.y END;

VAR n: troika;

BEGIN
    n.x := 3; n.y := 5;
    n.b;
    Out.Int( n.z );
END.
```

Here it will be very difficult to check whether a is needed or not. It is bound to the class *dvandva*, and *dvandva* is the base class of *troika*, of which there is one instantiation which is obviously used. So, do we require a?

Functional variables may also cause much grief:

```
-module(optimize).
-export([run/0]).

a( X, Y ) -> X + Y.
b( X, Y ) -> X * Y.

run() ->
    Func1 = a,
    Func2 = b,
    Func2( 3, 5 ).
```

How can we argue here for a being superfluous?

These are still unresolved issues, and in the end there is nothing like careful work and manual book-keeping. Remember this:

Artificial intelligence is no match for natural stupidity!

CHAPTER 13

Ordo Ab Chao: Sorting and Searching

The origins of the motto "Ordo Ab Chao" are as obscure as its meaning, which may range from "This will overcome the chaos" to "A structure as old as the world".

– ALEC MELLOR: "Rose Croix"

Relevance: Computers offer the advabtage of storing gigabytes or even terabytes of information. To be able to retrieve, among a vast body of data, those items that are of particular interest to us, it is mandatory to sort the data volume according to some criterion of interest.
Keywords: Sorting algorithms.

13.1 When is a group of data ordered?

One of the most common problems in informatics is the sorting of data. We are not going to discuss trees here; let our task be simply to arrange a sequence of elements in such a way that every element is "greater than or equal to" its predecessor, according to our chosen criterion.

How can we do this?

Basically, we need the following things.

- The data structure (or rather infrastructure) itself. This may sound trivial but already causes a pain in most lower- to mid-level languages. C and PASCAL, for example, simply do not possess lists, and they have to be implemented first, based on structures with references to each other. JAVA does not have them either but at least provides standard library functions, which are, however, verbose and cumbersome.
  In PYTHON or HASKELL, we use a list and call it quits.
- A splinter of code that does the following.
  (1) Pick two arbitrary items, $i[x]$ and $i[y]$.
  (2) Apply a criterion κ to $i[x]$ and $i[y]$ and see whether they are in the correct order.
  (3) If they are not, swap them.
  Picking the items is the easiest thing, if the infrastructure is well-designed; we get them by i[x] and i[y] (PYTHON) or i!!x and i!!y

(HASKELL).

Applying the criterion can be done very smoothly, e.g., by using a λ function.

As for swapping, PYTHON allows us to write simply: `i[x], i[y] = i[x], i[y]`.

- A framework that causes the elements to be compared and swapped.

13.2 Bubblesort

bubblesort

The simplest way of sorting consists in just comparing the items pairwise all along the list for several cycles. During each cycle one more element will arrive at its proper position. This is known as *bubblesort*, as the migration of elements to order has some similarity to the rising of bubbles in a glass of lemonade.

First cycle:

$$[\,\mathbf{6, 5},\, 7,\, 4,\, 9\,]$$
$$[\, 5,\, \mathbf{6, 7},\, 4,\, 9\,]$$
$$[\, 5,\, 6,\, \mathbf{7, 4},\, 9\,]$$
$$[\, 5,\, 6,\, 7,\, \mathbf{4, 9}\,]$$

Second cycle:

$$[\,\mathbf{5, 6},\, 7,\, 4,\, \mathit{9}\,]$$
$$[\, 5,\, \mathbf{6, 7},\, 4,\, \mathit{9}\,]$$
$$[\, 5,\, 6,\, \mathbf{7, 4},\, \mathit{9}\,] \qquad \text{(no further comparisons necessary)}$$

Third cycle:

$$[\,\mathbf{5, 6},\, 4,\, \mathit{7, 9}\,]$$
$$[\, 5,\, \mathbf{6, 4},\, \mathit{7, 9}\,] \qquad \text{(no further comparisons necessary)}$$

Fourth cycle:

$$[\,\mathbf{5, 4},\, \mathit{6,\, 7,\, 9}\,]$$

Final:

$$[\,\mathit{4,\, 5,\, 6,\, 7,\, 9}\,]$$

♠ *Write a bubblesort program that works on arbitrary lists. Provide for inversion (biggest first) by passing the criterion as a functional parameter. Hint: use the module **random** to facilitate the generation of raw lists.*

13.3 Insertsort

As the name implies, the *Insertsort* algorithm creates order by inserting items at the correct locations of a nascent list. The correct location is defined as one where the new element is placed between one that is less or equal and one that is greater or equal. The insertsort is a very straightforward algorithm, especially when using a language that possesses lists, and unlike the bubblesort, it is reasonably fast. Naturally, the insertsort is especially suitable for collecting data from external sources.

Here is a universally applicable implementation in PYTHON:

```python
class insertsort:
    "A derivative of the common list object whose components are sorted automagically upon creation or insertion. Uses an insertsort algorithm with linear searching."
    def __init__( self, contents = [] ):
        self.__core = []
        self.__add__(contents)
    def __str__(self): return str(self.__core)
    def __len__(self): return len(self.__core)
    def append(self,item):
        if len(self.__core) == 0: self.__core = [item]
        elif len(self.__core) >= 1:
            if item > self.__core[-1]: self.__core += [item]                    # > last
            elif item < self.__core[0]: self.__core = [item] + self.__core    # < first
            else:
                for i in range(len(self.__core)-1): # linear search
                    if self.__core[i] <= item <= self.__core[i+1]:
                        self.__core = self.__core[:i+1] + [item] + self.__core[i+1:]
                        return
    def __add__(self,ali):
        for x in ali: self.append(x)
        return self
    def __getitem__(self,n): return self.__core[n]
    def __getslice__(self,a,b): return self.__core[a:b]
```

This somewhat strange-looking class uses a technique known as *operator overloading*, which will be discussed in Vol. 2. This means that some of our familiar operators are redefined for objects of this class. The class *insertsort* redefines the addition/concatenation operator +, the index operator [a] and the slice operator [a:b] as well as the output in text form using the "print" command and the % operator. Objects of class *insertsort* can be created either without any data or using a list as initial data; otherwise, they can be used as any normal list would, including mappings, filterings, list comprehensions and modifications using the += operator:

```
>>> a=insertsort()
>>> print a
[]
```

```
>>> b=insertsort( [ 9, 7, 8, 5, 3, 10, 1, 2, 6, 4 ] )
>>> print b
[1, 2, 3, 4, 5, 6, 7, 8, 9, 10]
>>> print b[3]
4
>>> print b[5:7]
[6, 7]
>>> import math
>>> map(math.sqrt,b)
[1.0, 1.4142135623730951, 1.7320508075688772, 2.0, 2.2360679774997898,
2.4494897427831779, 2.6457513110645907, 2.8284271247461903, 3.0,
3.1622776601683795]
>>> print b
[1, 2, 3, 4, 4.5, 5, 6, 7, 8, 9, 10]
>>> b += [1.5, 7.5, 3.5]
>>> print b
[1, 1.5, 2, 3, 3.5, 4, 4.5, 5, 6, 7, 7.5, 8, 9, 10]
>>> [x for x in b if x > 4]
[4.5, 5, 6, 7, 7.5, 8, 9, 10]
>>> for x in b: print x,
...
1 1.5 2 3 3.5 4 4.5 5 6 7 7.5 8 9 10
>>> ("%s" % b).replace(" ","&")
'[1,&1.5,&2,&3,&3.5,&4,&4.5,&5,&6,&7,&7.5,&8,&9,&10]'
```

♦ *Currently, this uses* PYTHON*'s built-in comparison operators which excludes the use of more complex data as elements, as with "native" lists. Modify the algorithm to take a functional parameter as a sorting criterion, as you have seen it in the "Full Impact" chapter, p. 143.*

13.4 Quicksort

The fastest known sorting algorithm, appropriately termed Quicksort, is a recursive thing. It works like this.

Sort a list by taking the first item, placing all items which are lesser in a new list and all items which are greater in another new list. The result is the sorting of the first new list concatenated to the first item concatenated to the sorting of the second list.

Don't worry if you do not understand at first glance. It is, in fact, so simple that it is difficult to grasp. Maybe an example will help you:

▶ sort [6, 5, 7, 4, 9]

first item = 6, first new list = [5, 4], second new list = [7, 9]

    ▶ sort [5, 4]

    first item = 5, first new list = [4], second new list = []

▶ sort[4]

first item = 4, first new list = [], second new list = []

⇒ [] ++ [4] ++ [] ⇒ [4]

⇒ [4, 5]

▶ sort[7, 9]

first item = 7, first new list = [], second new list = [9]

▶ sort[9]

first item = 9, first new list = [], second new list = []

⇒ [] ++ [9] ++ [] ⇒ [9]

⇒ [7, 9]

⇒ [4, 5] ++ [6] ++ [7, 9] ⇒ [4, 5, 6, 7, 9]

HASKELL:

```
qsort []     = []
qsort (x:xs) = qsort elts_lt_x ++ [x] ++ qsort elts_greq_x
               where
                 elts_lt_x   = [y | y <- xs, y < x]
                 elts_greq_x = [y | y <- xs, y >= x]
```

Let us analyze this:

$f[] = []$ – the f of an empty list is an empty list.

$f(x:xs) = \ldots$ – the f of any (other) list which consists of the head x and the tail y is ...

$a ++ b$ – the concatenation of two lists a and b.

$[a]$ – a list containing only the item a.

$[element | element \leftarrow source, element < threshold]$ – "list comprehension". A list built of all *element*s taken from the *source* list for which the condition *element* < *threshold* is fulfilled (list comprehension; in PYTHON: `[y for y in xs if y < x]` or accordingly).

Now in order "to deter others from the like crimes" an implementation in C:

```
qsort( int *a, hi, lo ) {
  int h, l, p, t;
  if (lo < hi) {
    l = lo; h = hi; p = a[hi];
    do {
```

```
            while ((l < h) && (a[l] <= p)) l++;
            while ((h > l) && (a[h] >= p)) h--;
            if (l < h) {
                t = a[l]; a[l] = a[h]; a[h] = t;
            }
        } while (l < h);
        t = a[l]; a[l] = a[hi]; a[hi] = t;
        qsort( a, lo, l-1 );
        qsort( a, l+1, hi );
    }
}
```

In C, `int *a` means that a is a pointer to integers, which is equivalent to saying that it is an array of integers the size of which is not known. Therefore, the additional parameters hi and lo are required[1]. The C program differs from the HASKELL specimen in that it does not have to generate new lists every recursion level. It may thus run a bit faster and use less memory, but the HASKELL version is clearly superior in terms of legibility and intelligibility.

♠ Write a quicksort program, in PYTHON, of course(!), that works on arbitrary lists. Provide for inversion (biggest first) by passing the criterion as a functional parameter.

♠ Though taken from a paper intended to show the supremacy of HASKELL and functional programming in general, the HASKELL version is not ideal. Can you think of any way of improving performance? If so, implement this in PYTHON too – or in HASKELL, if you are really tough.

♠ Assess the different algorithms for efficiency depending on the size of the input list. Hint: look at the module *time* first. Compare this also to PYTHON's built-in sort function, and discuss the difference.

13.5 Comparison

On the aging Linux workstation *nyarlathotep* equipped with a G4, 1 GHz and 768 MB of memory, PYTHON implementations of the three algorithms were compared by sorting a list of n integers between 0 and 100 000, generated by using the *random.randint()* function. Generation of the list was linear, taking approximately 16 μsec per item. The results (averaged over $\frac{10^6}{n}$ runs) were:

Algorithm	10 items	100 items	1000 items	10 000 items
Bubblesort	0.5 msec	52.6 msec	22.381 sec	>1 h
Insertsort	0.15 msec	6.9 msec	487 msec	70.1 sec
Quicksort	0.01 msec	0.11 msec	1.4 msec	19.85 msec

---

[1] Some experiences are not really required. Contrary to my usual custom, I took this from a textbook and did not bother to check it.

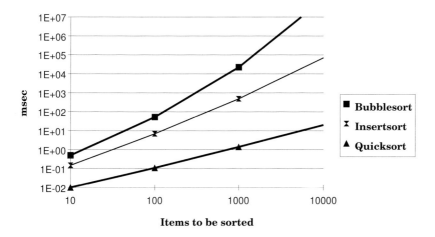

For a discussion of the general problem of speed and scalability, please see Vol. 2.

13.6 Others

There are numerous other sorting algorithms, e.g., shellsort and heapsort[2]. Most of them are specialized, designed for non-linear structures (e.g., trees), or simply older than quicksort (which was invented in the mid-1960s).

To iterate is human; to recurse, divine.

---

[2] Most recently, a method named "Burstsort" has been published, which is rumoured to be even faster than the quicksort algorithm.

CHAPTER 14

Welcome to the Library

For a while Casanunda gazed at the orang-outan, who was wearing a sweater with the word PONGO on it. "So you are the librarian, sir?"
"Ook", the orang-outan affirmed.
Amazed, Casanunda turned to the other visitors from the University. "But what about the stench?"
"Oh, he doesn't mind", the dean replied.

– Terry Pratchett: "Lords and Ladies"

Relevance: In practical application, much of the power of any programming language is a question of how many "building blocks" for frequent tasks are readily available to the programmer. In general, such building blocks are grouped, by functional relation, into "libraries" dedicated to peculiar themes.
Keywords: Library functions.

We have mentioned the PYTHON module system now and then, but we haven't used it much so far. So let us repeat once more.

- Built-in functions, or *BIFs*, are part of the core of the language. There is yet another jihad going on about how many of them are necessary. The extremes are COMMON LISP which has several hundreds of functions, all of them implemented as BIFs, and O'CAML, where there are theoretically no BIFs at all (the inevitable ones being assigned to an imaginary module named `Pervasives` which is loaded automatically).
- Library functions are contained in modules, which must be imported or loaded.

BIFs

Libraries are as old as operating systems themselves, but were not taken into account by programming language designers before the 1980s. For example, traditional C ("K&R C" because defined by Kernighan and Ritchie) does not support libraries or modules on the language level. Program parts are just compiled and then linked together with the necessary libraries – it is up to the linker to find out whether there is something missing, often resulting in error messages like "error at 0x220F310A: reference not satisfied" when a library was found missing. Much worse, the linker tried just to tether up things; when there was an error about parameters (e.g., invoking a library function with only

one parameter where two were required), there was no way to find that out –
until the program crashed. Happy debugging.[1]

The advent of graphical user interfaces has further stimulated interest in modularization, because it is simply not possible to do everything that is needed for complex window and menu systems via BIFs. In addition, the availability of more abstract type systems now allows the implementation of libraries for many purposes (from data trees to linear algebra).

The grouping together of related functions within a module is also more user-friendly as it is much easier to come to grips with a structured reference manual. One of the reasons why most lispers prefer SCHEME to COMMON LISP is that you really have to know the vast built-in functionality of COMMON LISP by heart, for otherwise you will hardly manage to find the routines you need.

PYTHON has relatively few BIFs, as most of its built-in functionality has assumed the shape of operators (i.e. list processing: `list[3:5]` instead of `slice( list, 3, 5 )`, or mathematics: `e ** (-x)` instead of `power( e, -x )`). Among the more important of these there are `print` and `input / raw_input`, which you have encountered before.

You may easily inspect the PYTHON internals by typing `dir()` at the toplevel prompt. The interpreter will answer: `['__builtins__', '__doc__', '__name__']`. By `dir( __builtins__ )` you may obtain a list of the entire built-in functionality excepting the command `print` which simply does not fit anywhere.

♠ *Try this.* Now.

Next, you may question the help system about any of these:

```
>>>help(reload)
Help on built-in function reload:
reload(...)
    reload(module) -> module
    Reload the module.  The module must have been successfully imported before
```

Now for the library functions. They must be imported, which can happen in two ways.

- `import module` loads a complete module into the interpreter's "name space", i.e., makes its contents available. Functions must be addressed in a "qualified" way: `y = module.function( x )`.
- Alternatively, single functions may be imported from a module by `from module import function`. They are then directly available: `y = function( x )`. This can be extended: `from module import *` makes *all* components of the module directly available.

---

[1] C-99, the most recent standard, offers some improvements on this but it is still possible to fool the system.

- As PYTHON is slightly functional, you may single out library functions (that is the third possibility): `fun = module.function` binds the library function to the new variable *fun*.

Modules may be nested, in which case the reference has to be written accordingly:

```
import Sound.Effects.echo
Sound.Effects.echo.echofilter(input, output, delay=0.7, atten=4)
```

♠ *The library reference is contained in the `pythondox.zip` archive which you can download from my web page: it is to be found under `lib/lib.html`. Can you say a few words about the modules `operator`, `linecache`, `pickle`, `marshal`, `string`, `math` and `time`?*

♠ *Pick one module and give a comprehensive description of its functions.*

♠ *Put your implementation of the `set` and `bag` datatypes into a module named `orcatypes`[2] and write a short program to demonstrate the correct use of `orcatypes`. The `orcatypes` module is to comprise the following functionality:*

- *creation of sets and bags*
- *addition of items to, and removal of items from sets and bags*
- *test for the presence of a member in a set or bag*
- *inspection of sets and bags: functions for determining the size of a set or bag, and for bags, the number of different elements contained therein (as opposed to the total number of elements)*
- *automatic generation of sets and bags: e.g., a method that inserts all characters from 'B' to 'T' or all numbers from '666' to '999' into a set or bag*
- *conversions of sets into bags and vice versa*
- *output as lists and strings.*

Of course, proper documentation is expected!

♠ ***This is very advanced stuff.*** *Implement a module named `arrows2D` which contains the functionality for working with vectors in two-dimensional space. A vector consists of a starting point and a direction. The starting point is just a pair[3] of coordinates; the direction may be given in terms of either the end point of direction (angle) or length. The following functions should be implemented:*

- *creation of vectors (assuming a length of 0 if only the starting point is given)*
- *mathematical addition of vectors (if possible, also of complete lists of vectors)*

---

[2] ORCA (a PASCAL derivative) is among the few languages implementing these types.

[3] More generally, an aggregate of n values in an n-dimensional space.

- scaling (multiplication of vector arrows with scalar values[4])
- intersection of vectors: do they intersect, and if so, what is their intersection point and angle?
- conversion from end point to angle/length representation and back
- a function calculating the area of a diamond whose borders are formed by two arrows

e.g., we would like to have:

```
>>>a = arrows2D.arrow( 5, 5 )
>>>a.length()
0.0
>>>b = arrows2D.arrow( 10, 10, 20, 20 )
>>>b.length()
22.360679774997898
>>>b.angle()
45.0
>>>c = arrows2D.arrow( 20, 10, 10, 20 )
>>>arrows2D.intersect( b, c )
(15.0, 15.0, 90.0)
>>>d = arrows2D.add( b, c )
>>>d.length()
20.0
>>>d.angle()
0.0
>>>arrows2D.diamond( b, c )
100.0
>>>e = arrows2D.add( a, d )
>>>e.endpoint()
(5.0, 25.0)
>>>f = e.scalmul( 2 )
>>>f.endpoint()
(5.0, 45.0)
>>>arrows2D.__doc__
'This module implements twodimensional vectors.'
```

The design is completely up to you.

♠ Using your previously acquired knowledge and a bit of common sense, try to understand and explain what the following OBERON program does. Pitfalls are the double use of the . operator (which may signify either module or structure access) and the various ENDs:

---

[4] "Scalar" is to be understood in the mathematical sense here, not as on pp 74ff.

```
MODULE Tartan;

IMPORT XY := XYplane, Random;
CONST W = XY.W; H = XY.H;
VAR x, y: INTEGER;

BEGIN
   XY.Open;
   LOOP
      x := Random.Roll(W);
      FOR y := 0 TO H-1 DO XY.Dot(x, y, XY.draw) END;
      y := Random.Roll(H);
      FOR x := 0 TO W-1 DO XY.Dot(x, y, XY.erase) END;
      CASE XY.Key() OF
      | 'q': EXIT
      | 'c': XY.Clear
      ELSE
      END
   END
END Tartan.
```

A couple of months in the laboratory can frequently save a couple of hours in the library.

Part 2

WATER

CHAPTER 15

♣ A Very Short Project: Trithemizing a File

"I would not be so confident if I were you. I think this miserable Gollum has been sneaking again."

– J. R. R. TOLKIEN: "The Lord of the Rings"

Relevance: Preventing data from being divulged to those who are not concerned is not only an important task in computing but also serves for illustrating how to deal with files and their contents.
Keywords: Cryptography algorithms: Caesarean, Trithemian and Vingenère Ciphers.

15.1 A brief history of coding

Many approaches have been made to make information inaccessible to those whom it does not concern. The best method is probably to use a "proprietary information format", e.g., Navajo-speaking telecommunications officers in the U.S. Navy during World War II, and the second best, to use multiple encodings, e.g., to represent each letter by a pair of numbers indicating a line and character number in the Mahābhārata. Due to the size of the key, it is possible to encode each character, even in a very long document, by a separate pair, so statistics cannot get a grip on this[1]. Next comes embedding your relevant information in a vast surplus of irrelevant information, e.g., hiding your text among the pixels of a bitmapped image. However, often these approaches are not feasible.

The ancient Greeks, especially the Spartans, used a method based on positional shuffling for the encryption of relevant military information. The words were written on a tablet or plate and read not row by row but column by column, the length and width of the tablet being the key. This was facilitated by the ancient habit of writing without spaces.

---

[1] The Roman emperor Gaius Iulius Octavianus, a.k.a. Augustus, is said to have used such a system based on the first hundred lines of Homer's Iliad.

Bioinformatics Programming in Python. Rüdiger-Marcus Flaig
Copyright © 2008 WILEY-VCH Verlag GmbH & Co. KGaA, Weinheim
ISBN: 978-3-527-32094-3

```
D E A R P A U
S A N I A S H
A V E Y O U E
V E R M E T A
G I R L B Y T
H E N A M E O
F A N N A X X
```

The result thus being

```
dsavghfeaveieaanerrnnriymlanpaoebmaasutyexuheatox
```

Looks great at first glance, but is easily deciphered once a certain coding method is suspected... as will certainly have been the case the very moment the first tablet fell into the wrong hands.

The simplest method, which does not involve positional shuffling, is just to replace characters by fancy glyphs, as in Edgar Allan Poe's narrative "The Gold Bug", or by other characters[2]:

```
def goldbug( str, key ):
    "Encode text in Captain Kidd style -- a tribute to E.A.Poe"
    r = ""
    for ch in str:   r += chr( ord( ch ) + key )
    return r

>>>print goldbug( "programming in cobol sucks", 1 )
qsthqbnnjmh!jo!dpcdm!tvdlt
```

This is most easily deciphered by statistical analysis if you only know the language of the message and the encoded message is above a certain size. Advanced statistics, however, are required if the coder uses "homophonic extension", that is to say, uses multiple encodings for the most frequent symbols. On the other hand, extending the key by setting an arbitrary encoding for every letter does not really increase security but makes writing and deciphering a lot more cumbersome.

Mary Queen of Scots

It may be interesting to know that *Mary Queen of Scots* used such a code to encode her private letters. The code was broken, the deciphered letters were used for testifying against her and thus had a strongly negative influence on her half-life.

---

[2] The latter method has been described by Gaius Iulius Caesar (100–44 BC) in "De bello Gallico" and is therefore often referred to as a "Caesar cipher". However, we reject this term as we do not feel that G. Iulius actively contributed to the development of this technology, even if a marked preference for steganography seems to have been running in the family (Caesar being the uncle of Octavianus).

15.2 Johannes von Trittheim (Trithemius) and his method of steganography

The German Benedictine monk (and later abbott) Johannes von Trittheim (a.k.a. *Trithemius*, 1462–1516 AD) seems to have been a typical Renaissance scholar – a jack of all trades and a master of the most remote ones[3]. Among other things, he wrote a book about encryption methods where he came up with the simple but powerful idea of changing the code for every character. This was the state of the art for more than four hundred years – the famous "*Enigma*" coding machine of WW2 was still based largely on this concept.

Trithemius

Enigma

It is very easy. Let the key be k. Begin by adding k to the 1st letter, $k+1$ to the second, ... $k+n-1$ to the nth. If the letter obtained is beyond the end of the alphabet, restart at the beginning. Trithemius himself did this by using a wheel of 26 characters, just like the cyclic lists of HASKELL, but it is just as easy to use the remainder of a division. So for a character c at position n, the new value is $c' = \frac{c+(k+n-1)}{26}$

For a full ASCII char set, we would not use 26 but 256.

Thus, we get

    hj rwfsh fo obzautl sjq ys ujh foyxxbe gcbeijhq wq mopp

as an encrypted version of

    hi osama we collect you at the airport tomorrow at noon

(provided, of course, that we skip the spaces and encrypt the letters only, which we do here just for the sake of demonstrating.)

Alternatively, we could use "char XOR (pos+key)" instead of this offsetting.

♠ *Are you familiar with the logical / bitwise operators AND, OR, NOT, XOR? What does XOR ($\oplus$) signify? Become familiar with the basics of Boolean algebra if you choose to implement your algorithm on the basis of bitwise inversion.*
♠ *To warm up, write a program that trithemizes a single string and then detrithemizes it again.*

15.3 Tools of the trade

The program is intended to be called from the command line (p. 70) by entering

    python trithem.py key sourcefile sinkfile

sourcefile is then read, encrypted using *key* (either by addition or by XOR, but please document which one you are using!)

---

[3] Except for boxing, tennis and car racing, which means that he was a nobody.

e.g.,

```
python trithem.py 22 foo.bar foobar.crypt -e
```

There are two instruments to which I would like to direct your attention.

First of all, we have to access the command line arguments: in this case, "22", "foo.bar", "foobar.crypt" and "-e".

sys.argv   The module `sys` provides a variable named `argv` (for *argument vector*). This is basically a simple list of strings, in which the first entry `sys.argv[ 0 ]` contains the name of the PYTHON script, whereas the others are the arguments in the order in which they were entered (p. 71). Thus,

```
['trithem.py', '22', 'foo.bar', 'foobar.crypt', '-e']
```

In order to use the key, convert it to a numeric value by writing `key = int( sys.argv[ 1 ] )`.

Second, we must open files. **Caution: opening a file for writing will overwrite its contents, if a file of this name was in existence before!**

A file is an object which is opened by simply constructing it:

```
filedescriptor = file( filename, mode )
```

e.g.,

```
infile = file( sys.argv[ 2 ], "rb" )
```

where "r" stands for read access, "w" for write access and "b" – combined with the others – for binary mode. We may then access it as described previously:

```
for l in infile:
    ...
```

before we proceed to close:

```
infile.close()
```

♡ *Become familiar with the* `sys` *and* `file` *modules first.*

There are a few other things you should know.

for
- PYTHON's *for* works on files, by reading them line by line.
- Alternatively, you may read the file *en bloc* by using the `read()` method of all file objects. This method will return a string which contains the complete contents of the file.
- Don't forget to open the files in binary mode with "rb" and "wb".

- When you have a string s, you may produce a substring from a to b with $s[a:b]$.
- PYTHON's *for* works on strings too, character by character. Alternatively, you may use MAP. Recursing on strings is theoretically feasible but not to be recommended in PYTHON (why?).
- The ASCII code[4] of any given character c is obtained by a = ord(c). Vice versa, the char corresponding to a given ASCII code is obtained by c = chr(a).
- For format conversions, there are str = hex(integer) and str = oct(integer). They convert a number into the correct hexedecimal or octal notation.
- The very astute may also use the % operator, described on pp 337ff.
- If you are really into S&M, regular expressions (p. 341) will be the right thing for you.

15.4 Your task

It is simple. Write the program. Take proper measures not to overwrite the file with itself. Test it by decoding the encoded files again. The program should expect four arguments: the key; the source file; the destination file; and an "option" specifying what is to be done: **-e** for encrypting and **-d** for deciphering. *For the really tough:* For the purpose of making hardcopies, the program should be able to dump the generated codes as hex numbers, dependent on the last argument (**-h** or **-x**):

```
python trithem.py 22 foo.bar foobar.crypt -h
```

should produce the following output for the "tomorrow at noon" input:

```
68 6a 20 72 77 66 73 68 20 66 6f 20 6f 62 6a 61 75 74 6c 20 73 6a
71 20 79 73 20 75 6a 68 20 66 6f 78 78 62 65 20 67 63 62 65 69 6a
68 71 20 77 71 20 6d 6f 70 70
```

---

[4] It is worthwhile mentioning here that ASCII is obsolescent, and this is at least in part due to Eastern Europe's rejoining the western world. The big problem with the original ASCII definition is that it does not make provision for special characters (English being the only language which uses a Latin alphabet without any special tokens or diacritic marks!). There are *de facto* standards for ASCII variants catering for most languages of Western Europe, but with Eastern languages becoming more important, this has become burdensome, as characters such as "ü", "ţ" and "ř" now have to be represented. Instead, a new format named Unicode is now being propagated that actually uses 31 bits, which means that it is capable of containing and representing any writing system ever devised by humans, from European accented or otherwise modified letters to CJK (Chinese, Japanese, Korean) glyphs, Thai and Indian Devanāgarī and things as remote as the Sequoia script for Algonkin languages. A derivative of Unicode is UTF-8, which uses 1 byte for common Latin letters, 2 bytes for European special characters, and 3 to 5 bytes for everything else. Of course, the recipient of your message will have to use the same character representation, otherwise things just will not work.

Of course, then you will also have to implement a proper decipering algorithm for this. But try... if you succeed, you will get quite a number of extra credit points (and a kissss from Anna, too).

♠ *Try to see yourself on the other side of the fence. Write a program that reads in a file, counts the occurence of each character, sorts the characters by frequency and produces a simple text-based histogram as output. Apply this to different files, especially longer text files and trithemized text files. What does this teach us?*

15.5 Professional cryptography now – random remarks

Kāmasūtram

On an amusing note, in the famous Indian book *Kāmasūtram* the knowledge of cryptography is placed prominently on the list of female virtues, though it is not quite clear for which purpose.

15.5.1 Pimp your code: the Vigenère Cipher.
Vigenère Cipher

A more powerful algorithm, known as the *Vigenère Cipher* and first published in 1586, can be obtained by using a multi-digit code for the Trithemius algorithm. That is to say, if our key is 1:4:7, this will mean that the first letter of the file is encoded by adding 1, the second by adding 4, the third 7; the fourth, 2, the fifth, 5, the sixth, 8; the seventh, 3, the eighth, 6, the ninth, 10; and so on. It is interesting to know that this code was used by the Confederates during the American Civil War, and its breaking may have contributed to the victory of the North.

In general, coding systems, where the length of the key is unknown, offer more security. Ideally, messages thus encoded cannot be deciphered without the key even if the algorithm is known. Another means of increasing security consists in superimposing several cryptography systems upon each other (*super-encryption*).

super-encryption

15.5.2 World War II: the solved enigma.
By using a key with as many letters as the message has and discarding the key after single usage, the Vigenère Cipher is indeed capable of guaranteeing perfect secrecy. The great strength of the *Enigma* coding machine was that it offered more than 10^{20} different keys, making it effectively impossible just trying to see whether decoding yielded anything sensible.

Enigma

The PYTHON module `rotor` contains a complete implementation of the "Enigma" machine, which baffled the skill of British cryptanalysts (led by no lesser a personage than Turing) even after they had got their hands on a working coding machine and were thus able to study the algorithm in full detail. Breaking of the "Enigma" code was supported by deliberately feeding characteristic expressions into the German information system, e.g., by bombing an innocent lighting buoy (an act silly enough to warrant its reporting) and waiting for the word *Leuchttonne* to appear in encoded texts. Nevertheless, by

Leuchttonne

modern standards even the "Enigma" system is considered as insecure, and its use is deprecated.

15.5.3 Frustrating Dan Browne: brute force *vs* smart brains.
One of the first novels by the now best-selling American thriller manufacturer Dan Browne is the slightly moronic hacker story *Digital Fortress*, which, in spite of all its imbecilities, makes one good point. Basically, any code may be broken simply by testing all possible keys. In the story, the National Security Agency (NSA) of the United States have built a monstrous computer named TRANSLTR, comprising three millions (!) of processors, for the sole purpose of decrypting all internet traffic. Though right now this may be dismissed easily, as it is currently much simpler to enlarge keys than to expand brute force deciphering machines (another Malthusian situation), the advent of some new technology such as quantum computing may easily shift this balance[5]. But a bit of common sense is sufficient to foil all cryptanalysis.

Digital Fortress

How is the "endpoint" of a brute force approach determined? A key is obviously correct if the result of deciphering makes some kind of sense. Thus, there are two obvious countermeasures.

- Incorporate nonsensical data into your message. This practice was widely used in the age of manual coding and codebreaking, such as the famous WW2 message THE TURKEY TROTS TO THE WATER RR WHERE IS SEVENTH FLEET RR THE WORLD WONDERS which caused bitter feeling among the commanders of the 7th American fleet because they did not realize that the last sentence was a dummy like the first one and thus considered the simple request for a report as a harsh criticism of their activities.
- Mix your message with structurally correct but irrelevant data so that a plurality of "obviously correct" keys can be identified by the brute force approach, each of which yields a different but "sensible" decoding. When the number of sensible alternatives is high enough, and in particular when the message itself is distributed over a number of differently encoded partial messages, decoding will simply not be enough to determine the proper message. Imagine we mix up four different messages, each with a code of its own:
    (1) hi osama, what is going on, down there, where you live
    (2) hi wladimir, we collect you, at the kremlin, next sunday
    (3) hi angela, enjoy a good meal, at the airport, whenever you go there
    (4) hi george, there's a new rapper in town, named trouble-you, tomorrow at noon
  with the proviso that at each interpunctation sign the reader is supposed to switch to the next message. Knowing the true codes *and their order*, the reader will not have any problems decoding the real message (see above). Any code-breaker would only obtain the above phrases, or, knowing the

---

[5] It will be remembered that in World War II, the Naval Enigma, the most sophisticated form of coding machine available, was finally broken by a combination of theoretical advances, treason, sound reasoning, educated guesses and the development of the bombe machine for testing the subset of likely keys.

principle but not the proper order, he might even run into misinterpretations such as `hi george, enjoy a good meal, at the kremlin, where you live`. Obviously, there is no way to reliably discriminate them. And human reasoning is easily tired...

15.5.4 One-time pads: the (almost?) perfect solution.

In fact, there *is* a code that can be demonstrated to be theoretically unbreakable – provided that some technical requirements are met, which is harder to do than it seems. The concept is amazingly simple.

(1) Apply a different encoding to every letter of your message (i.e., use a key the size of your message).
(2) Employ a random generator to generate your encoding sets.
(3) Transfer your code sets to the receiver by a secure route.
(4) Use every encoding set only once, then destroy it.

The full power of this approach, developed in the early 20th century, was realized only later, partly because of the high workload it entails, which makes it impractical without at least some sort of calculating machine. It was widely used by the Soviet Union in the aftermath of WW2, in particular during espionage activities. The Soviet codes consisted usually of single sheets or "pads" which were marked with serial numbers for identification of the code. An agent would receive a batch of these, use one at a time to encode his report prefixed with the serial to be used (of which the only existing duplicate was at home) and burn the *one-time pad*. However, in violation of the rules, pads actually were occasionally re-used (it seems there was a chronic shortage of these), and the generator was not fully random (as every "random" number generator is). Taken together, these two weaknesses allowed the American *Project Venona* to decipher a few percent of encoded messages – enough to identify a number of spies in critical areas such as the Manhattan Project or the White House itself.

♠ *Design an OTP system of your own, comprising a randomized code generator, an encoding and a decoding facility.*

15.5.5 Symmetric and asymmetric codes.

All codes described so far are "symmetric", that is to say, the key for encoding is identical to that used for deciphering. There are also "asymmetric" codes which use a deciphering key which is different from the one used for encoding. They can be used for establishing the authorship of a document. The deciphering key may be made publicly available; if this "public key" can be used to decipher the file, authenticity is guaranteed – or at least this is the theory. Currently, this is a widely used technique to distinguish executable binaries provided by trustworthy sources[6] from trojans.

---

[6] For example, a vendor of open software may even display the public key for his files on his web pages and thus prevent others from spreading doctored software in his name.

♡ Do not trust in any encryption method. Whatever has been encoded, may be deciphered – if only by stealing the key. You may assume that any mail you send is inspected on its way by several espionage programs (which are technically illegal but still there) screening for certain key words, some of them serving anti-terror institutions, others... If you are using a closed software system, you may also be sure that, no matter how much anti-espionage software you put in, all your files are routinely scanned for anything of interest. This may sound paranoid, but I swear this really happened. Once, my computer chanced to crash while it (it, not me!) was just doing some really clandestine business – and after restart there were enough files left on the hard disk to enable me to reconstruct that I had, by mere coincidence, caught it in the act of sending all files containing the words "parallel processing" and "evolutionary algorithms" to, you would not like to know whom! So probably the best thing, if you speak neither Upiq nor Quitchua or Zulu, is to write down all your confidential data by hand and scan them in as a graphics file. And last not least, never forget that – — — — — — — — — — — — — — censorship in a free country – — – — – — – — – — – very silly – — – — – — — — .

"The key's to. Given!" – James Joyce: "Finnegans Wake"

CHAPTER 16

Some Thoughts on Compression and Checksums

> Die Teufel fanden sich bei allzugroßer Hellung
> In sehr gedrängter, unbequemer Stellung.
>
> — JOHANN WOLFGANG OF FRANCKFURT, also called GOETHE: "Faust, der Tragoedie zweiter Theil"

Relevance: Bioinformatics involves the generation and processing of large amounts of data which must be stored efficiently while at the same time their integrity must be guaranteed.

Keywords: Malthusian laws of computing; lossless compression; checksums; the LZW and Adler32 algorithms.

16.1 Introduction

There seems to exist a Malthusian law[1] regarding the amount of data to be managed. The volume will always increase faster than the means of storing and handling it. From the first "personal computers" of the early 1980s, with soft magnetic disk drives ranging from 70 to 360 kB, capacity has risen to its current height, and there is still no end in sight. At the time of writing, the rewritable CD (650 MB capacity) is obsolescent, and the DVD (4.8 GB per disk) is already going to face severe competition from *Blu-Ray^{TM}* and competing technologies offering 50–80 GB of storage capacity. The reason is that the ongoing development of computers opens up new fields of applications not thought of before. Graphics and video processing immediately spring to mind. At the same time the sciences, and in particular the life sciences, are learning to harness the power of computers for entirely new dimensions of work, which generate vast amounts of data, from confocal fluorescence microscopy[2] to MPSS[3].

---

[1] Malthus postulated that population grows exponentially while food production increases only in a linear fashion. In computing, Wirth said that "software gets slower faster than hardware gets faster".

[2] In this technology, a three-dimensional image is constructed from optical measurements.

[3] Massive Parallel Signature Sequencing, developed by Lynx (Heidelberg), is a technology which basically allows one to monitor genetic activity at the level of detecting each RNA molecule within a tissue.

Bioinformatics Programming in Python. Rűdiger-Marcus Flaig
Copyright © 2008 WILEY-VCH Verlag GmbH & Co. KGaA, Weinheim
ISBN: 978-3-527-32094-3

However, the gigabytes thus produced are generally unlike the digits of π in that they possess a certain degree of redundancy and can be found, at closer scrutiny, to comprise certain intrinsic patterns. Taking advantage of this fact, the volume of data may be reduced, sometimes drastically, thereby effectively increasing the actually useful power of hardware.

Basically, there are two very different approaches to this problem, which will be discussed separately: lossy and lossless compression.

16.1.1 Lossy compression.

Lossy (not "lousy"!) compression tries to reduce data volume by discarding useless information. One might say that lossy compression is a *transformation of a subset of the information* contained in the primary data. Thus, lossy compression is essentially tantamount to *abstraction*; and it will be understood that the information once lost in compressing cannot be restored but replaced by something similar enough for all practical purposes.

The philosophical implications of this are highly interesting, as all understanding, and human language, may be basically regarded as a lossy compression algorithm. When describing a particular situation, we reduce the universal complexity around us to a small number of defined items in order to convey the essential meaning at the cost of sacrificing all the details. "A house with an oak tree beside it" is a precise description but not sufficiently detailed to allow a recognizable graphical rendering. Obviously, this kind of abstraction is still far beyond our contemporary information technology, even with neural networks and other cutting-edge approaches.

At a more limited and more easily achievable level, statistical analyses may be generally classified as methods for lossy compression. For example, traditional regression analysis tries to derive an abstract formula from a multiplicity of data items, thereby getting rid of the individual items' peculiarities. A very advanced algorithm might reduce a table of logarithmic values to the *ln* function. Thus, an lossy compression algorithm will always be a *hypothesis generator* and perform, to some degree, what Kant called "subsumption". Right now, however, this ability is still too superficial to be of practical relevance.

hypothesis generator

One of the best-known examples of lossy compression is that utilized in the JPEG (Joint Photographics Experts Group) graphics format, which is widely used, in particular on the internet, where efficient compression is more important than precise rendering of details. JPEG is optimized for photographic pictures, i.e., images which are rich in colour and may tolerate a certain degree of blurring.

Similar approaches can be used for most video and audio data.

However, in the life sciences there are many other types of data which are not amenable to lossy compression without severe detriment to their usefulness. A nucleic acid or peptide sequence, for example, does not tolerate any "averaging" or "blurring". Likewise, non-photographic images often require faithful rendering of each individual detail, even at the cost of requiring a reduction in colour. For these, the second kind of compression is highly preferable.

16.1.2 Lossless compression. By contrast, lossless compression is a *transformation of all the information* contained within the primary data (from a theoretical point of view, this might be called a *tensor*). Hence, each and any piece of information contained in the original data can be reconstructed from the compressed data. It is evident that this excludes all approximations and interpolations and restricts the means of compression to the identification of redundancies. Even so, tremendous savings are possible with certain types of data.

tensor

The format of a compressed file must needs be simple, since any administrative overhead would be counter-productive. Of course, this brings us to the "markup" problem (p. 45ff) of differentiating between data and meta-data. A possible solution is described below.

Apart from this, the art of lossless compression consists mostly in devising algorithms which are good at identifying, within an acceptable time frame[4], as many repetitive patterns as possible and replacing them with back-references to previous occurences.

It is interesting to note that, whereas ideal lossy compression would ultimately result in a hypothesis generator (e.g., derivation of the logarithm function from a table of logarithmic values) or even in a somewhat scary kind of genuine artificial intelligence (capable of compressing observations on thousands of peas into Mendel's laws...), ideal lossless compression would ultimately result, by elimination of all redundancy and detectable patterns, in a "white noise" indistinguishable from mere statics. In fact, some believers in extraterrestrial intelligence have suggested that the failure of the *SETI project* is caused by the fact that the "others", being faced with the same kind of Malthusian IT trouble as we are, namely insufficient physical bandwith for their communication, have optimized their compression technology to such a degree that their emissions cannot be identified as being artificial. Setting aside these amusing speculations, it is a fact that the potential of one of the Voyager space probes was indeed greatly enhanced by remotely reprogramming it to use compression for transmission of photos – an astounding feat, given the fact that the free memory of the probe's computer had less than 300 bytes (!) of free memory available.

SETI project

From this it will also become apparent that encoding and compressing are related arts. Moreover, it will also be understood that a compression algorithm may be easily modified to yield a powerful tool for the identification of regular patterns within nucleic or amino acid sequences.

---

[4] Anyone who has ever seen *gzip* or *bzip2* at work will have an idea of how laborious these algorithms are.

16.1.3 Aside: processor speed *vs* memory size. The fundamental rule of the suboptimum dictates that no capacity will ever be fully used, and also that the arrangement of a certain pile of data which allows for swiftest processing is generally of no more than intermediate efficiency with regard to memory use, and vice versa.

For example, it will be easily seen that in nucleic acids each base pair has an information content of 2 bit, whereas amino acids may be more than sufficiently represented using 5 bits ($2^5 = 32$) per residue. However, contemporary computers work with bytes of 8 bit each as the basic entity, and generally each nucleic or amino acid is represented by a full byte. For nucleic acids, this means that actually four times as much memory or disk space is occupied as would be strictly necessary. In order to work with 2-bit representations of amino acids, the processor would have to extract the relevant portion of a given byte (by ANDing it with $3, $C, $30 or $C0, respectively) in addition to the normal operations.

In general, in the life sciences processor speed is a more limited – and limiting – resource than memory capacity, the reason being simply that in the age of 64-bit processors, available memory is limited virtually only by available funds. If desired, one can get a computer with a terabyte (1024 gigabytes) of memory and a hundredfold the disk capacity; by contrast, processors in the tera*hertz* range are not available for any money and will probably never be, due to the intrinsic limitations of semiconductors – and the fact that scaling of processes to a multiprocessor system is a difficult task, which we will return to in the second volume.

That is to say, the need for data compression has to be balanced carefully against the need for reasonably fast processing. Nevertheless, for the efficiency of permanent storage, which is not supposed to be speed-critical, compression is essential.

16.2 A simple lossless compression algorithm

16.2.1 Theory. A very straightforward lossless compression algorithm simply tries to identify runs of identical bytes and replace them with a code signifying the number of such bytes. Primitive though this may be, this is amazingly efficient with 8-bit (256-colour) images, where each pixel is represented by a single byte and areas of a given colour will therefore necessarily comprise long runs of identical bytes. However, it is easily defeated by simple patterns[5].

Using PYTHON, it does not require extraordinary prowess to load a file into memory, scan it and identify runs of multiple bytes. The question raised above

---

[5] In the 1980s, this was one of the few algorithms which could easily be ported to "personal computers" and was therefore widely used in early painting programs such as Degas or STAD.

(p. 45ff) with regard to meta-data, however, still applies. In the end, we will have to write all our data, both the compressible and the incompressible, into a single file. As the input file may use any combination of bytes, how is it possible, easily and reliably to tell compressed from uncompressed data?

On p. 337, the technique of "escape processing"[6] will be discussed, which deals with essentially the same kind of thing. When certain delimiters, e.g., quotation marks, are required to denote a string, how is it possible to write a string that comprises such delimiters as glyphs? The C/PYTHON solution is very neat. The backslash, which is very rarely used in normal texts, is used as an "escape character" which signifies that the following character has a meaning different from its normal one – e.g., a backslash followed by a quotation mark means a quotation mark glyph rather than the end of the string, as a backslash followed by an "n" means a newline. However, how can we get backslash glyphs into our string? Very easily: two adjacent backslashs mean a backslash glyph – "escape from the escape".

Now this can be put to good use here. We define one byte out of the complement of 256 as our "escape character", and we further define that a particular byte following the escape character "un-escapes" it again. The most rational choice for the escape character will be the byte which occurs least frequently within the data chunk to be compressed. Of course, it will have to be saved together with the compressed data, otherwise we will not be able to uncompress again!

So any byte within our encoded file may be *either* a "normal" byte *or* the escape character. If it is normal, fine. If it is the escape character, the next byte may be *either* a zero byte *or* a byte indicating the number of bytes in the run which was identified. If it is zero, fine (in this case, the escape character is "unescaped" and actually represents a data byte of this value). If it is not, it indicates the number of bytes in the run which was identified and is consequently followed by a third byte giving the value of the bytes in the run.

16.2.2 Practice. Now let's try to implement this in PYTHON.

First of all, we should decide on a style. Obviously, the functional approach is very straightforward. Both the compressor and decompressor are functions that transform a set of data into a different set of data. Alternatively, an object-oriented approach makes a lot of sense. We define an object which may hold both the compressed and the uncompressed form of data, and which provides methods for conversion, for input and output and so on (or, in SMALLTALK lingo, is capable of reacting to messages telling it to do conversion, input and output and so on).

In order to demonstrate the practical use of classes and inheritance, the following is an OO solution. We begin by defining the central functionality:

---

[6] In the old days, the character "Escape" (ASCII 27) was used in the communication between computers and printers to indicate that a control sequence was to follow.

```python
class enc_dec_core:
    """This class implements simple compressor functionality suitable for
    256-colour images. It comprises only the functions required for
    processing of the data (i.e. compression and decompression, herein
    referred to as encoding and decoding). These functions are:

    - rarest_byte( S ): identifies the rarest byte and its frequency in string S
    - code( S ): compress string S
    - uncode( C, R ): uncompress compressed data chunk C assuming that R
                     was the least frequent byte in the string which C was
                     originally generated from.

    The sole purpose of this class is to provide these functions to
    classes which inherit from it.

    This class is hence not supposed to use a constructor, or any
    non-local variables, and it is not designed for instantiation. Of
    course, it *is* possible to create objects of this class, but they
    are not useful for anything particular.
    """

    # -- Section 1: Internal necessities --
    def rarest_byte( self, stringo, traceon=False ):
        """Identify the byte occurring least frequently within a chunk of
        data which is passed as a string.
        rarest_byte actually returns *two* values, the first of which is
        the least frequent byte and the second its frequency.
        Strategy:
        1. Create a dictionary ('table') where there is one entry for each byte.
        2. Set all these entries to zero.
        3. Scan the data chunk byte by byte, and for each byte, increase
           the corresponding dictionary entry by 1.
        Thus, for each byte the frequency will be obtained.
        4. Set the putative maximum frequency to chunk length and the
           putative least frequent byte to an arbitrary value.
        5. Scan the dictionary entry by entry; if the respective entry has
           a frequency which is less than that of the putative least frequent
           byte, set the putative least frequent byte to this entry and the
           putative maximum frequency to this frequency.
        6. Return the putative maximum frequency and the putative least
           frequent byte as being confirmed.
        """
        table, allchars, maxim = {}, map( chr, range( 0, 256 ) ), \
                                 len( stringo )  # step (1)
        for ch in allchars: table[ ch ] = 0       # step (2)
        for l in stringo: table[ l ] += 1         # step (3)
        rar_b, n_rar = chr(0x00), maxim           # step (4)
        for ch in allchars:                       # step (5)
            if table[ ch ] < n_rar: rar_b, n_rar = ch, table[ ch ]
        if traceon: # debugging output desired
            print "\nRarest byte is 0x%02x : %d occurences" % (ord(rar_b),n_rar)
        return rar_b, n_rar                       # step (6)
```

```
# -- Section 2: Encoding/decoding algorithm proper --
def code( self, text, traceon=False ):
    """Convert any chunk of data, handed over as a string, into its encoded
       (compressed) form, and return the encoded data and the least frequent
       byte identified within the data chunk.
    """
    my_code, siz, i = "", len( text ), 0 # initial values to start with
    if traceon: print "Searching rarest byte among %d bytes..." % len( text )
    rarest, n_of_rarest = self.rarest_byte( text, traceon )
                                            # get least frequent byte
    if traceon: print "Encoding..."
    # Processing the large bulk of data:
    while i < siz - 2:
     # The last 2 bytes are not considered here -- they are incompressible.
        if traceon and i % 100 == 0: print "\t%3.1f%%\r" % (100*float(i)/siz),
        ch = text[ i ]       # Look at current byte.
        if ch == rarest:     # Is it the ominous least frequent byte?
            my_code += rarest + chr( 0x00 ) # then "escape from escape"
            # It will be noted that a multiplicity of the least frequent byte
            # cannot be compressed.
        else:                # Otherwise it may be single or multiple.
            # We compress only if there are at least 3 identical bytes:
            if text[ i+1 ] == ch:
                if text[ i+2 ] == ch:
                    # Find the end of the run, which may be determined by
                    # - another byte; or
                    # - the end of the chunk; or
                    # - the number of 255 repetitions being exceeded.
                    j = i + 1
                    while j < i+0xFF and j < siz and text[ j ] == ch: j += 1
                    # Add repetition signal to the code:
                    my_code += rarest + chr( j-i ) + ch
                    i = j # Skip the repetitive run...
                    continue # ... and resume at next byte behind that.
                    # As we won't waste space by compressing two consecutive
                    # bytes  into a three-byte sequence, two situations will
                    # never occur in the code:
                    # ... - [least frequent byte] - [0x01] - ...
                    # ... - [least frequent byte] - [0x02] - ...
                    # So basically these two might be used for further sophistications,
                    # e.g. with [0x01] indicating that not a *byte* but a *word* will
                    # follow, enabling the use of 16-bit values where they are useful,
                    # thereby making it possible to compress runs of up to 65535 bytes.
                    # [0x02], by contrast, might be used to indicate a change in the
                    # escape character, the rationale being that different portions of
                    # a file, e.g. a low-colour image, may have differing "least frequent
                    # bytes".
            my_code += ch # less than three -- just ignore. Potentially SLOW!!
        i += 1 # and advance to next byte
    # And now for the last bytes, if there should be any left (they MAY have been
    # processed if they are a part of a run.):
    while i < siz: # As there can be no more than 2 bytes left, don't try to compress.
        ch = text[ i ] # Just get byte and determine whether it is a least frequent byte
        if ch == rarest: my_code += rarest + chr( 0x00 ) # and either add escape from escape
        else: my_code += ch # or the byte as it is
        i += 1              # and advance.
    if traceon: print "\nDone."
    return my_code, rarest
```

```python
def uncode( self, stringo, rar, traceon=False ):
    """Restore the original data from a chunk of compressed data ('stringo'), assuming
    that 'rar' was the least frequent byte in the string which C was originally
    generated from.
    """
    res, i, maxl = "", 0, len(stringo) # initial values to start with
    if traceon: print "Decoding %d bytes..." % maxl
    while i < maxl: # Iterate all over the compressed chunk, without 'buts' or 'ifs'.
        ch = stringo[ i ] # Look at current byte
        if traceon: print "%d -- Is 0x%02x (%c) == 0x%02x (%c) ?" % (i, ord(ch), ch, ord(rar), r
        if ch == rar: # Is it the one identified as escape character?
            if traceon: print "YES"
            if stringo[ i+1 ] == chr(0x00): # If 0 follows, this is escape from escape.
                if traceon: print "Here follows just a single char %c" % rar
                res += rar # Then we add the least frequent byte to the plain text
                i += 2        # and skip the escape character and the 0.
            else: # If anything else follows, this must be a compressed run.
                # We determine which character is meant, and how often:
                amount, encod = ord( stringo[ i+1 ] ), stringo[ i+2 ]
                if traceon: print "Here follow %d repetitions of 0x%02x(%c)" % (amount,ord(encod
                res += encod * amount # Reproduce the run and add it to the plain text.
                i += 3            # Skip the escape sequence.
                if traceon: print "Continue at %d" % i
        else: # No, it is just a normal character.
            if traceon: print "NO\t",
            res += ch     # add it to the plain text
            i += 1        # and proceed
    return res
```

Admittedly, this is not really OO yet. The class *enc_dec_core* is more of a library module than a genuine object class, since it comprises no data. However, by inheritance we may easily derive a fully-fledged object class, here named *turbocore*:

```python
class turbocore( enc_dec_core ):
    """Based on the functions provided by 'enc_dec_core', this class implements
    a type of object comprising the fiollowing members:
    (A) DATA:
        - A chunk of data for storing the plain text.
        - A chunk of data for storing the encoded text.
    (B) METHODS:
        - Set/get the plain text.
        - Set/get the compressed data (code).
        - Get rarest byte, compression ratio, &c.
        - Convert the object's plain text and code into each other.
    """

    # -- Section 1: Constructor --
    def __init__( self ):
        """The __init__ method is the constructor. It is invoked automagically
        by the PYTHON system for each new object as soon as it is created.
        """
        self.initialize_core()
```

16.2 A SIMPLE LOSSLESS COMPRESSION ALGORITHM

```
def initialize_core( self ):
    """The initialization proper has been outsourced to a
    public method which is called from within the constructor.
    WHY? Because we wish to do inheritance. In inheritance,
    methods are overridden; this also holds true for the
    constructor. Thus, when a class inherits from 'turbocore',
    creation of an object of the derived class will NOT cause
    the 'turbocore' constructor to be called again. This is a
    deliberate feature of PYTHON to allow for multiple inheritance:
    If a class inherits from more than one base class, which base
    class constructors were to be called, and in which order?
    Instead, there can be only one. However, by outsourcing the
    functionality, we may design the constructor of the derived
    class so as to properly execute the base class initialization
    too; see below.

    The initialization procedure is used here to set up the data
    attributes of the object.

    Data attributes (the term corresponds to 'instance variables'
    in Smalltalk and to 'data members' in C++) are things which
    are peculiar to any individual object, rather than the class
    as a whole; i.e., different objects of the same class may
    have different values here.

    Data attributes used here
    (the trailing __ indicates they are private):
    __text      - the plain text content of the object
    __code      - the code content of the object
    __reliable  - flag which indicates whether text and code agree
    __rarest    - the least frequent byte in the text portion

    All of these are accessible only via respective methods. PYTHON
    provides a good deal of encapsulation, by hiding private attributes
    not only from outsiders but also from the heirs of a class.
    """
    self.__text = self.__code = ""
    self.__reliable, self.__rarest = False, 0x00

# -- Section 2: Getting and setting the object's states --
# - 2.1: Setting attributes
def set_text( self, text ):
    """Set the plain text of a Turbo object to the string passed.
    At the same time, unset 'reliable' toggle.
    """
    self.__text, self.__reliable = text, False

def set_code( self, code, rarest=-1 ):
    """Set the code portion of a Turbo object to the string passed.
    At the same time, unset '__reliable' toggle.
    Optionally (if known), the '__rarest' attribute my be set.
    If set_code is invoked with a single attribute, __rarest
    will not be changed.
    self.__code, self.__reliable = code, False
```

```
            if rarest >= 0: self.__rarest = rarest
    # - 2.2: Getting attributes and other information about the object
    # These methods should be self-explanatory.
    def get_text( self ): return self.__text
    def get_code( self ): return self.__code
    def get_rarest( self ): return self.__rarest
    def is_reliable( self ): return self.__reliable
    def text_rarest_byte( self ): return self.rarest_byte( self.__text )
    def compression_percentage( self ): return (100.0*len(self.__code)) / len(self.__text)

    # - 2.3: Methods for inducing internal states changes
    def encode( self, traceon=False ):
        """Sending this message will cause the object to adjust its code
            to its plain text content. Relies on the method "code"
            in section 6 below.
        """
        self.__code, self.__rarest = self.code( self.__text, traceon )
        self.__reliable = True # after encoding, both are guaranteed to be matching

    def decode( self, traceon=False ):
        """Sending this message will cause the object to adjust its
            plain text to its code content. Relies on the method "uncode"
            in section 6 below.
        """
        self.__text = self.uncode( self.__code, self.__rarest, traceon )
        self.__reliable = True # after decoding, contents are guaranteed to be matching
```

Lovely. Now we have OO as it ought to be, with encapsulation of data together with the functionality required to handle them.

The astute reader will not have failed to notice that all our data members are locked in in such a way that they can be accessed only via the setter and getter methods. Normally, PYTHON's philosophy is based on the concept of intelligent people doing computer programming, and it does not possess the notorious interdicting features of, e.g., JAVA. However, here it makes sense not to rely on the programmer respecting the privacy of data members, as this also allows the object to take due note of any change to its state. For example, the flag __reliable will be unset whenever new data are fed into the object, and set again when compression or decompression is performed.

But in general, doing compression and decompression alone is not sufficient. We further need a layer of file access functionality. Thus, we derive a class *turbo* from the previous *turbocore* class:

```python
class turbo( turbocore ):
   # -- Section 1: Constants --
   # N.B.: PYTHON does not provide "genuine" constants. It rather relies on the programmer's
   # politeness not to change their values. In general, it is a good idea to establish some
   # convention for discriminating constants from variables. Here, following the C/C++ style,
   # identifiers written in capital letters are not supposed to be changed.
   __MAGIC = "\xAF\xFE\x01" # "magic" identification of compressed files
   __BINWRITE, __BINREAD = "wb", "rb" # Python file system constants

   # -- Section 2: Constructor --
   def __init__( self, traceon=False ):
      """As described above, the base class ('turbocore') must be initialized
         explicitly, by calling 'initialize_core', which, being public, is
         inherited and accessible from within all 'turbo' objects.

         Data attributes used in 'turbo' (the trailing __ indicates they are private):
         __header    - the prefix for the identification of a compressed file
         __traceon   - flag to indicate whether debug information is shown
      """
      self.initialize_core() # explicitly call turbocore constructor
      self.__header, self.___traceon = "", traceon # set data attributes

   # -- Section 2: File access methodology --
   def read_text_from_file( self, filename ):
      """Method to read the plain text of an object from a file whose name
         is passed as an argument.
         N.B.: This is a method, not a function -- it changes the state of
         the object in lieu of returning any value.
      """
      self.set_text( file( filename, self.__BINREAD ).read() )

   def write_text_to_file( self, filename ):
      """Method to write the plain text of an object to a file whose name
         is passed as an argument.
         N.B.: It is the object's content which is written, not an argument.
      """
      file( filename, self.__BINWRITE ).write( self.get_text() )

   def write_code_to_file( self, filename ):
      """Method to write the plain text of an object to a file whose name
         is passed as an argument. During this process, the header with
         the appropriate values is created and first written to the file
         (see below).
         N.B.: It is the object's content which is written, not an argument.
      """
      self.create_header()
      f, stuff = file( filename, self.__BINWRITE ), self.__header+self.get_code()
      f.write( stuff )
      f.close()
```

```
def read_code_from_file( self, filename ):
    """Method to read the code of an object from a file whose name
        is passed as an argument. This is more complex than reading the
        text, since the header has to be taken into account. We also
        use the header to perform a few primitive sanity checks.
        N.B.: This is a method, not a function. It changes the state of
        the object instead of returning the code as a value; instead it
        yields the two checksum values saved within the header to facilitate
        sanity checking. Moreover, it DOES NOT perform automatic decoding
        of the code portion but merely unsets the '__reliable' flag.

        Suggestion: This method might be private, as it is intended to
        be called from 'read_and'decode' only.
    """
    data = file( filename, self.__BINREAD ).read() # get file en bloc
    # Macerate the header and code:
    f_magic, f_crc_t, f_crc_c, f_rarest, code = data[:3], ord(data[3]), ord(data[4]), data[5], data[6:]
    if f_magic != self.__MAGIC: # file does not appear to be product of this script
        raise ("File magic incorrect: 0x%02x%02x-%02x rather than 0x%02x%02x-%02x"
 % (ord(f_magic[0]), ord(f_magic[1]), ord(f_magic[2]), ord(self.__MAGIC[0]),
ord(self.__MAGIC[1]), ord(self.__MAGIC[2])))
    self.set_code( code, f_rarest ) # store stuff in code portion, unset '__reliable'
    return f_crc_t, f_crc_c         # return the two checksum values

def read_and_decode( self, filename ):
    """Method to READ AND DECODE a file produced by this script:
        1. Invoke 'read_code_from_file' to verify magic, get code and checksums
        2. Invoke 'decode' of the base class 'enc_dec_core' (or whatever it has
           been overridden with) to perform a decoding and set the text portion of
           the object in accordance with the code portion.
        3. Calculate checksums for both the text and code portions of the object
           and compare them to what has been stored. If they are identical, we can
           be pretty sure about file integrity.
    """
    f_crc_t, f_crc_c = self.read_code_from_file( filename ) # Step (1)
    self.decode()                                           # Step (2)
    # Step (3):
    crc_t = crc_c = 0xFF
    for ch in self.get_code(): crc_c ^= ord(ch)
    if crc_c != f_crc_c: raise ("Checksum C incorrect (0x%02x:0x%02x)\
 -- file corrupt" % (crc_c, f_crc_c))
    for ch in self.get_text(): crc_t ^= ord(ch)
    if crc_t != f_crc_t: raise ("Checksum T incorrect (0x%02x:0x%02x)\
 -- file corrupt" % (crc_t, f_crc_t))

def create_header( self ):
    """Method to create the __header member of the object in accordance with its
        other contents.
        N.B.: This is a method, not a function -- it changes the state of
        the object in lieu of returning any value.
    """
    self.__header = self.create_any_header(
self.get_text(),self.get_code(),self.get_rarest() )
```

```
def create_any_header( self, textstr, codestr, rarest ):
    """Method to create a header for two chunks of data.
    The header is a 6-byte structure comprising the following information:
       - Byte #1: Magic 0xAF
       - Byte #2: Magic 0xFE
       - Byte #3: Magic 0x01 (version number)
       - Byte #4: checksum for text portion
       - Byte #5: checksum for code portion
       - Byte #6: Rarest byte, i.e. the one used as an escape code to denote
                  the beginning of compressed sequences
    "Checksum" is a bombast for this simple check, which basically consists of
    starting with 0xFF and XORing each byte in turn with this. However, this
    primitive mechanism will detect file corruption with a 99.6 % chance.
    With two of these values being used in parallel, the chances are almost
    nine-five. They might be even higher if we  additionally store the
    length of the chunks.
    N.B.: This is a function, not a method -- it returns the header sequence
          without changing anything about the object.
    Design issue: As this is a class function, where is it best placed?
    """
    crc_t = crc_c = 0xFF
    for ch in textstr: crc_t ^= ord(ch)
    for ch in codestr: crc_c ^= ord(ch)
    return self.__MAGIC + chr(crc_t) + chr(crc_c) + rarest
```

16.2.3 Rev it up. So far, so good. The class *turbo* may now be put to work as in the following manner, wherein an object is created, a file fed into it, encoded, the code written to a file, decoded and compared to the original; then the object is abandoned, a new one created and fed the code file created by the first, then decoded and compared.

```
# main program: just a few tests
def dump( n1, n2 ):
    def ddump(x):
        for a in x: print "%02x(%c) " % (ord(a),a),
        print
    print "Error -- something is wrong. %d : %d\n" % (len(n1), len(n2))
    ddump( n1 )
    print
    ddump( n2 )

from sys import argv
ar = [ "bitmap.bmp" ] if len( argv ) < 2 else argv[1:]
for fil in ar:
    t = turbo()
    t.read_text_from_file( fil )
    contrl = t.get_text()
    t.encode()
    t.write_code_to_file( fil+".cmp" )
    t.decode()
    print "Compression: %2.1f%%" % t.compression_percentage()
    if t.get_text() == contrl: print "Everything works well."
```

```
else: dump( t.get_text(), contrl )
del t
u = turbo()
u.read_and_decode( fil+".cmp" )
if u.get_text() == contrl: print "Everything works really well."
else: dump( u.get_text(), contrl )
u.write_text_to_file( "new-"+fil )
```

Of course, further classes may be built around this. It is also possible to override the methods presented here with more efficient ones. This will be discussed below.

16.2.4 Criticism and considerations. There are two-and-a-half weaknesses in this algorithm.

(1) It is limited to working on the byte level.
(2) It does not recognize any patterns but only adjacent elements of data; thus it will be unable to perform any compression on an image consisting of vertical one-dot lines in different colours. It is also unsuitable for dealing with text files or program code.
(3) As it is, the algorithm does not compress runs of more than 255 identical elements of data, requiring at least 3 to store them, thereby achieving a maximum compression of no more than approximately 1:80. This is not really a serious drawback as pictures are generally not in the gigabyte range, and more efficient compression agorithms actually pose a security threat. A file may be hand-crafted in such a way that an algorithm such as LZW, described below, creates a terabyte monstrosity when being unpacked – a common trick of hackers to blow up test programs for incoming email (*mailbomb*).

mailbomb

It will also be noted that, even though the algorithm performs no modification of the data unless at least a minimum of benefit can be achieved, still the addition of a few "meta" bytes to the compressed file is inevitable to store necessary information, so there may be files that actually increase in size. This is a general problem of compression algorithms. There cannot be any algorithm that will be able to reduce the size of every file. This can be argued for mathematically, but plain logic suffices. If there were such an algorithm, its output files could be compressed over and over again, until every primary file is compressed to the size of one byte. Obviously, this is nonsense.

A compression algorithm uses regularities to reduce file size. With completely chaotic data, no compression is possible, and vice versa. As has been pointed out during the SETI project, ideally compressed data show no regularities whatsoever.

16.3 The LZW compression algorithm

16.3.1 History. In 1978, Jacob Ziv and Abraham Lempel developed the *LZ78* compression algorithm (US Patent[7] 4,464,650), which was improved by Terry Welch, who published it in 1984. The improved algorithm[8] has since been known as the *LZW* algorithm and has gained widespread acceptance, as it is straightforward and does not require any statistical voodoo. On the downside, its analysis of data is incomplete, and its compression rate thus imperfect, even though it is universally applicable and capable of compressing texts in western languages to approximately one-half of their original size (and nucleotide sequences to less than one-third).

LZ78

LZW

16.3.2 "Work like an Egyptian". To understand the LZW algorithm, let us begin with a few considerations.

♠ *Information is measured in bits, wherein one bit corresponds to two alternatives. How much is the information content of a single base pair of DNA? And of the SV40 virus (5243 base pairs)?*

As we all know[9], the Latin script is traditionally considered to comprise 26 letters plus the space. Thus, each letter has an information content of $log_2 27 \approx 4.755$ bits; or, in other words, 4.755 bits are required to represent one letter. Now Anna has recently read a book on Middle Egyptian and found out that hieroglyphs may be either letters in our sense or *diglyphs*, i.e., tokens which have the meaning of two letters[10]. This, she supposes, might be used to save space. Or might it not?

diglyphs

This would require 27^2 individual glyphs, each thus requiring $log_2 27^2 \approx 9.510$ bits to be represented, while at the same time only requiring half the number of characters. However, $\frac{log_2 27^2}{2 \cdot log_2 26^2} \approx 1.011$, or, in other words, this would require about 1.1% *more* bits than before.

The key is that not all combinations of single letters are required (Middle Egyptian indeed employs only a few dozens of diglyphs for the most frequent combinations). So if we peruse our text carefully, we may find that of the 702 possible diglyphs maybe only about half are actually used, wheras combinations such as *yc* or *qx* will probably never appear[11]. In this case, if we build our own table

---

[7] The US are unique in granting patents for software, games and business methods, which most civilized nations consider as purely intellectual achievements and thus deny the technical nature which is a prerequisite for patentability. However, by now protection for the LZ* algorithms has expired (20 years after the filing date).

[8] US Patent 4814746 (June 1st, 1983) and US Patent 4558 302 (June 20th, 1983), both expired.

[9] If you do not: congratulations on having come so far.

[10] For example, *gb* is represented by a sign showing a goose, whereas *g* is represented by a jar holder and *b* by a leg.

[11] In a modern English text. They may appear in other languages, but in these, in turn, other combinations will be unused.

of diglyphs, we would need only $\frac{350}{2 \cdot log_2 26^2} \approx 0.899$ bits per letter of the original, or, in other words, we would achieve more than 10% net gain!

triglyphs

Now Anna learns that Middle Egyptian even used a couple of *triglyphs* for particularly frequent combinations of three letters[12]. So why should we restrict this system to diglyphs – or to any N-glyphs? Basically, any repetition may be represented by a glyph of its own. The fact that not all possible combinations are actually used is what yields a net benefit[13].

So this is basically how we do it. We scan our file to be compressed and build a table of N-glyphs, into which we enter only those that actually occur in the file. If any combination represented by an N-glyph does occur more than once, the second and all further occurences are replaced by the N-glyph. Proceeding in this way offers the advantage that the table, which may become quite bulky, does not have to be saved: the decompression algorithm may build it in the same way while reading the compressed file and then replace the N-glyphs with the letter combinations they represent.

How do we build a table of N-glyphs? Any two subsequent characters form a diglyph. Any diglyph plus the next character is a triglyph, and so on.

16.3.3 The algorithm.
The LZW algorithm itself is amazingly concise:

16.3.3.1 *Compressor algorithm.*

```
let w := None;                                          # begin with zero
initialize table with single characters;  # necessary for decompressing
while there is another character, which we call k, do
    if w+k exists in the table then let w := w+k; # more than a diglyph
    else
        add w+k to the table;                       # enter new diglyph
        output the code for w;
        let w := k;
    endif
done
output the code for w;                                  # flush
```

16.3.3.2 *Decompressor algorithm.*

```
read a character;
output this character;                 # the first one is never a diglyph
let w := this character;
while there is another character, which we call k, do
    let entry := table entry for k;  # single characters yield themselves
    output entry;
```

---

[12] Similar to the cyrillic letter for "shtsh", which Anna particularly likes.

[13] Of course, if the data set to be compressed is completely random, it will increase rather than decrease in size. Remember: there cannot be any algorithm that will be able to reduce the size of every file. Those who do not listen can stay after class.

```
        add w+(first character of entry) to the table;   # build table - again
        let w := entry;
done
```

It will be understood here that brevity does not necessarily imply simplicity, still less legibility. The iterative program structure is a formidable barrier to understanding; the only way is to simulate the process using paper and pen (which is strongly recommended). A purely functional version of the algorithm *might* be more intelligible...

♡ *Confused? Overwhelmed? Stunned? Welcome to the club. Comprehending other people's algorithms is an extremely laborious task, even to the experienced. Don't be fooled by the fact that some twenty lines of pseudocode are sufficient to express the entire thing: These twenty lines were considered as sufficiently inventive to warrant the granting of several patents. In stark contrast to the situation in, e.g., biology, where complexity is directly related to the amount of data available, size does not matter in IT. If this is little comfort to you, remember that the discovery that $E = mc^2$ was justly awarded a Nobel Prize.*

Before we continue, it should be mentioned that two widely used file formats involve the use of the LZW algorithm (and the chances are that you have worked with all three of them before, albeit unknowingly): *GIF* (Graphics Interchange Format) for lossless 8-bit graphics, e.g., in web applications, and *PDF* (Portable Document Format) for complex documents. Patent quarrelling over the use of LZW in GIF actually motivated the creation of the alternative *PNG* (Portable Networks Graphics) format, which employs a different compression algorithm.

GIF
PDF
PNG

16.3.4 A slightly silly example. The tongue-twister

```
these three thick brothers threw these three thick pieces of
                    these thick things
```

contains obvious candidates for di- and triglyphs: *th, the, thr, these* and *thick*, among others. So what happens if we scan this piece of data for LZW compression?

Let us represent the N-glyphs with numbers in square brackets:

#	Letter	Possible N-glyph	Table entry	Output
1	t	t		
2	h	th	[1]	t
3	e	he	[2]	h
4	s	es	[3]	e
5	e	se	[4]	s
6	_	e_	[5]	e
7	t	_t	[6]	-

#	Letter	Possible N-glyph	Table entry	Output
8	h	th		
9	r	thr	[7]	[1]
10	e	re	[8]	r
...				
77	n	_thin	[303]	[32]
78	g	ng	[304]	n
79	s	gs	[305]	gs

Thus, in the end the compressed version will be:

these [1]re[5][1]ick bro[1]ers[6]h[8]w[23][3][10][24][9][23][12][14]piec[3] of[27][4][32][13][32]ngs

Exactly what will happen during decompression? This is left as yet another exercise for the reader.

♠ *What are you still waiting for? Get yourself a pen and a piece of paper and decompress the above line manually.*

16.3.5 An exemplary implementation. As we all know, computers do not possess any common sense[14]; time will show whether this is really a disadvantage. Thus, many elements of an algorithm which are self-evident to the human reader still have to be written down explicitly. PYTHON is advantageous in that it offers at least built-in facilities for the management of a "table" like the one mentioned above. Dictionaries (p. 88) are excellent for this purpose. Nevertheless, lots of additional definitions are still required, and for this reason an actual implementation is considerably longer than the algorithm alone:

```
import string
def code( x ): return unichr( x ).encode( "utf8" )

def compress( uncompressed ):
    def fuse( mylist ): return string.join( map( lambda(x):"%d"%x, mylist ), "-" )
    chars, table, buff, result = 0x100, {}, [], ""
    for i in range( 0, chars ): table[ "%d" % i ] = i
    for i in map( ord, uncompressed ):
        if len( buff ) == 0: glyph = "%d" % i
        else: glyph = "%s-%d" % ( fuse( buff ), i )
        if table.has_key( glyph ): buff += [ i ]
        else:
            result += code( table[fuse( buff )] )
            table[ glyph ], buff = chars, [ i ]
            chars += 1
    if len(buff) != 0: result += code( table[fuse( buff )] )
    return result
```

---

[14] They only think they have.

```
def decompress( compressed ):
    chars, table, buff, chain, result = 0x100, {}, "", "", ""
    for i in range( 0, chars ): table[ i ] = code( i )
    decoded = compressed.decode( "utf8" )
    for i in map( ord, decoded ):
        if buff == "":
            buff = current = table[ i ]
            result += current
        else:
            if i < 0x100:
                current = table[ i ]
                result += current
                chain = buff + current
                table[ chars ] = chain
                chars += 1
                buff = current
            else:
                if table.has_key( i ): chain = table[ i ]
                else: chain = buff + buff[ 0 ]
                result += chain
                table[ chars ] = buff + chain[ 0 ]
                chars += 1
                buff = chain
    return result
```

Test this, e.g., with the following lines, and toy around with it until you feel that you have fully grasped the inner workings of the algorithm:

```
def test(x):
    print x
    y = compress(x)
    print y
    z = decompress(y)
    print z
    print len(x),len(y),len(z)
test("there these three thick brothers threw these three thick pieces of these
    thick things")
```

Language notes. You will have seen the extensive use of mappings here. In this piece of code, they are used to convert groups of a particular type into groups of a different type. With regard to the individual data item, conversion of a number into a string representation may be achieved conveniently by using the % operator, which will be discussed in more detail on p. 339.

♦ *What does the expression* string.join(map(lambda(x):"%d"%x, mylist), "-") *do? Hint:* mylist *is supposed to be a number of integers.*

There is a further subtlety to this implementation.

In the beginning (of computing), there was... the English language, which is unique among all languages using the Latin script in that it does not use diacritics or special characters. So initially, no provisions were made for umlauts, accents, ogoneks and the like. However, it was soon realized that successful

codepages

Unicode

UTF-8

marketing of computers to non-Anglo-Saxon countries required the availability of a full complement of special national characters. Initially, kludges were used, most amusing among these probably the little switch below the keyboard of the Apple IIe which turned square and curly braces and the like into umlauts. Later, so-called "*codepages*" were used which essentially do the same thing by software. However, by 1990 the need for a universal encoding was generally recognized, and a new 16-bit standard was agreed upon: *Unicode* – a standard that encompasses literally all writing systems ever devised on this planet (including Chinese logograms and even *rongorongo*, the script of Easter Island), with the notable exception of the ancient but still used Hungarian scarved glyphs known as *rovásirás* (suppressed for political reasons, it seems)[15].

However, in most cases a 16-bit encoding is simply overkill. The vast majority of all western letters *can* be represented in 8-bit form, and doubling the size of all text documents is not an attractive prospect. This problem was solved by creation of *UTF-8*, which is a hybrid encoding that manages to represent all basic Latin characters and numbers in 8-bit format but uses 16-bit[16] Unicode for everything that is non-basic Latin.

Thus, UTF-8 is the most efficient way to store LZW compression products. As we begin with a table wherein every 8-bit value is represented, and must, at least in theory, be able to guarantee that any couple of 8-bit values can be represented, essentially 16-bit values would be required. The use of UTF-8 enables us to use 16-bit only where it is really required and to cut back the representation to 8-bit otherwise, which offers a significant advantage.

♠ *This time, all comments and docstrings have been left out on purpose. Maybe this will help you to appreciate their usefulness, as your job is now to insert them. Please analyze the program carefully and make sure you have understood – and appropriately commented on – each individual line and also inserted proper docstrings. If you are not sure what is going on, modify lines and test-run the program. Insert as many control points as you like and play Big Brother with the program.*

16.3.6 LZW and molecular biology. As we have learned before, a compression algorithm uses regularities to reduce file size. This is tantamount to saying that a compression algorithm is a tool for discovering regularities in a run of data.

♠ *Contemplate how LZW, or a similar algorithm, might be modified to identify repetitive portions in nucleotide or amino acid sequences.*

---

[15] The rovásirás script is sometimes called "Hungarian runes" but is evidently older than, and may be the ancestor to, Nordic runes.

[16] There are tradeoffs for everything, and some particularly rare symbols are actually represented in 24-bit format; but never mind that.

16.3.7 LZW and Python. Among PYTHON's modules, there is one named "zlib", which provides very efficient implementations of LZW-style compression and decompression algorithms, CRC codes (see p. 309) and the like.

♠ *Using the documentation, familiarize yourself with the contents of the "zlib" module.*

♠ *Now this is fun. How would you design an LZW mailbomb? (Warning: Please don't try this at home.)*

The following lines create a replacement for the *turbo* class which overrides the basic functions with the "zlib" algorithms:

```
class turboplus(turbo):
    import zlib
    def code( self, text, dummy ): return zlib.compress( text ), chr( 0x00 )
    def uncode( self, stringo, rar, dummy ): return zlib.decompress( stringo )
```

Naturally, files produced with these two algorithms will be mutually incompatible, so one would also modify the magic numbers here. And of course, the "rarest" byte in the header will now be meaningless. But nevertheless, the simple sanity checks still work!

♠ *Redesign the program in such a way that proper CRC checksums are used.*

16.4 Digression: prototyping and extending

The advantages of using PYTHON here are easily seen. Fast development, clear structure, easy debugging and a lot of support on the language side (e.g., the fact that the programmer is disburdened of all memory management). When running the conversion methods of the *turbo* class, you will, however, see that the speed of execution is far from intoxicating[17]. So what can be done about this?

One possibility is to use PYTHON for "prototyping", i.e., development and testing of the algorithm, which is then transferred into a compiled language such as C^{++}. However, the usefulness of this approach seems to be much overrated, for the simple reason that what slows down the PYTHON script will also slow down the C^{++} program unless it is radically modified. Moreover, many of the more advanced features of PYTHON simply do not exist in C^{++}. (If you are not sure to what this alludes, just turn back to the LZW section.)

In the present case, the main culprit is the inconspicuous line my_code += ch, since its execution may require extensive realignment of memory and data.

---

[17] This holds true in particular on Windows^{TM} systems. Because of mysteries not to be penetrated into by mere mortals, the memory management scheme of the Windows^{TM} kernel is such that, after a fulminant beginning, PYTHON will quickly "tire" when strings or lists become too long. By contrast, the Linux, BSD and Solaris^{TM} kernels keep their pace quite steadily.

Keep in mind that in PYTHON a string is a dynamic structure. A C++ program using a corresponding dynamic structure would require the same amount of realignment of memory and data and hence be only marginally faster.

Here one should start by asking what assumptions can be made for the purpose of getting rid of such complex operations. In the present case, it is easy to see that the compressed data chunk can be guaranteed to be either smaller or exactly equal in size to the uncompressed block, but never larger. Thus, the C++ program could get rid of all dynamic structures by temporarily reserving for the compressed data a block of memory the size of which equals that of the uncompressed block. This given, the "inconspicuous line" mentioned above is simplified into plain memory access and hence requires not more than a handful of machine instructions which can be executed in a corresponding number of clock cycles.

But is it really necessary to rewrite the entire program in C++, discarding all the benefits of PYTHON even where speed is not critical, e.g., in file handling? Luckily, it is not. The PYTHON has an interface to C which allows one to easily implement PYTHON library functions in C (or any other low-level language with compatible calling conventions, sich as C++).

Thus, the optimal solution consists of replacing the *code* and *uncode* functions with C routines, which can be done at an arbitrary later stage of development by simply overriding them. So this, again, demonstrates the benefits of OO.

16.5 Preventing unnoticed tainting of data

The capacities and transmission rates of modern devices are vast and still increasing. One single *Blu-Ray*^{TM} disk can hold approximately as much as 50 000 floppy disks, each of which is good for several hundred pages of text; and there are networks available which can transfer all this within approximately ten minutes from one computer to another. The dimensions of time and space which are available for a single bit are thus minute, and the danger of loss or falsification of data during transfer or storage is correspondingly high. Depending on the nature of the data, a single corrupt bit can tarnish a database in the gigabyte range. This is a problem of particular concern in bioinformatics, where particularly vast amounts of data have to be handled; and it is, ironically, aggravated by the use of compression algorithms as described above, as a single bit in a compressed file may have a much greater "meaning".

Algorithms have been developed to verify data integrity, and many of them are implemented in hardware and form an integral part of common storage and transmission services – whenever a failure seems to have occurred, the device automatically requests repetition until untainted data are provided.

The most obvious way of checking data integrity consists in adding up the byte values of all the bytes in the data block of interest. Alternatively, they may be joined bitwise by "exclusive or". In both cases, a value is obtained that is characteristic for the byte set of the data block – the *checksum*. However, two complementing corruptions, or a change in the order rather than the value of the bytes, cannot be detected using this simple mechanism. There are many names that give *The Number of the Beast*!

checksum

The Number of the Beast

One of the most straightforward solutions is the *Adler-32* checksum algorithm (named after its inventor, Mark Adler). In fact, it is incredibly simple. We calculate *two* checksums, one for all the bytes and one for all the byte pairs in the block to be guarded. These checksums are 16 bit each and are combined to form the lower and the higher word, respectively, of a 32-bit return value.

Adler-32

```
def adler32( data ):
    a, b = 1, 0
    for i in data:
        a = (a + ord(i)) % 65521
        b = (b + a) % 65521
    return (b << 16) + a
```

65521 is used becaused it is the greatest prime number below 2^{16}, which has practical advantages.

This algorithm is widely used for maintaining larger data volumes. It was found, however, that it is not fully reliable for data blocks of less than 128 byte size. Such small packages, which are essential to the proper functioning of many networks, are therefore better entrusted to other algorithms such as the *CRC* (Cyclic Redundancy Check) algorithm.

CRC

To pythoneers, professional implementations of both Adler-32 and CRC are available as parts of PYTHON's "zlib" module.

"Now mark Time's finist joke: putting Allspace in a Notshall." – James Joyce: "Finnegans Wake"

CHAPTER 17

Dealing with Errors

> It is none else than here the Red Crosse Knight
> Who yonder slew that foul mishappen wight,
> Beastlie and deforme, that rightly ERROR hight.
>
> – EDMUND SPENSER: "The Færie Queene"

17.1 What are errors good for?

Basically, there are two kinds of error:

- errors which bring a program to an "undefined state" (i.e., a crash)
- errors which do not upset the apple cart but cause the program to yield wrong results.

From a naive point of view, the latter appear not quite as bad: "as long as the program still runs..." However, this is a mistaken analogy. In fact, errors which go unnoticed can lead to much worse – sometimes disastrous – results. Think of "mission critical" software for air traffic or nuclear facility control, or for the management financial transactions "by wire". Here the hardware is usually designed to go to "emergency stop" if control is lost, i.e., if the software crashes. With the program running wild, however, there is a good chance that nobody will notice that anything has gone wrong before it is too late, and lives may be lost – or even money, which seems to be held to be even worse by many people nowadays.

It is very difficult to do anything about logical errors. There are approaches to verification by formal methods, some of which are in turn accessible to computerization (automatic theorem provers); only these do have a fundamental disadvantage. *They do not work with imperative (including OO) programming.* Why? Because an imperative program is like a cooking recipe. Unless you set it to work, you will not find the results. Declarative programming (both λ- and π-based), on the other hand, is more like a bluescript. The program contains a definition, and by methods analogous to algebra, these may be transformed into something clearer. For example, our HASKELL quicksort (p. 265) can ultimately be transformed into a statement that every member of the list is greater than or equal to its predecessor[1]. Thus, the correctness has been formally verified. All very well, but it just does not work with imperative languages.

When using an imperative language, the only possible approach is to narrow down the bandwidth of possible states to those which are sensible. In our address file program (p. 32), the PASCAL version has limited postal codes to values between 0 and 99999. So if anybody, by mistake or out of a strange sense of humour, enters 123456 as the postal code, this will cause a *runtime error*[2]. Thus, the logical error, which would otherwise probably go unnoticed, is transformed into something that brings the program to an abrupt halt, usually an exit with (in PASCAL) an error message like "Runtime error in line 9873: Value P of type PLZ out of range".

PYTHON does not permit this, as this does not mix well with weak typing. Instead, a PYTHON script relies on "defensive programming" to avoid logical errors – like this:

```
def get_plz( prompt ):
    "Ask for a PLZ and ensure it being in the proper range"
    plz = input( prompt )
    if plz < 0 or plz > 99999:   return get_plz( prompt )
    else:   return plz
```

(Another use of recursion!) Alternatively, one might use a non-rejecting loop or, for the very rude, a "goto" statement to ensure a proper value:

```
int get_plz( char *prompt ) {
int plz;
    do {
        printf( "%s", prompt ); scanf( "%i", &plz );
    } while (plz < 0 || plz > 99999);
    return plz;
}
```

♠ *Try to implement this way of getting a proper postal code in* PYTHON *in a non-recursive fashion.*

Most noteworthy in this respect is EIFFEL, where the *requires* and *ensure* keywords can be used to limit the range of acceptable arguments or legal return values of a function or method.

---

[1] Admittedly, this is a science of its own, and for larger programs, finding the "right" criterion for correctness may be as difficult as writing the program.

[2] If anywhere in the program such an out-of-bounds value should be assigned directly to the postal code, this will be caught during compilation.

17.2 When the walls come tumbling down: errors leading to program termination

Besides these artificially contrived errors, a program may incur errors stemming from "natural sources" – operations which just cannot be completed. The simplest of these is an instruction which is unknown to the processor (which happens, for example, when the processor runs into data areas, or when code sections are overwritten with data). A division by zero, obviously, or an overflow also unhorses the processor, as does access to a memory address which is not available on this computer[3]. This latter error is the general bugbear of C and C++ programs. I would estimate that at least 90% of all program crashes are due to "address errors" and "bus errors". The overflow of internal stacks and the inability to satisfy memory requests are also frequent sources of trouble. Finally, multi-tasking OSes raise error conditions when internal access privileges are violated (i.e., when a program tries to interfere with another's allocated memory areas, it is bumped off at once)[4].

17.3 Error signalling by return value

There is a point which we have not yet emphasized, mostly in order to avoid confusion. A program[5] does not operate in "empty space". The famous quotation attributed to Isaac Newton is valid here as well: "If I have seen further, it was by standing on the shoulders of giants." That is to say, our PYTHON script operates using megabytes of subordinate software (the interpreter, its libraries and the different layers of the underlying OS). And whenever a function belonging to these lower strata is called, no matter how indirectly, it may always happen that, in spite of the program being perfectly error-free, it is not possible to satisfy the demand. Probably the most frequent case is an attempt to read a file which does not exist (or to write to a protected file – both cause the same kind of problem: the file may not be opened).

Here we have a condition under which execution of the program cannot and must not proceed as planned, for normally a file is opened with the purpose of accessing its contents. PASCAL is very rigorous about such a situation: Unless special measures are taken, the attempt to open a non-existent file via the

---

[3] Either because the address is beyond the limits of physical memory, because the address is actually occupied by some memory-mapped device, or because of some other mistake – in general, words, longwords and quadwords have to be "word-aligned", i.e., placed at even locations, and trying to perform a word or longword operation on an odd address will invariably crash the program.

[4] The reactions to such errors are extremely system-dependent and range from a simple system reset on older computers to a nice graphical box informing you about the problem and offering to send a detailed error report to the author via email. The all-time champion, however, was the AmigaOS, which notified the user about a "Guru Meditation Error" (*sic*) number so-and-so before requesting the system disk for restarting.

[5] At least the kind of program we are interested in. System programming, especially working on the OS kernel, is a different matter.

function `reset` causes an immediate runtime error and a *Ciao bello* from the program.

There is a more gentlemanly solution to this, and this is used by, e.g., C. The rationale is that opening a file will always yield a reference to that file which later serves to access the file. So why should we raise an error if we can just return a value that is immediately discernible as invalid? In C, the function `fopen( filename, mode )` returns a pointer to a structure of the type `FILE` describing the file. Nobody knows what is really contained within this structure, still less cares – but this pointer is passed to all functions operating on open files (*fread, fwrite, fseek, ftell* and finally *fclose*) in order to identify the file in question, and *they* do know. So what a kind of pointer does *fopen* return if it cannot open the file? The usual thing in C: a NULL pointer – a pointer set to 0x00000000 (32-bit systems) or 0x0000000000000000 (64-bit processors), implicitly equivalent to a Boolean FALSE. Thus, the C programmer will write (the declaration *vartype* *$*vp$ always meaning that *vp* is a pointer to a variable of type *vartype*):

```
FILE *fp = fopen( "file-which-is-needed.txt", "r" );
if (fp) {
    ... /* perform strange and wonderful operations here */
    fclose( fp );
} else printf( "Error:  could not open file" );
```

This works well with library calls in C, where pointers are a prominent feature and thus there is usually a value at hand which signifies "stop, something is wrong". But what about a simple integer[6] division? For x = y / z there is simply no "undefined value" result in the case when z is zero (nor is there for x = y + z and the similarly realistic case that the sum of y and z may exceed the maximum integer value, causing an overflow). From 0x00000000 to 0xFFFFFFFF, all 2^{32} values are legal. A possible way of dealing with this is restricting the effective integer width to 31 bits (allowing a range from – 1.07×10^9 to $+1.07 \times 10^9$, sufficient for most applications) and using the 32nd as "error flag" bit.

A more elegant solution is made possible by structured data types. Imagine we are supposed to write "mission-critical" software working on integers[7], where a breakdown of the program due to divide-by-zero or overflow errors is to be excluded at all costs. This is very easy. We define a structured data type, or an object class, named `airbagged_integer`, consisting of the integer value proper and an ancillary status value which may assume *Valid*, *DivideByZero*, *AdditionOverflow*, *SubtractionOverflow*, *MultiplicationOverflow*, *GeneralError* and maybe a few others. This data type is then bundled together with the

---

[6] Floats use to have more complex internal structures, among which an "undefined" value also exists.

[7] Financial transactions are usually calculated using integers only in order to avoid rounding problems.

necessary functions for adding, subtracting, multiplying and dividing values of type *airbagged_integer*[8] to form a module.

Elegant, but necessarily inefficient in terms of speed and memory usage. Moreover, although it allows very precise error handling, this is restricted to the immediate surroundings of the error zone. If we have a deep call hierarchy, the fact that an error has occured may have to be passed back upwards through several levels of calls until it is reported back to a branching where the program's reaction itself is determined. The latter point is of greater importance in most applications, as processor speed and memory size are nowadays more than a match for most real-life tasks.

17.4 Error signalling by exceptions

Amazingly, one of the first systems which implemented this modern method (albeit in a primitive fashion) was the Sharp pocket computer from 1984, with 8 kilobytes of RAM and a built-in BASIC which featured the construct ON ERROR GOSUB, meaning that you could specify a subroutine which was executed in case anything went wrong. This subroutine could then RESUME at any other line.

This illustrates the basic tenet of "signalling by exception". If a runtime error occurs, do not kill the program but instead look for some error trapping routine. (If there is none, we still have the option to kill the program in order to avoid worse damage.)

It is comparatively easy to implement this in a hierarchical fashion. If an error occurs and there is no trapping implemented in its vicinity, automatically report the error to the calling routine. If none is implemented there either, go one further step higher... until either the error *is* trapped somewhere or we come to the main routine of the program. If it is not trapped there, then finally use the default – dump an error message and *ciao bello*.

Such a system is built into the PYTHON interpreter, as it is built into JAVA, ERLANG and O'CAML. JAVA terminology[9] speaks of "throwing" and "catching" error reports, and each function must explicitly declare whether it may generate ("throw") such error reports, and if so, what type they are (if in doubt, Exception will be okay). In this case, the calling function must either "catch"

---

[8] In Vol. 2 we will discuss "overloading". Basically, you define operators for new data types so you don't have to write `airbag.intmul( x, y )` but can write `x + y` even if x and y are not of a built-in type but of type *airbagged_integer*

[9] The authors of JAVA make a great fuss about their language's error handling capacities, maybe to imply that it is very good at error handling, or perhaps just to conceal that they did not know how to get rid of all this boilerplate while preserving the quality of their compiler's code generation (not the authors' weakness but an intrinsic shortcoming of the imperative approach, which does not *explain* things). ERLANG's error handling is at least as good and much more slender in structure. The positive aspect of this complicated approach is that it unfolds the very last detail of the concept to the one who is willing to work his or her way through it.

the report or "throw" it again. The compiler will not have it otherwise. In the following example, the error may occur within the function *gupta* which is called by *foo* which is in turn called by the main program which is to deal with errors. *foo* does nothing (as far as the error is concerned) except for passing it back to the higher level.

JAVA:

```
private static int gupta( blah parameter ) throws Exception {
    ... // here things may go wrong
}

private static int foo( blah bar ) throws Exception {
    ...
    int chandra = gupta( bar );
    ...
}

public static void main( String[] args) {
    int f;
    ...
    try {
        f = foo( x );
    } catch (Exception e) {
        error_message( e );
        return;
    }
    display_result( f );
}
```

A similar mechanism for trapping errors exists in ERLANG (note that this yields a tuple consisting of either 'EXIT' and the error message or *value* and the result of the function):

```
gupta( blah ) ->
    ...  % here things may go wrong

foo( blah ) ->
    ...
    chandra = gupta( blah );
    ...

case catch foo( bar ) of
    {'EXIT', e} -> error_message( [e] );
    {value, f} -> display_result( [f] ).
```

That's it for RUBY:

```
def gupta( blah )
    ...  # here things may go wrong
```

17.4 ERROR SIGNALLING BY EXCEPTIONS

```
    end

    def foo( blah )
        ...
        chandra = gupta( blah )
        ...
    end

    begin
        f = foo( bar )
    rescue
        error_message( $!.message )   # $! is the error object
    ensure
        display_result( f )
```

CAML riders also love it:

```
    let gupta blah =
        ... (* here things may go wrong *)

    let foo blah =
        ...
        let chandra = gupta blah in ... ;;

    try let f = foo x in display_result f
    with e -> error_message e;;
```

Oddly enough, even COMAL had this:

```
     FUNC gupta( blah$ ) CLOSED
         ... // here things may go wrong
     ENDFUNC

     FUNC foo( blah$ ) CLOSED
         ...
         chandra = gupta( blah$ )
         ...
     ENDFUNC

     TRAP
         f = foo( bar$ )
         PRINT f
     HANDLER
         error'message( ERRTEXT$ ) // ERRTEXT$ is the system message
     ENDTRAP
```

And now here's how pythoneers do it:

```
    def gupta( blah ):
        ...   # here things may go wrong
```

```
def foo( blah ):
    ...
    chandra = gupta( blah )
    ...

try: f = foo( bar )
except Error, e:  error_message( e )
else: display_result( f )
```

In all three cases, we actually get a bit more information than just that something nasty has happened. Information about the nature of the error is generated and shoved into a variable (*e*). In ERLANG, a language which is definitely not OO[10], this is a tuple; in the OO languages, an object.

Keep this in mind. In PYTHON, exceptions are objects. Trouble which cannot be dealt with leads to the generation of a specialized object. Functions possess, so to say, two "interfaces" to the environment, like network connections on the hardware level. There is the "data wire" – the path which normal results take, as in mathematics – and there is the "control wire", which is used for signalling problems (in ERLANG, where everything is transparent, this is most easy to see: the first part of the "caught" tuple is the control wire, the second the data wire). This is less an analogy than a homology. In both cases it was found to make good sense to have "meta-information" about success and failure separate from the mainstream of information[11]. However, whereas a network connection won't work without the "control wire", this is optional in most programming languages which implement this mechanism. If no control wire is provided by the program, the interpreter or runtime system falls back on its standard defaults.

In general, a "try" attempt may be followed by several catches for different errors and, in PYTHON, by an "else" clause for the case that no error at all has occurred:

```
try: f = foo( bar )
except VerySillyError: print "Aren't you ashamed?"
except OutOfCheeseError, e:
    print e, " - Please reload from drive /dev/edamer"
except Error, e: error_message( e )
else: display_result( f )
```

And of course, error conditions may be generated deliberately; see p. 344.

---

[10] Joe Armstrong, one of the godfathers of ERLANG, once used the strange expression *Eppur si sugat* with reference to OO in general and C++ in particular. Later he explained it in a private mail to me: "My Latin/Italian is non-existent – I wanted to say "and yet it still sucketh..."

[11] In the section on markup languages (p. 45), this subject has been touched briefly.

♠ *Using* Python*'s* try:/except: *mechanism to catch errors during manipulation of native integers, implement the* `airbagged_integer` *data type and its allied functionality described above. Hint: compare airbagged integers to complex numbers before proceeding to devise a structure; make information accessible via methods rather than by directly accessing attributes – this is preferred, but why? Do we have to implement a full set of comparison functions too, or is there any way of employing the normal operators here (without overloading)? What about speed? Discuss.*

♠ *Objects are, as we know, black boxes. If exceptions are objects the exact description of which is not even known, how are we to make any sensible use of them? In particular, how may the very common formula* `except Error, e: print e` *work? (Hint: turn back to p. 54 for this especially useful trait of objects.)*

17.5 Assertions

We have mentioned before (p. 312) that a weakly typed language has to rely on "defensive programming" to guard against nonsensical arguments. In Python, this is supported by "assertions", a concept which is gradually leaking into other languages as well.

Basically, an asertion is simply a condition which must be fulfilled; otherwise a runtime error will be raised. Thus,

```
assert plz >= 0, plz <= 99999
```

will make sure that $0 \leq plz \leq 99999$.

In Python, there is an additional sophistication in this as the assertion is checked only when the global `__debug__` flag is set, which cannot be modified from inside a script but is set at startup when the debugging option (`-d`) is used[12].

17.6 Crash & Burn

On June 4th, 1996, the first flight of the European Ariane 5 expendable launch system (*Flight 501*) failed dramatically. The rocket veered off its flight path 37 seconds after launch and underwent self-destruction when high aerodynamic forces caused the core of the vehicle to disintegrate. The payload was completely lost – four Cluster mission spacecraft, resulting in a loss of more than $3.7 \cdot 10^8$ US-\$.

Flight 501

The background was that the Ariane 5 was capable of considerably higher acceleration than its predecessor, the Ariane 4, from which much of the control

---

[12] Furthermore, when a module is compiled to byte code (*.pyc*), all assertions contained therein are omitted. Thus, time-consuming checks are performed only during development.

software – written in ADA using one of the very few compilers certified for safety-critical tasks – had been re-used. This higher acceleration led to an overflow in a 16-bit (i.e., word, p. 75) size variable, which generated an error condition, namely an overflow exception, as discussed above. For this exception, no handling had been implemented, so the processor shut down, causing a cascade of problems ultimately resulting in the crash.

Moral. When programming is sloppy, a certified compiler will output nothing but certified trash. Don't rely on promises of fault tolerance unless you feel that you have 370 mill. $ too much in your pocket.

As far as we know, our computer has never had an undetected error.

CHAPTER 18

♣ A Real-life Project: Generating a Restriction Map and Making Simple Predictions

Now stop that reasoning, friend – it's not required anymore – just get your weapons.

– CHRISTIAN GRABBE: "Napoleon or the Hundred Days"

18.1 Meet Micro-Willy

There *are* people who succeed in life without knowing anything about computers. One of these you see pictured above[1]. You have already encountered this gentleman when he was collecting values on p. 129.

18.2 What is a restriction map?

18.2.1 Background: restriction enzymes. In the 1960s, it was realized that bacteria protect themselves against infection by viruses (bacteriophages)

---

[1] Others are professional boxers, tennis players or racing drivers.

by means of enzymes which are able to cut DNA. These enzymes, named "restriction endonucleases" because they *restrict* the infectivity of phages, recognize specific runs of nucleotides and cut the DNA at defined sites either within or close to the recognition sequence. Restriction enzymes are always coupled with methylases which are specific for the same sequence. Methylated DNA is not cut. Thus, the bacterium avoids destroying its own DNA. However, it may be infected by a phage assembled in a bacterial cell possessing the same restriction/methylation system.

There are several classes of restriction enzymes. By far the most valuable are those that bind to a short palindromic sequence[2] and cut both strands within the sequence (either staggered or not), generating either overhangs or blunt ends.

♠ *Why does a palindromic recognition sequence make cloning much easier?*

Most of these enzymes recognize hexameric sequences (sixcutters), but there are also tetrameric recognition sites (fourcutters) and octameric ones (eightcutters). There are even a few fivecutters and one or two sevencutters. Some enzymes allow for variable nucleotides within the sequence. Generally, each restriction enzyme is known by a name that alludes to the species from which it was isolated, e.g., *Hind*III from *Haemophilus influenzae* strain D, *Nhe*I from *Neisseria heidelbergensis* and *Xho*I from *Xanthomonas holcicola*.

DNA ends with complementary overhangs may be joined again by using another enzyme, ligase. This procedure is at the heart of all recombinant DNA work. It has become such an everyday method that it is hard to imagine that in the mid-1970s the building of "unnatural" DNA was a major political matter. In fact, without restriction enzymes molecular biology would not be possible.

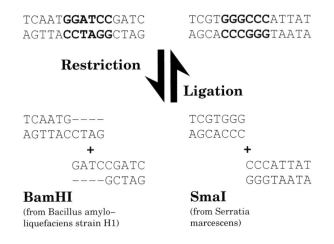

---

[2] A palindromic sequence is one that reads equally from both ends, such as "Hannah" or "Ein Neger mit Gazelle zagt im Regen nie".

18.2 WHAT IS A RESTRICTION MAP?

♠ *Assume the genomic DNA of the bacterium* Fubaromonas annae *comprises 4×10^6 nucleotides – contained within one single circular molecule – and contains equal amounts of GC and AT pairs. What is the average fragment size after treatment of the DNA with: a) the fourcutter* FooI *(GATC); b) the sixcutter* BarI *(GAATTC); c) the eightcutter* UrxI *(GCGGCCGC). In other words, how many restriction sites do you expect? What do you think about the actual distribution? Can you suggest any model for calculating the distribution?*

♠ *Do the same estimation for the semi-synthetic plasmid (small circular DNA construct capable of autonomous replication in a suitable host cell)* pFOObar$_{08}$15, *containing 4000 base pairs all in all, of which there are 60% ATs and 40% GCs.*

♠ *The biochemical analysis of different restriction enzymes has shown little, if any, similarity. They do not appear to be homologous. So what is the probable reason for so vast a majority recognizing palindromic hexamers?*

♠ *The endogenous restriction/methylation system* FanI *causes every G in the sequence CGNNCG (where N can be any nucleotide) to be methylated; unmethylated DNA will be cut right behind the first G (sometimes this is denoted as CG^NNCG). Do you think this may be a problem for any of the enzymes above?*

♠ *The plasmid* pFOObar$_{08}$15 *has a restriction site for* FanI, *which we have cloned, overexpressed and purified by now. The gene* ook *from* Chandra gupta *is contained within the common cloning plasmid* pUC19 *and may be excised using* FanI *– unfortunately we do not have the exact sequence, only a map of restriction sites. For some special purpose, e.g., expression in the schnorrovirus system, we need it in* pFOObar. *As* FanI *is not normally expressed in* E. coli *cells, we will have no problem with methylation. So, considering all these facts, is* FanI *a good choice for cloning under these circumstances, or should we rather resort to the tedious PCR? Explain your answer.*

18.2.2 Analytical digests. Apart from building new constructs, restriction enzymes are also useful for a "quick and dirty" characterization of DNA constructs. When "digested" with a restriction enzyme, constructs leave a characteristic "fingerprint" of pieces, which may be separated by electrophoresis and visualized. For many purposes, this can obviate the need for the tedious and expensive sequencing procedure[3].

18.2.3 Our exemplary application: working with the BLAH gene from *Pan troglodytes*. Micro-Willy plans to do some work on the BLAH protein, an orangylation-controlled fubarase suspected to play a role in cell cycle control. His chappie Dr. Meerkatz, from Affics Ltd., has kindly provided him with the DNA but has not taken the pains to add a description (in his

---

[3] During my first doctorate, I built a total of seventy-eight plasmids without doing a single sequencing reaction, all based on restriction digests only. The plasmids worked!

opinion everybody who does not know the sequence of *blah* by heart is a fool anyway) except that the DNA construct comprises "the bases from 3 to 1833 of the cDNA isolated in the lab of Babún and Rolloway".

From this description, it is easily possible to retrieve the correct part of the sequence from the international databases, but what can we do with it? It reads:

```
CTAAGGATCC GAGGATGCGG AAGGTCGGAA CCGAACTCAT GGCCCGGATC CTTATCATCC
TTCCAGGAGC TGCCTGGCTA GACTTGGCCA AGAGCATTAG CTCACATCGA TACGATCATC
TGGGTGCTTC ACCTACAGCA CGTCATCTAA TCAGTATTAT CCTTTCTTTC GATGCCTCCT
TGCAGTATGT AGGGGATATG AAGTATCTGC ATCCCGTTCC GTCCTCGAAG CCCGCCCTAA
TGTGTTCCGC GTCGCCGTGC GAGGAGTGGT CACGAGATCA GTACCTTATG GACCCGAGGG
GTCGCTGGAC ACGAGGACTA CACCTCTGGA CACCGACATC TATCTTTGTC ATGAAGAGTA
GTAGAACACA GAACGCGATC TCTATGACAC GAATAAGGGC CCATATCTCC TTCAGTCCAG
CGGTTTCAGC AAATTTGCTC CTTCCGCAGC GCAAAGTAAA GTCCTCGAAG ATATGTGTTC
CCCCCTGCCG TAGCTGGATG TCAGTCAGTC GAATATATGT CATAAGGCCT GTGAACTCTA
CCGAATTCGG TGCACACATC GGCGGCGAAT TGAAAGATGG CTCATTATAC CGTACAGACC
GCCGCGCTTT ACTATTCGAG ATCTTTCGCT GCTTATCCGA TGACTGTACA TTCACCTACC
CTCCACCAAA TAGAAGTTGG GGCGTTCGAG ATGTCGGTGT TACGAACCTC AGGGATCGAC
CAGAACGGGA GAGAATGAAC TACGCGACAA CGGTAGCTCC GGCCTATCGG CAGATAGTCT
ACCGAGTTCA GGGGACTTGG CGCCACGCTT ACCTCCGCCT GCCCTCAAAG AAGATCGGGC
ACTTAACATC TCTAGTCAGG CAGAATCGGT TGCTAGCACG CGCCCTTCGC CAGGTTCCCA
GCACATTGTT TATTACGCCG GGGAATAGTG AAATGCGGAG CCTTGCGGCA GCCACAGATA
TTTTAAGCA TGCGCCCTGG TGCTGGTGGT ATCATGAATC TGTCCTGTTG TCCCAGGGCC
GGCATGAACC CGCTCGTCTG GCGTATAACT CTCACCGTGC AACCGAGGCT CAATCGATCG
TCCGTCGATG GGCATTCATA CTCTTGTTAG TTATATTCGC CTTTTTTATA GCATTCGCAC
TGGCTATAGT CATTGTCTTA TTATTTCTGG CATTATTTTT GAGGAAGAGG AGGAAGAGGA
TTTTGGCCAC CTCGCCGGCA GCTCTGTGGC TAAGCAAGGC TAAGACACGG GGCCATAACG
GAGGATCTCA AATCGGGATG CCCTTTCCGT CGGTCGACAG GATTTGTCCT GTGCATTCTC
GCGCATGGAC TAGATGCTCT CAAGCATGTG AGGATTGTAC TCTCGTCACA GATCATAAGC
CAACTCGGAA TATATGTGTT TGCATGGAGA CCGGTACTCT GGTACGAATC CAGATAGCAT
GTTGGGATTG TTATACTGCC TACCTTCCGG CAGTTCGCCT GACCGTACGC AAGTACGCAC
GTGCACCTAT CTTTGGGGAT CAACTCCGTG TGGCATGTCT TGATTGTCCT CTGTTACCCG
GCTGCACCAG ATTTGGGCGT GTTGTCTGGA ACGACCCATC CATTCTTAAG TGTTCCACGG
AACGTGTATC TTGCTTTGAC TGTTTCACTT CCGTTACTAA TTATGATTCC TCGCGGGCAT
ACTGGGCGGG GACGGCATGT ATGGATTGTG CCATTCCTTT ACTTGAGAGC GTGTCTCACT
GTTTCTATCA GGTTACTCAA TTGCGGTCCG TGGGGTGATA CTCTGTTGAC AGGCTCCATA
AAAAAAAAAA AAAAAAAAAA GCTTGTAATC
```

The things we need are:

- the sites where common restriction enzymes cut
- the protein translations.

After some consideration and a number of bananas, Willy wisely decides to ask Anna for help.

18.3 Extracting data from a file

A GenBank file can be divided into two main parts: The *annotations* and the *sequence*. Here you see a sample of a typical entry:

```
LOCUS       L43967      580074 bp    DNA     circular BCT    17-MAY-1999
DEFINITION  Mycoplasma genitalium G37 complete genome.
ACCESSION   L43967
VERSION     L43967.1  GI:6626254
KEYWORDS    .
SOURCE      Mycoplasma genitalium.
  ORGANISM  Mycoplasma genitalium
```

```
            Bacteria; Firmicutes; Bacillus/Clostridium group; Mollicutes;
            Mycoplasmataceae; Mycoplasma.
REFERENCE   1  (bases 1 to 580074)
  AUTHORS   Fraser,C.M., Gocayne,J.D., White,O., Adams,M.D., Clayton,R.A.,
            Fleischmann,R.D., Bult,C.J., Kerlavage,A.R., Sutton,G.G.,
            Kelley,J.M., Fritchman,J.L., Weidman,J.F., Small,K.V., Sandusky,M.,
            Fuhrmann,J.L., Nguyen,D.T., Utterback,T., Saudek,D.M.,
            Phillips,C.A., Merrick,J.M., Tomb,J., Dougherty,B.A., Bott,K.F.,
            Hu,P.C., Lucier,T.S., Peterson,S.N., Smith,H.O. and Venter,J.C.
  TITLE     The minimal gene complement of Mycoplasma genitalium
  JOURNAL   Science 270 (5235), 397-403 (1995)
FEATURES             Location/Qualifiers
     source          1..580074
                     /organism="Mycoplasma genitalium"
                     /isolate="G37"
                     /db_xref="taxon:2097"
     gene            735..1829
                     /gene="MG001"
                     /db_xref="GenBank:3844619"
     CDS             735..1829
                     /gene="MG001"
                     /note="similar to GB:U00089 SP:Q50313 PID:1209517
                     PID:1673814 percent identity: 70.87; identified by
                     sequence similarity; putative"
                     /codon_start=1
                     /transl_table=4
                     /product="DNA polymerase III, subunit beta (dnaN)"
                     /protein_id="AAC71217.1"
                     /db_xref="GI:3844620"
                     /translation="MNNVIISNNKIKPHHSYFLIEAKEKEINFYANNEYFSVKCNLNK
                     NIDILEQGSLIVKGKIFNDLINGIKEEIITIQEKDQTLLVKTKKTSINLNTINVNEFP
                     RIRFNEKNDLSEFNQFKINYSLLVKGIKKIFHSVSNNREISSKFNGVNFNGSNGKEIF
                     LEASDTYKLSVFEIKQETEPFDFILESNLLSFINSFNPEEDKSIVFYYRKDNKDSFST
                     EMLISMDNFMISYTSVNEKFPEVNYFFEFEPETKIVVQKNELKDALQRIQTLAQNERT
                     FLCDMQINSSELKIRAIVNNIGNSLEEISCLKFEGYKLNISFNPSSLLDHIESFESNE
                     INFDFQGNSKYFLITSKSEPELKQILVPSR"
                     (...)
     gene            complement(579222..580031)
                     /gene="MG470"
                     /db_xref="GenBank:3845065"
     CDS             complement(579222..580031)
                     /gene="MG470"
                     /note="similar to GB:U00089 SP:Q50314 PID:1209518
                     PID:1673815 percent identity: 90.26; identified by
                     sequence similarity; putative"
                     /codon_start=1
                     /transl_table=4
                     /product="soj protein (soj)"
                     /protein_id="AAC72491.1"
                     /db_xref="GI:3845067"
                     /translation="MIISFVNNKGGVLKTTMATNVAGSLVKLCPERRKVILDLDGQGN
                     VSASFGQNPERLNNTLIDILLKVPKFSGSNNFIEIDDCLLSVYEGLDILPCNFELNFA
                     DIDISRKKYKASDIAEIVKQLAKRYEFVLLDTPPNMATLVSTAMSLSDVIVIPFEPDQ
                     YSMLGLMRIVETIDTFKEKNTNLKTILVPTKVNVRTRLHNEVIDLAKTKAKKNNVAFS
                     KNFVSLTSKSSAAVGYEKLPISLVSSPSKKYLNEYLEITKEILNLANYNVH"
BASE COUNT    200543 a   91524 c   92312 g  195695 t
ORIGIN
        1 taagttatta tttagttaat acttttaaca atattattaa ggtatttaaa aaatactatt
       61 atagtattta acatagttaa ataccttcct taatactgtt aaattatatt caatcaatac
                     (...)
   579961 gatcctgcaa cattagttgc cattgtagtt tttaatacgc cgcctttatt atttacaaaa
   580021 gaaatgatca tatatttaaa tgattataat atttctttaa tactaaaaaa atac
//
```

We shall discuss the annotation part in Vol. 2. For the moment it will be sufficient to say that there are not only lots of additional information contained in this, but also the "key" to the individual genes. For example, if we are looking for the Soj protein of *Mycoplasma genitalium*, we will have to find an entry containing `/product="soj protein (soj)"`. This leads us to base pairs 579222..580031, where the remark "complement" means that the other strand, not the one given here, encodes the protein (i.e., transcription and translation occur "upwards"). These base pairs can then be retrieved from the sequence part, which is delimited by the word `ORIGIN` on one side and a double slash on the other.

This, however, will be the subject of a later project, as it can be done more efficiently using regexes (p. 341) for scanning the input file. At this moment, we will put aside all AI and use our NI to get the sequence. That is to say, Micro-Willy is caused to load the file into an editor ("Shall I save this in M–t Word format?" – "No, Willy, use ASSSCII or 'Plain DOSSS' text!"), seek BLAH manually and copy the interesting part (the basepairs named by Dr. Meerkatz) manually. So we arrive at a file like the one displayed above.

The crude sequence consists of G's, A's, T's and C's, mingled with glyphs to be discarded – mostly whitespaces[4], but there may also be the numbers from the left margin. All of these have to be sorted out, yielding a plain string of nucleotides.

So the first subtask is: write the software to read-open the file, read in the contents, discard all non-nucleotides and store the result. This is to be done by a function

```
sequence = read_rough( filename )
```

and to take into account argument passing at script invocation.

Yet you are curious about how to parse the feature region? Okay... Here is a possible solution, straight from the forge...

```
type entry = { entry_isQuoted: bool; entry_cont: string; }

let getfile filename = (* read a file and return its contents as a single string *)
  try let channel = open_in filename in
  let chlength = in_channel_length channel in
  let data = String.create chlength in
    really_input channel data 0 chlength;
    close_in channel;
    data
  with _ -> "";;

let tokenize stri = (* split string into its components *)
  let toklist = ref [] in
  let ll = String.length stri in
```

---

[4] The term "whitespace" covers normal spaces, line endings (bot CR and LF), tabulators and form feeds.

```
        if ll <= 0 then [] else
          let isvalid ch = (ch >= '0' && ch <= '~') || (ch = '/') || (ch = '.') in
          let next x =
            if x mod 100 == 0 then Printf.fprintf stderr "%d k\t\r" (x / 1000); x + 1 in
          let cc = ref 0 in
            while !cc < ll do
              let c = stri.[!cc] in
                if c == '\"' then ( (* read up to closing quote *)
                  cc := next !cc;
                  let quoted = ref "" in
                    while stri.[!cc] <> '\"' && !cc < ll do
                      quoted := !quoted ^ String.make 1 stri.[!cc];
                      cc := next !cc;
                    done;
                    cc := next !cc; (* discard final quotation mark *)
                    let restok = { entry_isQuoted = true; entry_cont = !quoted; } in
                      toklist := restok :: !toklist; (* WAAY faster than @ *)
                ) else if isvalid c then ( (* just a token *)
                  let toked = ref "" in
                    while !cc < ll && isvalid stri.[!cc] do
                      toked := !toked ^ String.make 1 stri.[!cc];
                      cc := next !cc;
                    done;
                    let restok = { entry_isQuoted = false; entry_cont = !toked; } in
                      toklist := restok :: !toklist;
                ) else ( cc := next !cc; ) (* whitespace etc. *)
            done;
            List.rev !toklist;;

let istoken delinq what = if delinq.entry_isQuoted then false else delinq.entry_cont = what;;

let rec featureregion = function (* get the feature region of tokenstream only *)
  | [] -> []
  | head::tail -> let rec nosequence = function
      | [] -> []
      | h::t -> if istoken h "BASE" then [] else h::(nosequence t);
    in if istoken head "FEATURES" then nosequence tail else featureregion tail;;

(* Group a tokenstream into genes. Unfortunately, the GeneBank format is rather erratic and
 * not easily accessible to rigid functional analysis. We therefore take the purely pragmatic
 * way - iterate over the list of tokens generated by lexical analysis, take the token "CDS"
 * to indicate a new description and all /type entries subordinate to this. Thus we use an
 * accumulator which is grafted on the list whenever a new one is initiated; and all subordinate
 * entries are added to this accumulator one by one.
 *)
let hierarchy flatlist =
  let hlist = ref [] in
  let accum = ref [] in
    for x = 0 to (List.length flatlist)-2 do
      let tok_x, tok_y_text = List.nth flatlist x, (List.nth flatlist (x+1)).entry_cont in
        if not tok_x.entry_isQuoted then match tok_x.entry_cont with
          | "CDS" -> if !accum != [] then hlist := accum :: !hlist; accum := [ ( "CDS", tok_y_text ) ];
          | "/gene="          -> accum := ( "Gene name",   tok_y_text ) :: !accum;
```

```
              | "/product="       -> accum := ( "Gene product", tok_y_text ) :: !accum;
              | "/note="          -> accum := ( "Comment",      tok_y_text ) :: !accum;
              | "/codon_start="   -> accum := ( "Start codon",  tok_y_text ) :: !accum;
              | "/translation="   -> accum := ( "Amino acids",  tok_y_text ) :: !accum;
              | "/db_xref="       -> accum := ( "Accession",    tok_y_text ) :: !accum;
              | _ -> ();
      done;
    !hlist;;

let parse_gbfile filename = (* do the job *)
  let gb_data = getfile filename in
    let gb_tokens = tokenize gb_data in
      let gb_header = featuregion gb_tokens in
        let gb_genes = hierarchy gb_header in
          Printf.printf "\n%d genes read.\n" (List.length gb_genes);
          let prifu (a,b) = Printf.printf "%s : %s\n" a b in
            let prili l = List.map prifu (List.rev !l);
                      Printf.printf "================================================\n" in
              List.map prili gb_genes;;

let usage n =
  Printf.eprintf "Error: please pass only the filenames, not %i arguments!\n" n; exit 1;;

let main () =
  let argc = Array.length Sys.argv in
    if argc == 2 then try parse_gbfile Sys.argv.(1) with
      | Sys_error( _ ) -> Printf.eprintf "Error opening file %s\n" Sys.argv.(1); exit 2
      | _ -> Printf.eprintf "BUG: Something is very wrong!\n"; exit 3
    else begin usage( argc ); exit 1 end;;
main ();
```

♡ *An(n)alyze this.*
You will recognize the language: it's O'CAML. *This is a functional language, in which everything is evaluated to yield a value. Thus, an expression which reads simply* var *means in fact: "return the value of* var *as the result of the function." Unlike most other functional languages,* O'CAML *also supports iterative features such as loops, and to this purpose it has mutable variables in addition to its normal immutable ones – mutable variables are declared using the keyword* ref *followed by the initial value, accessed via the* ! *operator and modified using* :=. *Arrays and strings are eo ipso mutable structures; string elements are accessed by* x.[i] *and modified through the* ← *operator. The* type *keyword introduces the definition of a compound type, and a double semicolon ends a function definition. List functions are contained in a separate module, and they comprise* hd *and* tl *for head and tail of a list. Recursive functions are marked with the* rec *keyword. For more details, e.g., on the* ∧ *and* :: *operators, please consult your* O'CAML *documentation. And of course, have another look at pattern matching on p. 140 and prettyprinting (*Printf.printf*) on p. 339. One of the extraordinary things about* O'CAML *is that arbitrarily complex data types may be used without naming them explicitly, even though* O'CAML *is a very strongly typed language – from* f x *alone we cannot deduce easily whether*

x is supposed to be a simple integer or some wild list of structures. So the first step towards understanding this is to identify the data structures which are used for the passing to and fro of values. Focus on the two pivotal functions: tokenize and hierarchy. What kind of data do they produce? How does parse_gfile do its job?

18.4 Finding palindromes

Far from all commercially available restriction enzymes have palindromic restriction sites, but as a rule of thumb, recognizing palindromic hexamers and octamers will be sufficient to find most potentially interesting sites. The alternative consists in retyping the recognition sequences of all commercially available restriction enzymes – yawn.

♠ *What kind of* PYTHON *data structure would be most suitable to contain a comprehensive list of restriction enzymes?*

The exact definition of a palindromic DNA of length n is that each element at position e is the complement of the element at position $n - e$ (if we start numbering with 0).

♠ *This definition must be extended a bit to account for palindromes with an odd number of elements (what about the central nuke?). Do so.*

So we may simply walk the sequence base by base and apply our gauge at every step. Do the following 5, 6, 7 or 8 nucleotides form a palindrome?

♠ *Utter care must be taken at the end of the sequence, lest you try to read beyond the end of the sequence and cause a runtime error. Have you any idea how to handle this?*

It will be best to implement this as a function of that style:

```
sitelist = get_palindromes( sequence, position, length )
```

As we are using PYTHON, we are not restricted to a plain binary answer. We can design the function so that it has two possible return values.

(1) If there is no palindrome, simply `None`.
(2) If there is one, return the sequence.

Thus, if there is a palindrome, we may create a tuple containing position and sequence (and maybe the length if we think this will speed things up but it is not really necessary, since it is implicitly contained in the sequence of the palindrome). This tuple can then be stored away. We simply add it to a list.

So in the end, we will have a list like this:

```
[(533, 'CCATGG'), (712, 'GTGCAC'), (1066, 'GAAATTTC')]
```

Typically, this list is going to be considerably shorter (by several orders of magnitude) than the initial data pile. This is a common feature of data processing systems: reduction of the data load by selection and abstraction. (Pretty much as in natural perception.)

♠ *Do we have to scan both strands for recognition sites, or is one going to be enough? Explain your answer.*

It is easy to simply display this list. An alternative mode of output requires a bit more work but is also feasible (and I would recommend to implement both): printing the sequence with the palindromes in uppercase and the remaining bases in lowercase letters[5].

18.5 Naming the sites

You will find a list of palindromic sites in any supplier's catalogue or web site. For each of these sites, you may or may not find a corresponding restriction enzyme. So use the listing of restriction enzymes to generate a second list which comprises the position and the name of the corresponding restriction enzyme (or `None`) if there is none available.

Put this into a function

```
restlist = restrictions( sitelist )
```

Write a few DNA sequences to test this, e.g., `caggcggatcccgattcgggatcccggatccgctaatcg`. Your program will generate something like

```
[(6, "BamHI"), (9, "BamHI"), (26, "BamHI")]
```

This is correct, of course, but not easy to comprehend for human beings, especially when there are longer sequences and wildly mixed restriction sites.

So now you have to walk this list and generate a third one, which consists of tuples of three components each:

- the name of the restriction enzyme (sites for which there are no enzymes available may be dropped quietly)
- the number of sites found for this enzyme (optional – it is contained within the next)
- a list containing all sites.

And that's the way you do it. Begin with an empty summary list. Iterate (or recurse – that's up to you) over the whole of the restriction list. For each item of the restriction list, test whether the enzyme has been entered into the

---

[5] There are the string methods `upper()` and `lower()` to this end.

summary list before. If so, add the site to this enzyme's list of sites (the last component of the enzyme's tuple in the summary list). If not, enter the enzyme and the site as a new tuple.

It will be a smart move to implement this as a function

```
summary = compact( restlist )
```

and have the resulting summary list sorted.

So you end up with something like:

```
[("BamHI", 3, [6,9,26])]
```

Don't forget to add a "negative list" containing the enzymes which do not cut.

Basically, that's it! You may add some prettyprinting of the summary list if you want to impress Anna, but the essential information is contained in the list, no matter how it is displayed.

♡ *Can you imagine how much trouble it must be to implement this in C, where there are no lists or dictionaries but only pre-defined records and arrays of fixed size?*

18.6 Generating the protein translations

Now for something which is comparatively easy. For cloning and modification, we need not only the restriction sites but also the protein translations – two directions multiplied with three possible reading frames. Usually, the correct reading frame will be visible at once because it is going to be the only one that is not interrupted by stop codons.

♠ *Have a look at the genetic code. For DNA produced by some random process, e.g., chemical synthesis or telomerase activity, what would be the average length of peptides if this were transcribed and translated?*

So you need the DNA sequence (which you have) and the sequence of the complementary strand (which you have to deduce). Convert each of these into three lists of triplets:

```
rf1, rf2, rf3 = froggify( sequence )
rf4, rf5, rf6 = froggify( complementary_strand( sequence ) )
```

The optimal representation is certainly a list of strings like ['atg', 'cac', 'atg']. The method `string.split()` may be very helpful here.

After froggification[6] you may easily apply a projection function to each of these:

---

[6] This unofficial term alludes to the sci-fi evergreen "Starship Orion" where the "Frogs" are aliens whose communication is based on signal triplets.

```
pep1 = map( codon2aa, rf1 )
```

This will give you something like ['Met', 'His', 'Met'].

18.7 Making the map

This piece of the work is something for the aesthetically minded. By now, you have virtually all the information that is required: restriction sites and protein translations. Arrange this in such a way that the output consists not of a number of rough-hewn lists but a nice synopsis of sequence, translations and restriction sites. You are absolutely free to arrange this.

♡ *This may be a bit tricky. Refer to p. 337 for details on* PYTHON *output. You don't have to implement this if you think it's too much for you, but I will appreciate it if you try. And at any rate, I would like you to at least think about it.*

18.8 An artificial digest

Micro-Willy is not satisfied yet. He would like to have a prediction of what an agarose gel is going to look like after digestion of the *blah* DNA with a certain enzyme.

It is up to you whether to base this on a request mechanism ("what enzyme do you want to use for digestion?") or have the program spew out the complete set of possible restriction digests (you bought it, you get it).

At any rate, you will have to check what fragments are produced by digestion of the sequence with a given enyzme. Again, a list is the structure of choice for results:

```
fragmentsizes = digest( seq, enzyme )
```

Consider what kind of input is the best. When starting with the nucleotide sequence, just call some of the previously defined functions to obtain the sites. From these pieces of data, you may calculate the fragment sizes.

Now for the output. Theoretically, this could be written directly to a file for a graphics program. The Unix program `xfig`, for example, uses a plain text format for the representation of its objects, so this would need just a few lines of typing. However, we are content to output a text table containing the *migration width* and the *intensity* of the respective band:

Size in bp	Migration in % of loading dye	Intensity in % of original band
3300	30	50
2000	53	30
1200	70	20

♡ *The intensity of a band's fluorescence is directly proportional to the amount of DNA, and this is again proportional to the length. Migration, however, is controlled by a logarithmic relationship. Consult your method books for details. The sums of the visible fragments will not always correlate exactly to what should be. Very large fragments have difficulty in entering the gel at all, whereas very small ones just slide out. Also note that the loading dye itself may quench the fluorescence of the bands.*

If the enzyme does not cut at all, the output is to emphasize this fact.

18.9 Predicting problems with star activity

Micro-Willy is clueless[7]. He has done a digest with one of the predicted enzymes which should, in theory, yield one single band. However, there are three of them! When looking at the table above, he realizes that two of them, the smaller pieces of DNA, do not fit into the expected scheme. Scratching his head, he begins to realize that it was probably not a particularly smart idea to digest 10 μg of DNA with 100 units of *Uff*I overnight[8]...

The matter is that some restriction enzymes[9] are not absolutely site-specific, at least *in vitro*, but may recognize and cut, albeit at a low rate, "degenerate" sites which differ from the canonical site at one nucleotide.

Write the functions required to find such secondary sites for a given enzyme.

Now you have the basics of the system. There are two more aspects to be addressed. I expect that you will by now be able to find a good approach for yourself.

18.10 Profit motives – motive profit

It has been suspected that the fubarase is controlled by orangylation, but nobody has ever examined this thoroughly (not even Dr. Meerkatz himself), as the proof is quite difficult. Luckily, the last edition of KNOWLEDGE contains a communication by Bonobeau and Le Mur where the general motif ACxDC is shown to be the crucial motive for orangylation, which occurs at the aspartic acid residue within the sequence. Micro-Willy wants to generate mutations in which some or all of these sites have been deleted in order to do some activity tests on the mutants.

Write the function required to find the orangylation motives and correlate them with the restriction pattern. What advice could you give to Micro-Willy

---

[7] As usual.

[8] Under standard conditions, one unit of enzyme will digest 1 μg of DNA in 1 hour, so this was at least an "eightyfold overdigest". Bang the head that does not bang.

[9] In real life, *Kpn*I and *Pst*I seem to be especially inclined to such misdemeanour, but this is probably largely a matter of buffer composition.

concerning the planning of his experiments? Do you think he will have to do many PCRs?

18.11 A glance at transmembrane regions

We are going to deal with the KYTE–DOOLITTLE plot for depicting the hydrophobicity of a protein, in the second part of this course. Here we are only going to do the most primitive pre-screening to suggest whether the protein contains a putative transmembrane region.

As we all know, transmembrane regions consist of a run of 15–30 extremely hydrophobic residues: Val, Leu, Ile, Ala and Phe, among which no more than one really hydrophilic residue may be interspersed. If such a hydrophilic residue exists (usually a lone Ser), this is strongly indicative that the protein may dimerize (pairing of two transmembrane regions can cloak the energetically unfavourable hydrophilic residue[10] and thus change its state of activity). Very often this hydrophobic region is enclosed between two strongly hydrophilic regions which may not enter the membrane and thus hold the protein in place. On the extracellular site, this is less obvious because there glycosylation is often used, but on the intracellular side, we often find a run of 5–7 arginines and lysines[11].

Try to assess this. Write a function that is able to judge whether there is a putative transmembrane region contained within the protein and, if so, where it is located and whether it may be involved in dimerization of the fubarase.

18.12 Addressing a project

The general pattern for adressing such problems is as follows.

(1) Analyze the problem.
(2) Reduce it to its components.
(3) Try to find solutions for the components.
(4) Test them.
(5) Write the program.
(6) Test it.
(7) Document it.

---

[10] It also works the other way: "Leucine zippers" lead to dimerization of proteins in the aqueous phase.

[11] In my personal notes, I used to refer to this as a "tsuba" – the small handguard of the Japanese two-handed sword whose main purpose is to prevent the right hand from slipping off the handle and getting hurt by the blade.

Good luck!

CHAPTER 19

Advanced Techniques in Python

19.1 Escape processing and prettyprinting

One of the more intelligent features of C, which has diffused into most modern programming languages including PYTHON, is the way special characters are dealt with. It is completely based on the \ sign, also known as "backslash". In fact, PYTHON's runtime system relies heavily on the C libraries, so these two languages may be discussed together here.

A backslash means that the next character is to be treated differently. In programmers' slang, this is referred to as "escape processing". These are the possible combinations:

Escape code	ASCII code (decimal)	Name	Meaning
\n	13	Line feed	(see below)
\r	10	Carriage return	(see below)
\f	12	Form feed	eject paper, begin new page
\t	9	Tabulator	advance cursor to next column
\07	7	Beep	ring the bell
\"	32	Quote	place a " within normal text
\\	92	Backslash	
\0n	0n	any ASCII character (octal)	
\0xn	0xn	any ASCII character (hex)	

Line feed and carriage return. This is one of the most annoying system differences. Unix systems use a single \n for the end of a line of text, pre-OSX Macs a single \r, and MS-DOS[TM]/Windows[TM] boxes a combined \r\n sequence, the first char of which, when printed to the console, sets the cursor back to the beginning of the current line (X = 0), the second moves it one line downward (Y += 1). For this reason, we have to differentiate between "binary mode" and "text mode" when opening any file and also when transferring files across the network, e.g., via FTP. For printing, the \n is sufficient, as PYTHON comes from the Unix world. On Windows[TM] systems, the interpreter will do what is appropriate to be done in this case.

<small>Line feed and carriage return</small>

So you can print a table quite easily in one line:

```
print "Name\tFunction\nPeinemann\tpresident\nMbumba\ttreasurer\nFlaig\tbully"
```

will display

```
Name            Function
Peinemann       president
Mbumba          treasurer
Flaig           bully
```

upon the screen.

At the end of each "print" statement, PYTHON implicitly adds a line feed. This can be suppressed by adding a comma to the print statement:

```
print "Anna"
print "the hannah"
```

will yield

```
Anna
the hannah
```

whereas

```
print "Anna",
print "the hannah"
```

will give us

```
Anna the hannah
```

Note that the comma will always lead to insertion of a space! Thus,

```
print "According to our treasurer,",
print name, "owes", money, "EUR to the club."
```

gives the output

```
According to our treasurer, Hermann Holzbock owes 45.778 EUR to the club.
```

This can be undesired:

```
print "Meeting:", day, ".", month, ". at", hours, ".", min, ". At the serpentariate."
```

prints out

```
Meeting: 13 . 10 . at 19 . 15 . At the serpentariate.
```

In this case, there are two possibilities:

- Concatenation of the output strings with +. This is similar to JAVA, where the printing functions are ordinary procedures which accept only single parameters. However, in JAVA any object will automatically be

19.1 ESCAPE PROCESSING AND PRETTYPRINTING

marshalled into a string while in PYTHON, you *must* convert it to a string before! Thus,

```
print "Meeting:", day, "."+month+". at", hours+"."+min+". At the serpentariate."
```

will generate something you are not going to like:

```
Traceback (most recent call last):
  File "<stdin>", line 1, in ?
Type error: cannot concatenate 'str' and 'int' objects
```

You must write instead:

```
print "Next meeting:", str(day)+"."+str(month)+". at",
print str(hours)+"."+str(min)+". At the serpentariate."
```

to get:

```
Next meeting: 13.10. at 19.15. At the serpentariate.
```

- Use formatted output, as discussed below.

There is another very clever thing borrowed from C. In traditional C, the only method of console output is by means of a function named `printf`. This function takes a variable number of arguments, the first of which must be a so-called *format string*. This format string may contain escape characters and also placeholders for variables, which the following parameters will be sorted into, formatting them accordingly:

```
printf( "Dear %s, our treasurer regrets to report", firstname );
printf( "that you owe \n\t%3.2f EUR.\n", debt );
printf( "We'll meet again on %d.%d.\n", day, time );
```

will produce the following output:

```
Dear Hermann, our treasurer regrets to report that you owe
        45.78 EUR.
We'll meet again on 13.10.
```

Important placeholders are:

- **%d** : %d
  insert an integer variable which is printed in decimal notation.
- **%3d** : %3d
  insert an integer variable which is printed in decimal notation formatted to a field length of 3.
- **%03d** : %03d
  insert an integer variable which is printed in decimal notation formatted to a field length of 3 with leading zeros.
- **%x** : %x
  insert an integer variable which is printed in hexdecimal notation.
- **%o** : %o
  insert an integer variable which is printed in octal notation.
- **%f** : %f
  insert a floating-point variable.
- **%3.3f** : %3.3f
  insert an integer variable which is formatted to 3 digits before and 3 behind the decimal point.

- **%03.03f :**
  same with additional leading and trailing zeroes.
- **%s :**
  insert a string.
- **%12s :**
  insert a string formatted with leading spaces to fill a field of 12 characters.
- **%c :**
  insert a character.

PYTHON uses exactly the same tokens[1]. The only difference is that this is also done using the normal "print" command, hence the relation between format string and arguments must be clarified by inserting a % in between:

```
print "Meeting: %d.%d. at %d.%d at the serpentariate." % (day, month, hours, min)
```

prints out

```
Meeting: 13.10. at 19.15 at the serpentariate.
```

All these may be combined, of course:

```
print "Non-member %10s %c. lives at address\t0x%08x" % (fname, lname[0], addr)
```

will yield:

```
Non-member      Ronny R. lives at address            0x0000029c
```

(Normally you would use a form like 0x%08x for the dumping of pointers or other *memory* addresses – in other words, when you are interfacing PYTHON with C.)

A very nice feature is that formatting strings may be used independently of print:

```
>>>t = "%02d:%02d" % (3,5)
>>>t
'03:05'
```

This has been used in our implementation of the LZW algorithm and will become important with the `Curses` module (see below).

You will have noted that the second argument to the % operator is a tuple. This is not just a formalism. Indeed tuples may be passed directly to the formatting function.

```
reptiles = [("Anna","hannah",16,2,"Shakespeare"),
("Bodo","boa",18,5,"Molière"), ("Sammy","slowworm",1,6,"Pig Brother")]
```

---

[1] With the exception that C has a few more specifiers for data types which do not exist in PYTHON, e.g., %ud for unsigned integers.

```
form = "%s the %s is %d ft %d in long and likes %s."

for r in reptiles:  print form % r
```

This will yield:

```
Anna the hannah is 16 ft 2 in long and likes Shakespeare.
Bodo the boa is 18 ft 5 in long and likes Molière.
Sammy the slowworm is 1 ft 6 in long and likes Pig Brother.
```

The origins of the formatting string are as obscure as the invention of the "foo bar". LISP had it, it appears, well before C, using the tilde instead of the percentage sign, and of all the more modern languages, only ERLANG has followed LISP (because in ERLANG the percentage sign denotes a single-line comment). O'CAML has the `Printf.printf`, and even many BASIC dialects know the PRINT USING feature. The Algol languages alone have successfully refused this feature so far.

Variants of `printf` are `fprintf` which prints the output into a file, and `sprintf` which deposes the result into a string variable. Thus,

```
char announ[ 80 ];
sprintf( announ, "Meeting: %d.%d. at %d.%d", day, month, hours,
  min );
```

will cause the string *announ* to be filled with

```
Meeting: 13.10. at 19.15
```

These variants are implemented in some other languages, but not in PYTHON as they are simply not needed, thanks to the greater flexibility of the % operator.

19.2 Regular expressions

In C, the `printf` procedure has a counterpart named `scanf` (including `fscanf` and `sscanf`). It takes a formatting string and pointer to variables:

```
char *anything = "08.15";
float valu;
sscanf( anything, "%f", &valu );
```

extracts the float value from the string and assigns 8.15 to *valu*.

However, when compared to `printf`, this is strangely primitive. The following will not work:

```
char *anno = "Meeting: 13.10. at 19.15", action[ 20 ];
int day, month, hour, minut;
sscanf( anno, "%s: %d.%d at %d.%d", action, &day, &month, &hour, &minut );
```

What happens is that `scanf` first detects a %s in the format string and tries to read a string. As `scanf` is *greedy*, it will not stop at the colon but continue until the final null byte has been read. So there will be no further input to scan into the four integer variables[2].

Why are there no alternatives? This is what one Larry Wall wondered about. It inspired him to create PERL, a language which is very controversial[3], but everybody agrees that it introduced something really fantastic: regular expressions. Regular expressions are exactly that kind of formatting description which `scanf` lacks. At first glance, one sees that they are much more complicated than printf descriptions, mainly because there are many more ways of dealing with input than with output.

PYTHON also comprises regular expressions. They are implemented in the module `re`; for a description, please see file `lib/module-re.html` in your PYTHON documentation. It is too much material, and requires too many details and too few concepts into the bargain, to repeat it here. Guido van Rossum personally recommends: *"Mastering Regular Expressions* a book on regular expressions by Jeffrey Friedl, published by O'Reilly. The PYTHON material in this book dates from before the `re` module, but it covers writing good regular expression patterns in great detail."

The essential functions for extraction of values are `re.match` and `re.group`. The following example illustrates the power of regular expressions, although it will be completely unintelligible without a good look at the `re` documentation[4]:

```
>>>import re
>>>def annalyze( source ):
...     pattern = "(?P<maction>.*?):\s*?(?P<mday>\d*?).\s*?(?P<mmonth>\d*?)"
...     pattern += ".\s*?at\s*(?P<mhour>\d*)\D*(?P<mminute>\d*)"
..      cmp = re.match(pattern, source)
...     print "Action=", cmp.group("maction")
...     print "Day, month=", cmp.group("mday"), ",", cmp.group("mmonth")
...     print "Hour, minute=", cmp.group("mhour"), ",", cmp.group("mminute")
...
>>>annalyze( "Meeting: 13. 10. at 19 h 15" )
```

---

[2] This is so common an error about `scanf` that it is not surprising to see that this has slipped into Section 4.2.6 of the official PYTHON documentation. The statement "To extract the filename and numbers from a string like `/usr/sbin/sendmail - 0 errors, 4 warnings` you would use a scanf() format like `%s - %d errors, %d warnings`; the equivalent regular expression would be `([^\s]+) - (\d+) errors, (\d+) warnings`" is plainly wrong as far as the format string is concerned.

[3] Taste is a matter of taste. This holds true even if you have none.

[4] Actually I suggest that you print these pages separately to have them ready at hand.

```
Action= Meeting
Day, month= 13 , 10
Hour, minute= 19 , 15
>>>annalyze( "Orgy and chess:7.5.at 3:20 at Tiffany's" )
Action= Orgy and chess
Day, month= 7 , 5
Hour, minute= 3 , 20
```

19.3 File handling

All direct file access is done using the built-in module `file`. Its functions are always available. Do not try to `import file` as this will cause a runtime error. There is also the `pickle` module for converting objects into a notation suitable for storage and back. This format is not intended for human use!

Guido van Rossum defines the purpose and contents of the `pickle` module as follows:

> The pickle module implements a fundamental, but powerful algorithm for serializing and de-serializing a Python object structure. "Pickling" is the process whereby a Python object hierarchy is converted into a byte stream, and "unpickling" is the inverse operation, whereby a byte stream is converted back into an object hierarchy. Pickling (and unpickling) is alternatively known as "serialization", "marshalling," or "flattening", however the preferred term used here is "pickling" and "unpickling" to avoid confusion.

There are two modules to this end – `pickle`, implemented in PYTHON itself, and `cPickle`, whose implementation is in C, which leads to greater speed but some restrictions concerning the PYTHON class system.

19.4 More file handling: persistent dictionaries

We have discussed dictionaries before. The module `shelve` implements a very convenient mechanism for storing objects in a file. Basically, any such file consists of a sequence of pairs. The one component of each pair is a string that serves as a key, the other may be any legal PYTHON value. By applying the key, the associated value may be retrieved.

As this is stored in a file, the shelf is persistent and may even be transferred to other computers (provided they also run PYTHON with the `shelve` module). The low-level data format used for storing information on disk does not matter.

♠ *Homework.* Study the documentation on the "pickles" module and consider how the functionality contained therein might be employed in any of our previous examples.

19.5 String handling

Every PYTHON string is – implicitly – a string object, an instantiation of the *string* class. So the following methods are available with any string variable or constant, but you may also import the "string" module and call them as functions.

♠ *Load the "string" module into your PYTHON interpreter, consult the help function to find out which functionality is included and fiddle around with the string methods provided. Then write your own documentation with regard to the variables and functions contained therein. (N.B. Do not forget the variables. On many occasions, they are as useful as the functions.) Next, try to design your own implementations of some of the more interesting functions. Is the "string" module comprehensive? Which further functions would you like to see implemented? Devise a derived class "stringplus" which comprises this additional functionality.*

19.6 Raising exceptions

Sometimes it is useful to cause some routine to cause an error condition of its own, which can then be caught by a try:/except: mechanism somewhere else[5].

```
def oh_no( n ):
    if n < 0: raise RuntimeError, "shit happens"
    print n
    oh_no( n - 1 )

def oh_yeah():
    print "Begin."
    try: oh_no( 3 )
    except RuntimeError, e: print e
    print "Completed."

oh_yeah()
```

This will produce the following output:

```
[ophis@zacalbalam:~] python oh_no.py
Begin.
3
2
1
0
shit happens
Completed.
```

---

[5] This is the only kind of "long-range signalling" available in PYTHON, and it should be used with care and restraint.

19.7 "Curses": the functions for advanced text output

Unix was originally based on the client/server architecture, and this is still reflected in the way higher level I/O is done. As PYTHON comes from the Unix world, this is a part of the normal functionality which must be emulated somehow on other systems.

As the X11 windowing system runs on a "server" and opens its windows on a "client" which may or may not be identical to the server, so the Curses module runs on a "server" and communicates with a text-based "client". This may be another text console (at startup, Linux creates no less than six of them...), a terminal program such as *xterm* or *eterm*[6] or another computer connected via Telnet – this makes little difference. There will always be an official baud rate for the connection; even where this can be assumed to be extremely high – when the "client" is in fact the same computer – still the mechanism is that output happens in two steps, as it was sensible to do with low-baud terminals.

(1) Send your data to the buffer connected with the terminal.
(2) Only when all changes have been done, force a redraw of the screen (because this may be slow).

In PYTHON, such "virtual devices" are implemented as objects, of course. They are provided by the module `curses`.

Of a multitude of functions, the following four are by far the most important:

- **vc = curses.initscr():**
  opens a virtual console for output
- **curses.endwin():**
  shuts down the virtual console system
- **vc.addstr(y, x, textstring):**
  writes a text string to a defined position of the console (invisibly at first)
- **vc.refresh():**
  causes changes to become visible

vc = curses.initscr()
curses.endwin()
vc.addstr(y, x, textstring)
vc.refresh()

Use format strings to knock the output into proper shape before entrusting *addstr* with it.

This is how it works:

```
import curses, time
w = curses.initscr() # open curses terminal
w.addstr(3,3,"Of all the things I lost,")
w.addstr(5,12,"I miss my brain the most.")
w.refresh() # cause curses output to be displayed on the console
```

---

[6] Inspired by the file manager Konqueror, many KDE programs have little terminal boxes inside their normal user interface. The Curses library accepts these as well.

```
t0=time.clock()
while (time.clock()-t0)<3.0: pass # wait 3 sec
curses.endwin() # close curses terminal
```

Try it.

19.8 Module "os": the interface to all operating systems

os.system

PYTHON comprises a module aptly named *os*, which provides a standarized access to most of the functionality of the underlying operating system, independent of its make. Probably the most interesting function here is the *os.system* function, briefly mentioned before (p. 68). Basically, it takes a line of text that is executed by the shell as if it were a command line, with arguments, redirection and all the goodies possible. Thus, PYTHON scripts may use *os.system* to invoke each other, or non-PYTHON programs.

19.9 Utile et iucundum

Some other modules are so useful that their existence should be mentioned explicitly.

- *time* provides access to timer and timing functions.
- *poplib* contains the functionality required for downloading mail from a POP account.
- *smtplib* contains the functionality required for sending mail via an SMTP account.
- *zlib* offers compression of data using the famous *zip* algorithm as well as checksum generation for verifying data integrity.
- *xml**: these modules contain a complete XML parser.
- *math*: higher mathematical functions.

You will find their descriptions in your documentation.

Any teaching that can be put in a nutshell belongs there.

CHAPTER 20

♣ The Third Project: Python goes PCR

Yeah, Doctor Stein grows funny creatures and lets them run into the night;
They're now becoming politicians when the time is right.
Sometimes when he's feeling bored he's calling it a day;
He's got his computers and they do it their own way.
They mix some DNA, oil and a certain spray, you can watch it on the screen,
And the fellow's yellow and black, or sometimes red and green.

– Helloween: "Dr. Stein"

20.1 Monkey business

Willy decides to produce a number of fusion proteins where the C-terminus of the fubarase is linked to *something else*. Alas, he is not quite sure yet what this "something" will be. Fluorescent proteins, probably; but what is the best choice: blue, cyan, green, yellow or red FP? And maybe it would be nice to have the option for a Six-His-Tag for purification... or for an isoleucine zipper for enforced oligomerization... not to mention several other components of signalling cascades.

♡ *Do you know anything about the fluorescent proteins from Æquorea victoria? What other "reporter genes" are you familiar with?*

So the best strategy is obviously to design a "platform": build a derivative of the fubarase gene which has two unique and generally rarely cutting restriction sites at the 3' end of the ORF, just upstream of the stop codon; then add suitable restriction sites to your extension fragment by PCR. In general, useful gadgets are below 1 kbp in size, so they can be PCRed without major problems.

As for the restriction enzymes, it is not difficult to make a choice. There are only a few eightcutters, and of these, *Asc*I and *Not*I are not only by far the cheapest but also have the nice property of cutting sequences consisting exclusively of Gs and Cs, which are the least frequent ones in mammalian DNA. So it can be safely assumed that virtually any other gene may be cloned into these sites – provided there are no *Asc*I and *Not*I sites in the fubarase, which we have to make sure of first, of course. What we will also have to make sure of is that the reading frames of the fubarase and the extension match.

Bioinformatics Programming in Python. Rüdiger-Marcus Flaig
Copyright © 2008 WILEY-VCH Verlag GmbH & Co. KGaA, Weinheim
ISBN: 978-3-527-32094-3

20.2 Polymerase chain reaction: the boiling hell of Dr. Mullis

20.2.1 The polymerase chain reaction. The PCR (polymerase chain reaction) has been mentioned several times before. Briefly, this is a procedure for amplifying a piece of DNA between two defined sequences. In doing so, these sequences themselves may be modified or extended.

The process, developed by one Kary Mullis, is based on the fact that the DNA consists of two complementary strands which have a physical affinity to each other. G and C may form three hydrogen bonds with each other, A and T two. So above a certain length, two complementary strands will stick together up to a temperature depending on length and composition[1].

Another important premise is that DNA polymerases do not do their work *de novo* but only extend existing duplexes in the 5'→3' direction.

Having mixed the following ingredients:

- the so-called "template DNA" comprising (among other parts) the sequence to be amplified
- two short (15...30) synthetic oligonucleotides corresponding to the ends of the sequence of interest
- desoynucleotide triphosphates for building new DNA
- a thermostable DNA polymerase

we may proceed as follows.

(1) By heating the mixture to almost 100°C, the strands of the template DNA are separated.
(2) By cooling to approximately 60°C, a re-annealing is made possible. Here the synthetic oligonucleotides, by virtue of their greater abundance, have an advantage and bind faster.
(3) By warming again to 75°C, we set the polymerase into motion. It begins to extend the short duplexes and thus produces long primary transcripts which begin at one of the oligonucleotide sequences but still have an undefined end.
(4) This is heated again to 100°C.
(5) When cooling again to 60°C, the same thing happens as before. However, the oligos may now also anneal to the primary transcripts.
(6) At 75°C, the polymerase now produces either primary transcripts (where oligos have annealed to the template) or secondary ones (where they are bound to primary transcripts). The secondary transcripts extend exactly from oligonucleotide sequence to oligonucleotide sequence.

---

[1] Actually, supercoiling – the "twistedness" of the DNA – also plays an important role. Thermophilic archaea are able to keep their DNA in working order above 100°C only by rigorously supercoiling it. However, plasmids extracted from *E. coli* are not that strongly supercoiled, and linear DNA has no supercoiling, of course, so this parameter may be neglected.

Production of primary transcripts occurs at a linear rate, whereas secondary transcripts accumulate exponentially (since they may serve as templates themselves). So after some 30 cycles, per molecule of template we have about 30 primary transcripts but approximately 1 billion secondary transcripts (or as many as polymerization speed, oligo and desoxynucleotide supply will allow).

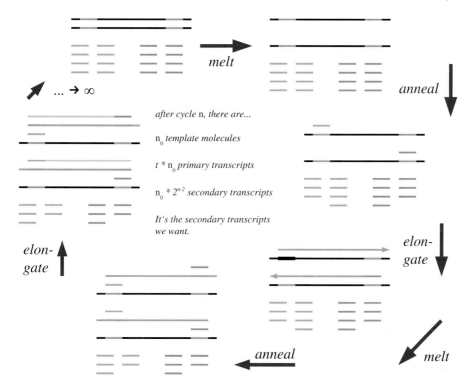

after cycle n, there are...

n_0 template molecules

$t * n_0$ primary transcripts

$n_0 * 2^{n-2}$ secondary transcripts

It's the secondary transcripts we want.

Like so many strokes of genius, this is conceptionally simple but difficult in detail. In particular, things tend to become more difficult with longer DNA stretches; using specialized polymerases and protocols, up to 30 kbp of DNA are reported to have been amplified successfully. However, most PCRing is done for pieces beneath 2 kbp.

20.2.2 Modifying the ends. What is the point of stubbornly amplifying DNA segments? Obviously, there are few applications for this[2]. However, the 5' ends of the oligonucleotides do not have to fit exactly (the 3' ends, where elongations begin, have to). Thus, we may add *things* to both ends of the DNA sequence – for example, new restriction sites, tags for purification (Six-His or hemagglutinine), organelle targeting or secretion sequences, ... The limiting factor is the maximum length of DNA that can be synthesized error-free and at a rational price – below 100 bp.

---

[2] Forensic "DNA detection" is among them.

Appending restriction sites is especially widely used, since this allows us to "whet" the ends so they fit into a given plasmid. Basically any restriction enzyme may be used which does not cut within the amplified sequence itself. However, some of them are cranky about cutting at the ends, so it is highly recommended to add not only the restriction sequence itself but a few base pairs more (their sequence does not matter as they are going to be removed anyway). A run of three Gs has proven effective in most cases.

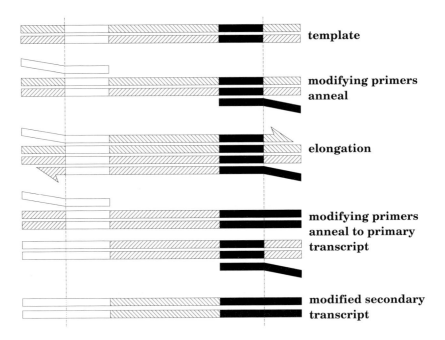

As this figure illustrates, modifying the ends of the primers may even improve the PCR as the primers bind more easily to secondary transcripts than to primary ones.

20.2.3 Rules of engagement. Chemical parameters (pH, salt, etc.) are generally optimized by the manufacturers of PCR enzymes, and concentrations are rather robust parameters (template concentrations, for example, may be varied over a range of more than 5 magnitudes with little effect on the outcome).

The most important chemical factor is [Mg^{++}], the effect of which is not quite clear; some workers report that a higher magnesium concentration increases specificity and amplification fidelity, whereas others have found the opposite. It is, however, absolutely necessary as a cofactor for all polymerases, and chelators such as EDTA are deadly to any PCR. Unfortunately, most degrading enzymes are also dependent on Mg^{++}, and therefore the ideal medium for long-term storage of DNA contains EDTA[3].

Temperatures and time, however, partly remain to be worked out.

	Time	Temperature
Melting	30"	95°C
Annealing	30"	*variable*
Elongation	1' / 1000 bp	74°C

It is never a good idea to use oligonucleotides ("primers") with widely differing annealing temperatures if you can help it.

Keep this in mind.

- **Higher working temperature:** more specificity
- **Lower working temperature:** more fault-tolerance.

Higher working temperature

Lower working temperature

♡ *Become familiar with terms like "hotstart PCR" and "touchdown PCR". This is not a course in molecular biology – therefore this is only a hint and not a* ♠ *– but you ought to know what molecular biology is about.*

In a modifying PCR, we actually have to consider two pairs of annealing temperatures; the one for the annealing of the primers to the template and the primary transcript; the other for the annealing of the primers to the secondary transcript. The latter will, of course, be higher – sometimes considerably. As the specificity of this part of the reaction is also increased, we may disregard this and focus on the annealing of the primers to the template.

In this reaction only the non-modifying section of the primers take part in the initial binding. So the melting temperature has to be designed with respect to these, not to the full primer sequence.

♠ *Explain this.*

20.3 Determining the melting point of DNA

Micro-Willy has a very straightforward way of determining the "melting point" of DNA – the temperature where the strands fall apart. He simply causes his

---

[3] In every lab working with DNA we find a buffer named "1xTE" – 10 mM Tris as a buffer (to avoid autocatalytic hydrolysis) and 1 mM EDTA for the inactivation of DNAses. Common procedures of DNA extraction such as the Birnboim–Doly preparation sometimes fail to completely inactivate the native DNAses of the bacteria. However, by now low-DNAs strains and advanced preparation methods including protein denaturation by guanidine hydrochloride and silica column chromatography have largely obviated the need for 1xTE.

EMBOSS

hiwi[4] to do it for him. The square-minded *hiwi* in turn starts the `dan` program from the *EMBOSS* package[5] (http://www.emboss.org), or he logs into the WWW pages of some supplier and uses their functions, never questioning their methods, treating their output as a revelation by some higher intelligence.

By `dan -help` we get the parameters of this program:

```
ophis@cirith-ungol:~> dan -help
  Mandatory qualifiers (* if not always prompted):
  [-sequence]          seqall     Sequence database USA
   -windowsize         integer    The values of melting point and other
                                  thermodynamic properties of the sequence are
                                  determined by taking a short length of
                                  sequence known as a window and determining
                                  the properties of the sequence in that
                                  window. The window is incrementally moved
                                  along the sequence with the properties being
                                  calculated at each new position.
   -shiftincrement     integer    This is the amount by which the window is
                                  moved at each increment in order to find the
                                  melting point and other properties along
                                  the sequence.
   -dnaconc            float      Enter DNA concentration (nM)
   -saltconc           float      Enter salt concentration (mM)
*  -formamide          float      This specifies the percent formamide to be
                                  used in calculations (it is ignored unless
                                  -product is used).
*  -mismatch           float      This specifies the percent mismatch to be
                                  used in calculations (it is ignored unless
                                  -product is used).
*  -prodlen            integer    This specifies the product length to be used
                                  in calculations (it is ignored unless
                                  -product is used).
*  -mintemp            float      Enter a minimum value for the temperature
                                  scale (y-axis) of the plot.
*  -graph              xygraph    Graph type
*  -outfile            report     If a plot is not being produced then data on
                                  the melting point etc. in each window along
                                  the sequence is output to the file.

  Optional qualifiers (* if not always prompted):
*  -temperature        float      If -thermo has been specified then this
                                  specifies the temperature at which to
                                  calculate the DeltaG, DeltaH and DeltaS
                                  values.

  Advanced qualifiers:
   -rna                boolean    This specifies that the sequence is an RNA
                                  sequnce and not a DNA sequence.
   -product            boolean    This prompts for percent formamide, percent
                                  of mismatches allowed and product length.
   -thermo             boolean    Output the DeltaG, DeltaH and DeltaS values
                                  of the sequence windows to the output data
                                  file.
   -plot               boolean    If this is not specified then the file of
                                  output data is produced, else a plot of the
                                  melting point along the sequence is
                                  produced.
```

---

[4] "Hilfswissenschaftler", which roughly translates to "specialist for everything which nobody else would do".

[5] This is a piece of Open Source software which represents a more recent version of the original GCG package. I suggest that you download and install it, provided your computer is capable of running it.

```
General qualifiers:
 -help              boolean     Report command line options. More
                                information on associated and general
                                qualifiers can be found with -help -verbose

ophis@cirith-ungol:~>
```

The implementation is an extremely ugly agglomerate of C code, some 500 lines in size, at the core of which we find a derivative of the formula given in "the Maniatis":

$$T = 81.5 + 16.6\log_{10} k + 0.41t - \frac{675}{l} - m$$

where k is [K$^+$] in mM, t the percentage of GCs in the sequence, l the length of the sequence and m the percentage of mismatches in the sequence. The **dan** program also takes into account the concentration of formamide, a potent inducer of DNA melting. Obviously, this was designed for Northern/Southern blotting rather than for PCR – nobody would use formamide in a PCR.

This formula was derived from the works of Bolton, McCarthy and Baldino, spanning more than two decades, and is not undisputed. Some researchers think that $\frac{500}{l}$ is actually closer to the truth than $\frac{675}{l}$. At any rate, the whole thing is a purely empirical affair without any foundation in theory. This reveals a rarely-thought-of bottleneck in bioinformatics.

♠ *Implement a program that reads a sequence file, extracts the DNA from it – you have done this before; simply re-use the code – and calculates the melting temperature of this strand.*

♡ *Note that this equation is valid only for "regular" DNA, that is to say, from 30 to 70% DNA and for sequences between 10 and 100 bp. Extremely GC-rich sequences may have an actual melting temperature which is higher than the predicted one. At any rate, do not expect overmuch from these calculations. They are largely a rule-of-thumb matter, and different implementations use slightly different formulas.*

♡ *Do you know what Northern, Southern and Western blotting are?*

20.4 ...and still my PCR doesn't work yet!

For the sequence of the fubarase gene, please turn back to p. 324.

The start and stop codons are easy to locate:

```
CTAAGGATCC GAGGATGCGG AAGGTCGGAA CCGAACTCAT
(...)
GGTTACTCAA TTGCGGTCCG TGGGGTGATA CTCTGTTGAC AGGCTCCATA AAAAAAAAAA
AAAAAAAAAA GCTTGTAATC
```

The upstream region is going to be left alone. It is ideal for working with, as it contains a *Bam*HI site. We simply use the same bases as primer: CTAAG-GATCC GAGG*ATG*CGG AAGGTCGGAA GGT...

By using dan or a similar tool, we may estimate CTA AGG ATC CGA GGA TGC GGA AGG TCG GAA GGT to have a melting temperature of $67.0°C^6$. That's fine, because Willy has learned it "in America" that one should set the PCR machine to 2...3 K below the estimated value, and the antediluvian PCR machine in Willy's lab (which still uses punch cards) has a program preset to 65°C and 2' elongation time.

Now the other side. We may use all the bases down to the final TGA which must be omitted if we wish to build a fusion protein. The sequence GCG GTC CGT GGG GTG ATA CTC TGT (that is to say, its complement) has a melting temperature of 67.1°C. Splendid. So we design the following extension:

```
5'-GCG GTC CGT GGG GTG ATA CTC TGT-CCg cgg ccg cgg gcc cag gcg cgc ccc
3'-CGC CAG GCA CCC CAC TAT GAG ACA-GGc gcc ggc gcc cgg gtc cgc gcg ggg
         +---------- overlap -----------+   +- NotI --++-SmaI-+ +-- AscI -+
                                             P   R   P   R   A   Q   A   R
```

*Not*I and *Asc*I have been explained before; it is advisable to place *Asc*I on the distal site because *Not*I is somewhat squeamish when it comes to cutting near the end. In addition, we have inserted a *Sma*I site for ligation to blunt ends, if this should be desirable, and also to keep the other two sites a bit apart. The translation looks quite nice too – hydrophilic and with two helix-breaking prolines.

We will also have to make sure that there is no homology between the two primers.

♡ *It might be a good idea here to become familiar with the 20 natural amino acids, their codes and their essential properties, as well as with the structure of the genetic code.*

So the PCR is eventually run:

$$\begin{pmatrix} 94°C & 30'' \\ 65°C & 30'' \\ 75°C & 120'' \end{pmatrix}_{30}$$

and put to the gel. What's the result, clearly visible on a 1% agarose gel?

Nothing.

Upon repetition of the experiment, strange things happen. Now and then there appears a pale band of the correct size, but obviously of very low quantity; it is not possible to purify and clone this. What is worse, this seems to be something

---

[6] At $[K^+]$ = 100 mM, as recommended by New England Biolabs for their polymerases.

stochastic. The appearance of this band does not fit into any logical pattern, and repeating a successful synthesis with exactly the same parameters generally fails. Addition of more Mg^{++} and fiddling around with the reaction time have no effect, neither has using the proofreading *Pfu* polymerase instead of the cheaper *Taq* standard[7].

What is really eerie is the total lack of reproducibility, both in the positive and in the negative sense. It almost seems as if there were something at work bent on defying logic and reason.

♡ *At this point, take a break, especially if you are a life scientist, and imagine yourself in poor Micro-Willy's shoes. What would you do?*

At closer scrutiny, *something* seems to have happened. In some of the samples, there is an amorphous smear between the little cloud of the primers and the place where the product appears, if it appears. It seems there has been some DNA synthesis, but pretty much out of control. So the notion that there was some chemical inhibitor of the polymerase (e.g., clandestine EDTA in one of the buffers of the kit used for isolation of the plasmids) can be discarded.

A bit of experience in the handling of DNA teaches us that in all probability this is single-stranded DNA – in other words, a primary transcript. So the polymerase must have worked, at least partially, but not managed to get the full PCR done. In other words, it seems that only one of the two primers has been elongated.

♡ *Become acquainted with the principles of electrophoretic separation. Why is it much easier to get clear separation of double-stranded nucleic acids than of single-stranded ones? How do proteins fare by comparison? What measures would you suggest if you could not help having to separate single-stranded DNA on a gel?*

Casus clarus, Willy thinks. Something is wrong with the other primer – it does not anneal properly. So let's try lower annealing temperatures until we get the other one bound too. This will be at the cost of specificity, but this is of minor importance here. We are not doing an amplification of a single stretch of genomic DNA to identify a criminal, but we are trying to copy a part of a purified plasmid.

Annealing temperature	Outcome
62°C	No product.
59°C	Pale band of the correct size.
56°C	No product.
53°C	Something of about 70 bp, hardly to be discerned from the oligos.
51°C	No product.
48°C	Blondie interrupts the run because she suspects an error in the settings.

---

[7] Ordering a new pair of primers with the same sequence in case synthesis *and* QA at the supplier's have failed at the same time also does not change anything.

Willy is flabbergasted. The thing at 53°C is really weird. There is no explanation for this either. After organizing some chalk and carefully drawing several pentagrams on the threshold of the lab door and around the PCR machine[8], he tries to make something out of that. Of course, any attempt to reproduce this fails, but now and then the 70 bp band turns up at other temperatures. Still no pattern visible.

♡ *Well, what's your opinion on this 70 bp band?*

20.5 Is there a hairpin structure?

At this point, something snaps in Willy's little mind. What if the problem is caused by self-annealing or hairpin formation of one of the two primers? If two primers of some 50 bases contain a sequence of approximately 30 bases that allows them to bind to each other, it is most likely that we may end up with a staggered duplex which the polymerase will gladly extend into a fully double-stranded DNA of approximately 70 bp length. As the 3' end is vital for a PCR, these short dsDNA pieces will no longer be available.

So maybe this is the solution? It is easy to understand that the initial binding to template or primary transcript is more difficult than that to the secondary transcript, where the primer can bind with its full length. In other words, there may be a kind of competition going on between the proper PCR and the "trash reactions", and it is mostly a matter of chance which one dominates. Therefore results may not be predicted.

♠ *Explain why the binding to the template is more critical.*

What are the prerequisites for formation of homoduplexes? As one partner has to be a mirror image of the other, there must be a longer palindromic sequence.

♠ *Explain this. How long will such a palindromic sequence have to be in order to cause trouble? What about the normal recognition sites for restriction enzymes, which are palindromic in their own right?*

Things may be even worse. When there is a palindromic sequence, the primer may muddle things up quite on its own by formation of a hairpin structure.

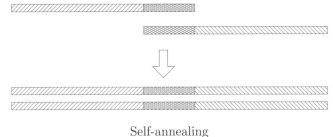

Self-annealing

---

[8] When all rational approaches have failed, feel free to try the irrational ones!

20.5 IS THERE A HAIRPIN STRUCTURE?

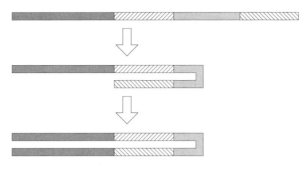

Hairpin formation

Lab folklore has the following formula for the thermal stability of DNA hairpins within a given sequence:

$$T = 81.5 + 0.41t - \frac{500}{l} - m$$

where t is the percentage of GCs in the sequence, l the length of the sequence and m the percentage of mismatches in the sequence.

Note that this is yet another approximation which does not do full justice to "stem-loop" or "panhandle" structures. In a sequence like GGGGGGGGGGAAAAAAAAAAAAAACCCCCCCCCC, the actual stability will be determined by the interaction between the G_{10} and the C_{10}, with the A_{14} folding itself quite comfortably into a "bubble" which will not have much influence on the stability of the structure.

Now **this is your task**.

Write a program that identifies potential self-annealing / hairpin structures, apply it to Willy's fubarase primers and consider the results. Make a suggestion on how to modify the experiment. Do the primers have to be redesigned, or will running the PCR at one certain temperature do it?

And this is how to do it.

Create a list of all the possible subsequences of the sequence, (e.g., of GATC, these are: GAT, GA, ATC, AT and TC). For each of these subsequences, calculate the thermal stability of the hairpin this sequence may form. It may be a good idea to form a tuple of the subsequence, with the hairpin-contributing nucleotides in capital and the other nucleotides in lowercase letters, its begin, end and melting temperature, and build a list from all these tuples. Then select tuples which are included within others. In the example above, if the melting temperature of AT is lower than that of GAT, of ATC or of GATC, it may be discarded. Then print the results in "human-legible" form.

20.6 Addressing a project

Refer to p. 334 for further tips.

"The Way is not about winning fights but about improving oneself. If I should ever leave the Way of self-discipline, I will be as good as dead, even if I should emerge victorious from all future encounters." – EIJI YOSHIKAWA

CHAPTER 21

The Wizards' Sabbath: A Gathering of Languages

> A mouse, taking her children for a walk, suddenly found herself facing a hungry cat. The mouse, however, immediately barked like a dog, and the cat fled. Turning to her children, the mouse said: "Now you see how important it is to learn foreign languages."
>
> – James Thurber

21.1 Considerable considerations

In this chapter, we will endeavour to make a comparison of a number of interesting modern programming languages based on very diferent paradigms.

The task is familiar from p. 243, where its relevance as a model problem has also been discussed, i.e., identify prime numbers by ascertaining for any number of interest whether it can be divided by any smaller prime number. We have compared this to the identification of open reading frames within nucleotide sequences in that both yield a previously unknown amount of data, which must be accommodated. However, in the case of prime numbers, some *assumptions* may be made, which can be used to modify the algorithm. Evidently, the number of primes found cannot be larger than the number of integers investigated. Thus, in the present case we may safely assume an upper limit. Depending on the programming languages, it may be preferable to rely on this assumption and use *static storage*, i.e., an array, rather than *dynamic storage*, i.e., a list.

<small>assumptions</small>

<small>static storage</small>
<small>dynamic storage</small>

♠ *Please turn back and make sure you have understood the basic principles. What is a list and what is an array? Which of these two forms is more efficient in terms of speed? What are the advantages of the other form? How does a stack fare by comparison, and to which of the two does it correspond?*

The most limiting factor in dynamic storage is the need to allocate and release memory upon demand. The operating system kernel has to keep track of the availability of any byte of RAM, for all processes at the same time, and also to make sure that none may trespass into any other's realm. Considering the complexity of modern multitasking operating systems, this requires a degree of administration that would scare Kafka. The positive aspect is that it is the top

---

Bioinformatics Programming in Python. Rüdiger-Marcus Flaig
Copyright © 2008 WILEY-VCH Verlag GmbH & Co. KGaA, Weinheim
ISBN: 978-3-527-32094-3

experts who are working on operating system kernels to provide you with the best solutions. Nevertheless, there are things that cannot be optimized away. At any rate, even the best solution will require finite time to complete. This time drops to zero if we do not use dynamic storage.

Apart from memory administration, access is an issue. Due to the structure of a list, the only way to access element n is to begin at the beginning and move from each member to its successor $n-1$ times; you will remember that a lists consists of elements, each of which comprises at least its value and a pointer to its successor, i.e., the address in memory where the next element is stored.

In an array, by contrast, the locaton of item n in memory is found simply by calculating $\alpha + n \cdot \sigma$, where α is the address of the first element and σ the size of each element. If we move on in linear fashion, we can even omit the multiplication and just add σ to the address of any given element to find its successor.

Thus, with static storage things may turn out as simple as the following piece of FOOBAR assembly code:

```
0001    ;;;; Calculation of prime numbers up to 30\,000
0002            load    0x0001  $03             ; R03 will be used for "1"
0003            load    0x0000  $02             ; and R02 for zero
0004            load    0x0003  $31             ; outer counter (I)
0005            load    0x7530  $32             ; max for outer loop: 30\,000
0006            load    0x0001  $37             ; primes found so far (HI)
0007            load    0xFFFF  $51             ; R51 is our pointer for array access
0008            load    0x0002  $55             ; our first prime is 2
0009            put     $55     $51             ; let list be [2], with R55 pointing to last
0010    :outerloop
0011            isgeq   $31     $32             ; is I>=max?
0012            jift    :outerend               ; Jump IF True: if so, quit outer loop
0013            load    0x0001  $40             ; inner loop counter (J)
0014            load    0xFFFF  $51             ; set R51 to begin of array
0015    :innerloop
0016            isgtr   $40     $37             ; is J>HI?
0017            jift    :innerend               ; Jump IF True: if so, quit inner loop
0018            get     $51     $52             ; list member into R52 for calculation
0019            mod     $31     $52     $28     ; R28 now contains modulo
0020            isequ   $28     $02             ; is it zero?
0021            jift    :noprime                ; Jump IF True: in this case break the loop
0022            add     $03     $40     $40     ; J=J+1
0023            sub     $51     $03     $51     ; decrease pointer for array access
0024            jump    :innerloop              ; go to "innerloop" anyway
0025    :innerend
0026            add     $03     $37     $37     ; HI=HI+1
0027            load    0xFFFF  $51             ; set R51 to begin of array
0028            sub     $51     $37     $55     ; index to newest one
0029            add     $55     $03     $55     ; we begin at zero...
0030            put     $31     $55             ; and save the new prime to the array
0031    :noprime
0032            add     $03     $31     $31     ; I=I+1
0033            jump    :outerloop              ; go to "outerloop" anyway
0034    :outerend
```

All the work is done in lines 0010–0034, which comprises a total of no more than 20 machine instructions, which might amount to a comparable number of clock cycles. Such code can be hand-crafted or generated by C and FORTRAN compilers.

However, the great drawback of static storage is that you have to allocate a single memory chunk of sufficient size, which must be available and is then no longer available for anything else until completion of the program. As discussed before, this may be problematic. Lists are much more economic in terms of memory, in particular if you cannot predict how much output will be generated.

We will now take a look at the following languages in addition to PYTHON, all of which should at least sound familiar to you by now.

- **Fortran**
  is not only the oldest of all programming languages and has a particular "tough guy" flair but is still widely used for engineering purposes and, since further revision is being scheduled up to FORTRAN-2008, is unlikely to be dethroned any time. Having undergone a considerable number of revisions during its fifty years of existence, its features for handling numerical data are still unsurpassed and in its more recent incarnations (in particular FORTRAN-95) supported by a full complement of modern structures, including object-orientation and functional programming (FORTRAN-90 and later being unique among imperative languages in that the keyword PURE allows one to restrict a function definition to genuine functional style). Above all, its whole array assignments and operations, such as the WHERE and FORALL statements and the ability to select in "slice" fashion any portion of a multi-dimensional array, have no equal (except perhaps in the esoteric field of dedicated array languages) and allow the compiler to generate *very* efficient code for "number crunching" and in particular to automatically produce code that takes advantage of multiprocessor hardware (p. 42). It is also capable, like C, of being combined with scripting languages such as PYTHON or RUBY. Recently, the FORTRAN–PYTHON combo has become fashionable for neuronal networks, with FORTRAN supplying the low-level infrastructure and PYTHON defining the high-level architecture of the net.
- **Oberon**
  is the grandchild of PASCAL and may thus be considered as the highest developmental step of the ALGOL line. Designed by no lesser personage than Nikolaus Wirth himself, it exemplifies imperative and object-oriented programming with outstanding clarity. For this reason, it is the second language taught at Oxford University (after the purely functional language HASKELL).
- **Lisp**
  was developed in the 1950s, but from the very beginning it incorporated such crucial concepts that it is still said that "those who do not study LISP are doomed to reinvent it – poorly." Some even claim that

LISP is defined by a certain cluster of features: head→tail lists, recursion, λ functions and the like, and any language possessing all of these is a dialect of LISP, which definition would classify all functional and declarative languages as LISP.

ERLANG

- **Erlang**
  is a functional language designed for real-life applications, in particular for distributed (multi-processor) computing, and (unlike LISP) characterized by clear and expressive syntax. Its unique blend of high-level data structures, built-in support for concurrency and excellent error handling combines with its system independence (virtual machine approach) and the genuinely Scandinavian quality of the libraries, to make ERLANG the system of choice for very large networking applications.

RUBY

- **Ruby**
  is a strictly OO language characterized by a more mainstream syntax than SMALLTALK, string processing features to rival PERL, and extreme conciseness without the orkish scrawls of C. It is remarkable in possessing a powerful system for guaranteeing security in open systems (internet servers), a flexible library and package management system

Ruby Gems
Ruby on Rails

  (*Ruby Gems*) and recently also an excellent web application framework (*Ruby on Rails*).

LUA

- **Lua**
  was designed with the intention of providing a powerful yet "lightweight" language that can be used with limited hardware resources and

extending
embedding

  is suitable both for *extending* (i.e., incorporation of "custom" features) and *embedding* (i.e., making a larger software project programmable, thereby serving effectively as a very powerful macro language). It is elegant and widely used "behind the scenes" and a large portion of the entertainment software currently sold comprises a LUA interpreter which runs the game proper and is amalgamated with an imaging machine implemented in a low-level language. For the same reasons, it is an obvious choice for the control ("scenario generator") of high-throughput scientific applications.

C++

- **C++**
  is the current *de facto* standard for application programming and, apart from the object system, very similar to its predecessor C, which is still *the* language for system programming. It is the language in which all of MICROSOFT OFFICE™, OPENOFFICE and the KDE Project are written.

JAVA

- **Java**
  has become famous for the "write once, run anywhere" concept[1]: It is

---

[1] Actually, this was first implemented in the 1970s with UCSD-PASCAL (which is often quoted as one of the main influences on the design of JAVA), but the JAVA system is enormously advanced both with regard to the structures of the language and to the system functions it provides.

much more than a programming language; its virtual machine, ported to almost any hardware by now, provides all facilities imaginable, from file access to graphics and multitasking, and thus guarantees that these functions are identically available on any computer. Thus, JAVA has gained particular importance in the field of *embedded devices* such as smartphones and palmtops, where it enjoys almost a monopoly. Characteristically, even other languages such as PYTHON and LISP have been ported to the JAVA virtual machine to be able to exploit its advantages.

embedded devices

Further programming languages (from ALGOL and LOGO to MERCURY and OCCAM) will be presented in the appendix.

♠ *What do you know about these languages and the paradigms they are based upon?*

21.2 Fortran

21.2.1 Description. Imperative (but possessing optional functional features and OO since FORTRAN-95), strongly typed, compiled. Originally developed by John Backus and coworkers at IBM in 1957, the "*For*mula *Tran*slator System" alias FORTRAN-I was the first of all "higher" programming languages. Code optimization was an integral part from the very beginning so as to be able to compete with hand-crafted assembler code.

21.2.2 Advantages/Disadvantages. FORTRAN is particularly apt at handling numerical data and especially arrays of such, for the handling of which the language design allows one to generate extremely efficient code. The GNU Fortran compiler is capable of implicitly translating array access into pointer access, and the forthcoming FORTRAN-2008 standard will define multiprocessor code generation. On the downside, FORTRAN has always suffered from the image of being an "engineering only" language lacking flexibility and expressiveness; Dijkstra calling it an "infantile disease" of computing.

21.2.3 Strategy. The program reserves spaces for an array of 250 000 integers and manages this by means of an integer containing the current count of registered prime numbers. Everything is done using two nested loops; if a number is found to be divisable by any previously registered prime number, the rest of the loop, comprising the registration, is simply skipped. In combination with FORTRAN's excellent code generation, this simple approach results in the fastest program possible when using the GNU compiler.

21.2.4 Listing. (using FORTRAN-77 – newer revisions offer considerably more sophistication!)

```
        PROGRAM PRIME
        INTEGER P, MAX, R, HI
        COMMON P(250000)
        MAX=250000
        HI=0
        DO I=3,MAX,1
            DO J=1,HI,1
                R=MOD(I, P(J))
                IF (R .EQ. 0) GOTO 200
            END DO
            HI=HI+1
            P(HI)=I
200     END DO
        DO I=1,HI,1
            PRINT *,P(I)
        END DO
        END
```

21.3 Oberon

21.3.1 Description. Imperative (object-oriented), strongly typed, compiled. OBERON is the fourth generation of ALGOL languages, created by Wirth in the late 1980s and named, whimsically, after the moon of Uranus. Wirth was fascinated with the precision and reliability of the Voyager space probe which was then just passing Oberon. Like ADA and unlike PASCAL and MODULA-2, it offers true OO facilities. However, Wirth tried to keep the language as simple as possible and just developed records into a very elegant mechanism. A record may be derived from another (inheritance), and procedures may be bound to record types (thereby becoming methods). Thus, classes are made available just as an extension of PASCAL's normal type system.

21.3.2 Advantages/Disadvantages. Among the imperative languages, OBERON is outstanding for clarity and expressiveness. However, it has never really left the experimental stage, so implementations are rare and libraries very limited.

21.3.3 Strategy. We use iteration on homespun lists, as this is quite instructive with regard to the internal workings of lists. Mechanisms very much like the one presented here are at the core of all list-processing languages and libraries.

pointer

The type *inode* is defined as a structure comprising (α) a numeric value ("valu") and (β) a reference (*pointer*) to another inode ("next").

↑

The NEW function allocates memory to a pointer, thereby creating a new element of the type referred to, and the circumflex operator (better represented as ↑ or →) gives the element to which a pointer refers. Thus, x↑.next is the successor of x or *NULL* if x is the last element, and x↑.next↑.valu is the value of the successor of x.

As OBERON is an OO language, everything is neatly wrapped up into the class *Primsieve*, comprising a pointer to the list of prime numbers and the methods *OOCE* (emergency stop in case of memory overflow), *add2list* (append new item to end of list), *printlist* (dump the whole list), *divisable* (check whether a given number may be divided by any number in the list) and *determinePrimeNumbers* (initialize, perform search and dump results).

You will note that a much faster implementation of *add2list* might be possible if we forego appending new items to the end, and instead graft them on the very top. This would save us from moving through the entire list. In fact, this is the *cons* function underlying LISP, which is also the basis of ERLANG's $[h|t]$ mechanism, where h is the first element of a list and t all the rest (see below).

cons

21.3.4 Listing.

```
MODULE prime;

IMPORT Args, Conv, Out;

TYPE inodeptr = POINTER TO inode;
     inode    = RECORD
                    valu : INTEGER;
                    next : inodeptr;
                END;

     PrimsievePtr = POINTER TO Primsieve;
     Primsieve = RECORD
                    list: inodeptr;
                 END;

PROCEDURE (VAR ps:Primsieve)OOCE; (* emergency stop *)
BEGIN
   Out.String( 'Out of cheese error - program terminated' ); Out.Ln;
   HALT(1);
END OOCE;

PROCEDURE (VAR ps:Primsieve)add2list(newval: INTEGER); (* add int to list *)
VAR rover : inodeptr;
BEGIN
   (* go to end of list *)
   rover := ps.list; WHILE rover^.next # NIL DO rover := rover^.next END;
   (* append new value *)
   NEW( rover^.next );
   IF rover^.next = NIL THEN ps^.OOCE ELSE
      rover^.next^.next := NIL;
      rover^.next^.valu := newval;
   END;
END add2list;
```

```
PROCEDURE (VAR ps:Primsieve)printlist;  (* show list *)
VAR rover :  inodeptr;
BEGIN
   rover := ps.list; WHILE rover # NIL DO
     Out.Int( rover^.valu, 1 ); Out.String( ' ' );
     rover := rover^.next
   END;
END printlist;

PROCEDURE (VAR ps:Primsieve)divisable(i : INTEGER) : BOOLEAN;
VAR c : inodeptr;
BEGIN
   c := ps.list; WHILE c # NIL DO
     IF i MOD c^.valu = 0 THEN RETURN TRUE END;
     c := c^.next;
   END;
   RETURN FALSE;
END divisable;

PROCEDURE (VAR ps:Primsieve)determinePrimeNumbers(n: INTEGER);
VAR x   : INTEGER;
BEGIN
   NEW( ps.list ); IF ps.list # NIL THEN
     (* create list anchor *)
     ps.list^.valu := 2; ps.list^.next := NIL;
     (* do the search *)
     FOR x := 2 TO n DO
        IF ps.divisable( x ) # TRUE THEN ps.add2list( x ) END
     END;
   ELSE ps.OOCE END;
END determinePrimeNumbers;

VAR buf         : ARRAY 10 OF CHAR;
primeCollection: PrimsievePtr;

BEGIN
  Args.GetArg(1, buf);
  NEW( primeCollection );
  primeCollection^.determinePrimeNumbers( Conv.IntVal(buf) );
  primeCollection^.printlist;
END prime.
```

21.4 Lisp

21.4.1 Description. Functional, weakly typed, interpreted or compiled. LISP came into existence as a spin-off from the FORTRAN project when it became clear that recursion and lists were not going to be incorporated into FORTRAN-I. It can therefore be said to be the second oldest of all languages.

LISP exists in two versions, COMMON LISP and SCHEME, each of which comes in a great number of implementations.

Some people use a different approach and define LISP as any language that offers a number of facilities for list processing, independent of its syntax. Using this definition, all functional languages are dialects of LISP.

21.4.2 Advantages/Disadvantages. LISP is the archetype of all languages designed for processing ill-defined aggregations of data by recursion, and it possesses the full set of features required to this end. It is very straightforward in its design, but its excessive parentheses make it very difficult to write (and still more difficult to read for humans).

21.4.3 Strategy. We recurse on lists, which form a prominent part of the language.

21.4.4 Language details. LISP has the simplest grammar of all languages. Everything is a list, each of whose elements may be another list or an "atom" (i.e., variable, numeric constant or string). A list is evaluated by considering the first element as the name of a function and the rest as its parameters, unless the list is "quoted" with an apostrophe, in which case it is simply itself. For example, *(+ x y)* returns $x + y$. A function is defined using the function *define name content*, where content is usually defined as a lambda function: *lambda parameters what-the-function-does*. The *cond* function is used for multiple decisions; (cond (a b) (c d) (else z)) means: if condition a is fulfilled, evaluate b; else if condition c is fulfilled, evaluate d; if no condition is fulfilled, evaluate z. *Car* returns the first element (head), *cdr* the rest (tail) of a list, and *cons* adds a single value to a list.

21.4.5 Listing.

```
(define divisable (lambda (i l)
  (cond ((null? l) #f)                  ; terminal: list empty
        ((zero? (modulo i (car l))) #t) ; head divisable?
        (else (divisable i (cdr l))))))  ; otherwise recurse

(define primelist (lambda (n max plist)
  (cond ((>= n max) plist)              ; maximum reached?
        ((divisable n plist) (primelist (+ n 1) max plist))
        (else (primelist (+ n 1) max (cons n plist))))))

(define prime (lambda (n) (primelist 2 n '() )))
(write (prime 50000))
```

21.5 Erlang

21.5.1 Description. Purely functional, weakly typed, compiled. Using strict and eager evaluation (including an almost-imperative use of local variables), ERLANG definitely stands out among functional languages (CLEAN, MIRANDA, HASKELL). It was developed in the 1990s at Ericsson Telecommunications for the special purpose of controlling networks, and only later became a fully-fledged, all-purpose programming language.

21.5.2 Advantages/Disadvantages. ERLANG was designed for real-life projects, and its central strength is its absolute reliability, surpassing even that of ADA. Its intrinsic parallelism is its most prominent feature; together with its weak typing, this is a very powerful approach for large ($> 10^5$ lines of code) projects. The ERLANG compiler generates excellent bytecode, but number crunching is not what ERLANG was built for, as its only aggregate data structure is the list. Large projects (the first real test for ERLANG was an actual telephony application comprising 190 000 lines of code) can be handled easily in ERLANG, provided one has a full understanding of the possibilities opened by the complementary concepts of message passing, pattern matching (p. 140) and weak typing.

21.5.3 Strategy. As for LISP.

21.5.4 Language details. Function definition and invocation in ERLANG is much closer to "normal" languages, with parentheses and commata. Lists are enclosed in square brackets, and the vertical bar separates head and tail of a list: $[h|t]$. Alternative function definitions (separated by semicolons, whereas different clauses of the same function are separated by commas and different functions by periods) may use *guards* such as the expression *when*, and the result is always announced by the arrow operator. Each program module starts with a header specifiying name, functions required from and functions provided for other modules.

guards

21.5.5 Listing.

```
-module(prime).
-export([main/0, main/1, run/1, primetime/1]).
-import(io, [fwrite/2]).

% a few paraphernalia to run the program in different environments:
main() -> run( 50000 ).
main( Arg ) ->              % process command line parameters
    [Tmp] = Arg, run( list_to_integer(atom_to_list(Tmp)) ).
run( X ) ->                 % use this to run from the Erlang shell
    display_list( primetime( X ) ), halt( 0 ).

% yield list of prime numbers up to n:
primetime( N ) -> enum( 2, N, [] ).
```

```
% count from alpha to omega and collect prime numbers:
enum( Alpha, Omega, Prime1 ) when Alpha > Omega ->
     Prime1;    % terminal condition: maximum reached
enum( Alpha, Omega, Prime1 )   ->
     case divisable( Alpha, Prime1 ) of
     true -> enum( Alpha+1, Omega, Prime1 );          % divisable
     _    -> enum( Alpha+1, Omega, Prime1++[Alpha] ) % otherwise
     end.

% test whether a number may be divided by any in a list
divisable( _, [] )   -> false;            % empty list yields false
divisable( N, [H|T] ) when N rem H == 0 -> true;           % yes
divisable( N, [H|T] ) -> divisable( N, T ).           % recurse on

% show a list
display_list( [] ) -> [];
display_list( [H|T] ) -> io:fwrite( "~w ", [H] ), display_list( T ).
```

21.6 Ruby

21.6.1 Description. RUBY was designed as a deliberate antithesis. Its creator, Yukihiro Matsumoto (a.k.a. *Matz*), considered PYTHON as overdone, unwieldy and semantically inflexible (e.g., in its use of significant whitespace). Most of all, he wanted a fully object-oriented language like SMALLTALK, which is still the "gold standard" for OO design, but recognized the need for a more mainstream syntax, as for highly developed string processing features. Thus, RUBY is sometimes considered as a "SMALLTALK-derived language for doing the work of PERL that has incorporated the lessons learned from PYTHON".

Matz

Having ousted PYTHON in Japan almost completely, RUBY's acceptance in the West is slowed down by a lack of non-Japanese literature. Nevertheless APACHE-2 features PERL, PYTHON and RUBY as equivalent languages, and there is now also a BioRuby package.

21.6.2 Advantages/Disadvantages. As for PYTHON. You will notice the extreme shortness of the program: The RUBY interpreter not only comprises all the stuff necessary for the handling of dynamic data structures such as lists, but also all the error handling as serious mistakes lead, not to a program crash, but to the raising of an error condition.

Currently (versions up to 1.9) RUBY is notorious for its slowness. However, a complete redesign of the executive mechanism is in progress and has shown very promising results so far.

21.6.3 Strategy. We use lists in an iterative fashion.

21.6.4 Language details.
For RUBY's OO-based loops, please see p. 126. Using braces instead of significant whitespace allows the use of "lambdoid" code blocks. Note that RUBY should be called "type-free" rather than "weakly typed", as there is only one type of data – the object.

21.6.5 Listing.

```
def divisable( i, l )
  l.each { |c| return true if i % c == 0 }
  return false
end

def prime( n )
  p = [2]
  (3..n).each { |x| p += [x] unless divisable( x, p ) }
  return p
end

puts prime( ARGV[0].to_i )
```

21.7 Lua

21.7.1 Description.
LUA is another scripting language of the PYTHON/RUBY assemblage, with a syntax very close to that of ALGOL or PASCAL. As a scripting language, it is (of course) interpreted, weakly typed, procedural and optionally object-oriented[2]. It is remarkable for the clarity of its syntax as well as for the compactness and portability of its interpreter (which does not break records but is swift enough for all practical purposes). This is about one-twentieth (!) the size of the PYTHON interpreter and written in pure ANSI-C without any system-specific extensions. Taken together, these features make LUA an ideal choice for embedding into larger applications. Products incorporating LUA currently range from *World Of Warcraft* to *Adobe Photoshop Lightroom*. People who emphasize the political aspects of software (a tradition begun by Richard M. Stallman) will probably not fail to point out that, so far, LUA is virtually the only computer language ever developed in a third-world country[3].

World Of Warcraft
Adobe Photoshop Lightroom

21.7.2 Advantages/Disadvantages.
The libraries are adequately small, leaving much to be done for the programmer (for example, there is no regular expressions module, no TCL/TK binding and no "BioLua" package). Like

---

[2] Strictly speaking, its design is *not* object-oriented. However, the versatility of LUA's data structures, together with its powerful function handling, effectively result in the availability of all OO features.

[3] Bringing Brazil's score to one more than Germany's. But we have better boxers, tennis players and racing-drivers to boast of.

Ruby, Lua is little known outside its home country, and thus most of the documentation is in Portuguese – unintelligible for people whose classical education is as low as the current standard[4].

21.7.3 Strategy. As for Python or Ruby.

21.7.4 Listing.

```
function divisable( n, l, f )
   local j
   for j = 1, f do
      if n / primes[ j ] == math.floor(n / primes[ j ]) then
         return true
      end
   end
   return false
end

max = 50000
primes = { 2 }
found = 1
for i = 3, max do
   if not divisable( i, primes, found ) then
      found = found + 1
      primes[ found ] = i
   end
end
for j = 1, found do
   print( primes[ j ] )
end
```

21.8 C++

21.8.1 Description. C++ can be understood only from a historical perspective.

C was developed by Kernighan and Ritchie ("*K&R*") about 1970[5] as an integral part of the Unix operating system. It was designed as a low-level language primarily intended for system (kernel) programming and allowed extremely efficient implementations of speed-critical routines. Debugging facilities were virtually non-existent; in the original K&R definition, not even type checking was performed (this became much better with the creation of the "ANSI C" standard in 1989). The current standard is C-99. Among other things, Python and Ruby are both implemented in C.

K&R

The importance, and impact, of C can hardly be overestimated. When in the 1960s "artificial intelligence" was the rage, not only were operating systems

---
[4] Who needs Latin for boxing, tennis or racing?
[5] 1967: BCPL as subset of CPL (in turn derived from Algol); 1970: creation of B by Ken Thompson.

written in LISP and for LISP, but even hardware for these machines was designed with LISP in mind. Similarly, the Inmos Transputer was designed as hardware basis for OCCAM, and Apple's Newton was optimized for FORTH. As these systems were sometimes referred to as "the LISP machine" or "the OCCAM OS", so all contemporary systems are optimized for C. x86, PPC, SPARC and MIPS are all designed for C, and all Unix derivatives as well as the 32-bit versions of Windows^{TM} may be called *"the C operating system"* – and in fact there is the so-called *Posix* standard established for them.

<small>"the C operating system"</small>
<small>Posix</small>

However, it was strongly felt that C's design was inadequate for managing large projects, and development into an OO language was proposed. Most straightforward in this respect was OBJECTIVE C, which was the foundation for the, then amazing, operating system of the NEXT computer and is still widely used by Apple developers, but has never found wide usage. Interestingly, OBJECTIVE C is a superset of C, i.e., an OBJECTIVE C compiler may readily digest a C-99 source as well.

Originally developed by Bjarne Stroustrup as another object-oriented extension to C, C^{++}, breaking backward compatibility with C, has by now become the language for most system and application programming. However, debugging and stability are still critical issues, and the concept of OO is difficult to reconcile with the generally low abstraction level of C.

21.8.2 Advantages/Disadvantages. The first thing a C/C^{++} neophyte will notice is the excessive use of cryptic tokens and expressions to the tune of `(*x++)->(*y--)&=*++z`. However, this is a typical "love it or hate it" situation[6]; in fact the notorious crypticity is easy to comprehend.

No other non-assembler language (with the possible exception only of FORTH) is capable of controlling the processor as precisely as C/C^{++}, or generating more efficient code for well-defined tasks. By using `unsigned long` as our data type of choice, we may work with numbers up to $2^{64} = 1.8 \times 10^{19}$. On the other hand, there are no higher abstractions, and although the code is extremely fast (if it is well-written), it is also extremely dangerous – checks are not just missing, they are simply impossible.

If everything works well, C^{++} combines the power of OO with the efficiency of C. However, higher abstractions (such as variable-size data strings and lists) are still mostly lacking, and so are checks and debugging features. For some unknown reason, the *gcc* compiler will not optimize C^{++} code as throughly as plain C, although it takes a lot more time to compile. And to add a real champion's opinion, here is Linus Torvalds on the idea of system programming in C^{++}: "It sucks. The fact is, C++ compilers are not trustworthy."

---

[6] "C^{++} is an insult to the human brain." – Wirth

21.8.3 Strategy.
In order to store our prime numbers up to n, we simply reserve a coherent chunk of memory large enough to accommodate n elements. This is done by a single call to the function `malloc`. Afterwards, the memory has to be de-allocated again by `free`. Access to this chunk is done using pointers, which will be explained in detail in Volume 2. Pointer access is the fastest way of dealing with data, considerably swifter even than array access (because it does not even require arithmetics). However, it must be admitted that it is really a hack, because no checks whatever are possible.

The entire thing is encapsulated into a class named `primsieve`. When this class is instantiated, it does the calculation; when the instance is subsequently deleted, it prints out the values. Admittedly, the task does not really lend itself readily to OO (unless one would, for whatever reason, wish to hold more than one prime number table in memory at the same time), but it is easy to see how things might be extended in the case when a prime number table is required by another project. In the program, both construction and destruction are performed within a single line in the main() function.

21.8.4 Language details.
The "class definition" actually contains no more than this – an index of all the methods and attributes of a class. All methods must have return values (`void` if no genuine value is returned), excepting the constructor and destructor, which have the same name as the class. The destructor is denoted by the $\sim$ sign. Later the actual code is written in the form of `resulttype classname::methodname(parameters)`.

21.8.5 Listing.

```
#include <stdio.h>
/* C++ does not have built-in list functions either. Therefore, we
 * do not use a chained list but a simple chunk of (maximum) elements,
 * which can be accessed at lightning speed with a pointer.
 * However, the chunk and the pointer are safely hidden within the objects.
 */
#define true 1
#define false 0

//--- Declaration
// This contains only the names and types of the members.
// The code of the methods is to follow later.

class primsieve {
public:   // These members are visible from without.
  primsieve( unsigned long n );  // constructor includes calculation
  ~primsieve();                  // destructor includes printing
private:  // These are not.
  unsigned long *chunk, *top;    // chunk and pointer
  void ooce();                   // emergency stop
  void printlist( unsigned long *from, unsigned long *to );
  int divisable( unsigned long n, unsigned long *base, unsigned long *beyond );
};
```

```
//--- Implementation
// The double colon (::) assigns the methods to the class.
// Unlike all other methods, the constructor and destructor do not have return values.

primsieve::primsieve( unsigned long n ) {
  unsigned long x;
  top = chunk = new unsigned long[ n ]; // no more "malloc" in C++
  if (chunk) {
    *top++ = 2;
    for ( x = 3; x <= n; x++ )
      if (!divisable( x, chunk, top )) *top++ = x; // add new prime number to chunk
  } else ooce();
}
primsieve::~primsieve() { printlist( chunk, top ); delete chunk; } // print and release

void primsieve::ooce() {
  printf("Out of cheese error\n");
  // something should be here to stop the program
}

void primsieve::printlist( unsigned long *from, unsigned long *to ) {
  unsigned long *rover;
  if (from && to)
    for ( rover = from; rover < to; rover++ ) printf( "%d ", *rover );
  printf( "\n" );
}

int primsieve::divisable( unsigned long n, unsigned long *base, unsigned long *beyond )
  register unsigned long *c;
  for ( c = base; c < beyond; c++ ) // scan the chunk
    if (n % (*c) == 0) return true;  // found something
  return false;                      // found nothing
}

//--- Invocation

int main( int argc, char *argv[] ) {
  int max;
  if (argc == 2) {
    sscanf( argv[1], "%d", &max );
    primsieve ps( (unsigned long) max ); // this line does all the work
  } else printf( "Error: we need 1 argument, not %d\n", argc );
}
```

21.9 Java

21.9.1 Description. Object-oriented, strongly typed, compiled (system-independent byte code). Produced in the early 1990s at Sun Microsystems for developing portable graphical applications, based upon C^{++}, with the deliberate aim of reducing the complexity of that language, while at the same time offering higher stability and better debugging facilities.

According to its devotees, JAVA is simply *the* language. According to SMALLTALK mujaheddin, "Java is a lightly-statically-typed, simple version of Smalltalk with the syntax of the 'C' family".

21.9.2 Advantages/Disadvantages. The JAVA standard comprises an exhaustive (some say exhausting) library, including classes for higher-level data structures, threading, exception handling, networking and graphics, and JAVA byte code will run on any machine equipped with a JAVA VM, although not half as fast as native code (although there are native code compilers available for many architectures). On the other hand, JAVA combines much of the ugliness of C with the verbosity of VISUAL BASIC, and the type system with its "wrapper classes" for primitives is just a hotchpotch.

21.9.3 Strategy. We use lists here. Although lists do not form an integral part of the language, they are implemented in a module named *Vector*. Alternatively, arrays might be used.

21.9.4 Language details. JAVA can hardly be mistaken for anything but a member of the C group of languages, and most of its structures (e.g., loops and branchings) are actually identical to those of C – not necessarily to JAVA's advantage, as this tends to obscure the differences between the two languages. Classes and their members can (and must) be defined very precisely, with expressions such as `private static void`. As it is the custom for OO languages, every class may contain or not contain a core which is executed when the class is run as a program – in this case, it must be implemented as a method named, simply enough, `public static void main(String args[])`. Most of JAVA's functionality is in fact within its extensive libraries, e.g., list handling (see below). As in EIFFEL, this does not really make programs look pretty.

21.9.5 Listing.

```
import java.lang.*;
import java.io.*;
import java.util.*;

public class prime {

    private static boolean divisable( int i, Vector l ) {
        // check whether i can be divided by any element of l:
        for ( int c = 0; c < l.size(); c++ ) {
            if (i % ((Integer)l.get( c )).intValue() == 0)
                return true;     // yes, it can
        }
        // seems it cannot:
        return false;
    }
```

```
    private static void primenumbers( int n ) {
        // initialize list
        Vector p = new Vector(); p.add( new Integer( 2 ) );
        // find prime numbers
        for ( int x = 3; x <= n; x++ ) {
            if (!divisable( x, p )) p.add( new Integer( x ) );
        }
        // print them
        for ( int cc = 0; cc < p.size(); cc++ )
            System.out.print( p.get( cc ).toString()+" " );
    }

    public static void main(String args[]) {
        primenumbers( Integer.parseInt(args[0]) );
        System.out.println();
    }
}
```

Some languages manage to absorb change but resist progress.

CHAPTER 22

Facing up to Python-3000

> Maybe man will be able to fly some day. In all things we are still at the very beginning.
>
> – Bertolt Brecht: "Life of Galilei"

PYTHON is still evolving. Over the years, hundreds of suggestions for improvement have been made and partially incorporated. An overview of the "Python Enhancement Proposals", abbreviated *PEPs*, can be found at http://www.python.org. By now, however, some modifications have been found necessary which cannot be reconciliated with all pre-established conventions. Thus, PYTHON is in for a major overhaul, with PYTHON-3000, the next version, being expected to take over in autumn 2008. The following does not strive to be an extensive list of changes, PYTHON-3000 now still being unfinished, but rather to provide some hints for what has to be expected[1]. It should be pointed out that the majority of changes relate to more arcane aspects of PYTHON, which will be discussed in the second volume.

PEPs

22.1 Functional "print"

22.1.1 Apocryphal parameters.

Relevance: So far, only such functions have been considered which show a one-to-one correspondence between parameters and arguments to be supplied – this should be the rule in any well-written program. However, in some cases, notably for system interfaces, variable argument sets may be advantageous. The new printing function of Python-3000 is a prime example for such a function.
Keywords: Keyword parameters and functions with variable argument sets.

From its very beginnings, PYTHON has allowed optional or *keyword parameters*. Such parameters are defined with a default value; if no suitable arguments are supplied, the keyword parameters are set to these default values. If more than one keyword parameter is present, the arguments are assigned sequentially; alternatively, upon invoking the function the parameters which are "meant" can be named explicitly. The following primitive example makes this clearer:

keyword parameters

---

[1] Vice versa, some things described here have already been incorporated into the most recent versions of PYTHON. Thus, from the fact that any particular feature is named here it should not be inferred that it is not present in PYTHON 2.x.

```
>>> def funz(a,b=10,c=11): print a,b,c
...
>>> funz(6)
6 10 11
>>> funz(6,7)
6 7 11
>>> funz(6,c=7)
6 10 7
>>> funz(6,c=7,b=5)
6 5 7
```

A second mechanism consists in parameters which take multiple arguments – all the arguments that cannot be assigned to any other parameter are packed into a tuple which is then stored in the "ragman" parameter. This must naturally be the last parameter in the parameter, and it is labelled with an asterisk preceding the variable name:

```
>>> def vunz(*x): print x
...
>>> vunz()
()
>>> vunz(1)
(1,)
>>> vunz(1,2)
(1, 2)
>>> vunz(1,2,3)
(1, 2, 3)
```

This mechanism can be freely combined with keyword parameters:

```
>>> def grunz(a,*x): print a,x
...
>>> grunz()
Traceback (most recent call last):
  File "<stdin>", line 1, in <module>
TypeError: grunz() takes at least 1 argument (0 given)
>>> grunz(1)
1 ()
>>> grunz(1,2,3)
1 (2, 3)
```

In PYTHON-3000 (and in PYTHON-3000 only), the need for the "ragman" to be the last parameter has been eliminated. There may be others following the asterisk identifier, but in this case it is mandatory that the arguments are named (*keyword-only parameters*):

keyword-only parameters

```
>>> def kunz(*foo,bar): print(foo,bar)
...
>>> kunz(1,2,3,bar=4)
(1, 2, 3) 4
```

```
>>> kunz(1,2,3,bar=4,5,6)
  File "<stdin>", line 1
SyntaxError: non-keyword arg after keyword arg
>>> def ounz(*foo,bar=665.99): print(foo,bar)
...
>>> ounz(1,2,3,4)
(1, 2, 3, 4) 665.99
>>> ounz(1,2,3,bar=4)
(1, 2, 3) 4
```

22.1.2 The new way of printing. In all previous versions of PYTHON, output to the console or a file was essentially done using the `print` statement with its peculiar syntax. However, the fundamental design principle of PYTHON has always been that "exceptions are not exceptional enough to break the rule". It was thus found appropriate to replace the `print` statement with an arbitrary-parameter function that uses keyword arguments, as described above, for flexibility:

Old	New	What it does
print "The answer is", 2*2	print("The answer is", 2*2)	
print a,":",b,":",c	print(a,b,c,sep=" : ")	Prints with separators.
print x	print(x, end=" ")	Appends a space, suppresses line feed.
print	print()	Produces line feed.
print (x, y)	print((x, y))	Prints the tuple (x, y).
print >>sys.stderr, "fatal error"	print("fatal error", file=sys.stderr)	Writes the words "fatal error" to standard error channel.

Concerning the *standard error* channel: The notion of channels, namely standard input and standard output, was discussed before (p. 67). The standard error channel ("StdErr") is a channel that is similar to the standard output channel but cannot be redirected.

standard error

What is this supposed to be good for? When a text interface is used (as in the majority of bioinformatics application, which are intended to simply transform a set of data into another set of data), writing the error message to the console is not always as straightforward as might be expected – because there is such a thing as standard input/output redirection (p. 67). Evidently, it might lead to embarassing results if an error message were deflected into some file instead of appearing on the screen. To avoid this, there exists a third standard channel which is never redirected. This is the standard error channel (*StdErr*). In PYTHON, this[2] is provided by the *sys* module, namely *sys.stderr*.

StdErr
sys.stderr

It will be seen that this, while eliminating a cumbersome statement from the core language, also permits a larger number of options to be passed to the printing function, thereby making it more powerful by customization.

---

[2] Together with *sys.stdin* and *sys.stdout* – which are, however, rarely used explicitly.

22.2 Standardized unicode support

22.2.1 Unicode in itself. It is worthwhile mentioning here that ASCII as mentioned before is obsolescent, and this is at least in part due to Eastern Europe's rejoining the western world. The big problem with the original ASCII definition is that it does not make provisions for special characters (English begin the only language which uses a Latin alphabet without any special tokens or diacritic marks!). There are de facto standards for ASCII variants catering most languages of Western Europe, but with Eastern languages becoming more important, this has become burdensome, as characters such as "ű", "ţ" and "ř" now have to be represented. Instead, a new format named *unicode* is now being propagated that actually uses 31 bits, which means that it is capable of containg and representing any writing system ever devised by humans, from European accented or otherwise modified letters to CJK (Chinese, Japanes, Korean) glyphs, Thai and Indian Devanāgāri and things as remote as the Sequoia script for Algonkin languages. A derivative of unicode is UTF-8, which uses 1 byte for common Latin letters, 2 bytes for European special characters, and 3 to 5 bytes for everything else. Of course, the recipient of your message will have to use the same character representation, otherwise things just won't do.

Obviously, a programming language will have to deal with the special character problem in two aspects:

(1) Special characters in source code – even though identifiers such as variable names are generally restricted to alphanumeric glyphs and thus the ASCII set, in an international context the dialogue with the user often requires string constants with special characters.

(2) Special characters in files to be processed – it goes without saying that a program should be able to deal with special characters.

Here a very good program suffers from the fact that some losing library does not understand unicode.

So far, PYTHON could be "made to" understand unicode in source texts, e.g. by using the magic formula

`# encoding=utf8`

as the first or second line of the program. How it dealt with unicode in files, however, was dependent on the underlying C library functions, i.e. the options PYTHON was compiled with. PYTHON-3000 now standardizes this:

(1) Source code is always interpreted as unicode unless another encoding is explicitly stated.
(2) For processing files, PYTHON-3000 now uses two different modes, one of which is for unicode and the other for plain bytes (binary files). Actually, this it not really new for the denizens of the Microsoft OS world, where a non-standard line ending (comprising two characters instead of one) has always necessitated discriminating between binary files and text files.

Of course, string constants as provided by the `string` module must now be either bytes or strings, too.

22.2.2 Accessing binary and unicode files. In opening files, PYTHON has always closely mimicked the situation in C: The first argument to the opening function is the file name, the second a chiffre indicating whether read ("r") or read/write ("w") access is desired. This chiffre can be prefixed with a "b" to indicate binary mode (thus "br" and "bw"). On Microsoft operating systems, this "b" traditionally indicates suppression of the interpretation of the double character line end as a single line end, which is required for handling binary files; on unixish systems, it was ignored. Now, however, it serves for discriminating text unicode files which will generally use the UTF-8 encoding (or any other encoding; this may be specified explicitly) from plain binary files.

22.2.3 Binary data in the times of unicode. In PYTHON-3000, strings are likewise unicode by default. However, there are also byte sequences which are not interpreted as unicode, and can thus be used for processing binary data. They are tagged by a prefixed "b". Thus:

- `"size"` will denote a unicode string;
- `b"size"` will denote a run of four bytes, namely 115, 105, 122 and 101;
- `"Größe in Å"` will denote another unicode string;
- `b"Größe in Å"` will be illegal.

22.3 Goodbye λ, it was nice to know you

PYTHON-3000 has shed several functional traits. First of all, there is no λ notation any more. Moreover, the `map` and `filter` functions still exist but in lieu of lists return generators (p. 392) now – objects which dole out values sequentially.

It is expressly intended that the use of `map` and `filter` with λ is to be replaced by list comprehensions (p. 128). Thus, prepare to use

```
y = [math.sqrt(x) for x in a if x>=0]
```

instead of

```
y = map( math.sqrt, filter( lambda n:n>=0, x ) )
```

range()
xrange()

When the function to be applied has side effects (e.g. influences other data structures or performs I/O), the proper structure to be applied is a loop. Thus it is only consequent that the *range()* function now does what was previously the job of *xrange()*: "Instead of returning a list, returns an object that generates the numbers in the range on demand. For looping, this is slightly faster than *range()* and more memory efficient."

As it has always been our custom so far to compare programming languages, a look at the functional languages is not amiss here. Basically, in processing large volumes of data they have the same problem which may be summarized like this: How can we avoid generating huge excesses of data as intermediates in a multi-step "pipeline" of data processing; for example, if we want to, for the purposes of sequence analysis, extract the nucleotide sequence from a GenBank file, achieve a processing which consists of (1) switching all letters in the raw data chunk to uppercase (function f_1) and (2) filtering out all letters which do not denote nucleotides (function f_2), without having an intermediate after (1) in the size of the original data chunk, which may easily amount to several gigabytes? In the pure mathematical notation $y = f_2(f_1(x))$ the intermediate – the result of f_1 – is obviously inevitable. Mathematical formalisms are not designed for disclosing efficient implementations.

lazy evaluation

The functional languages solve this dilemma elegantly by *lazy evaluation*, which is compulsory in the hardcore FP languages (MIRANDA, CLEAN, HASKELL) and optional in the others (LISP, O'CAML)[3]. "Lazy evaluation" in fact means that evaluation of a given expression is deferred until the result is needed. It will be understood that this is quite a complex task to be solved, and code produced by hardcore FP compilers is incredibly bloated because it is not always easy to determine when any particular value will be needed. PYTHON's iterators – which are, of course, impossible by design in a purely functional language – clearly present a much simpler solution.

Likewise, handling of dictionaries (p. 88) is now more on the object-oriented side.

---

[3] ERLANG will be missed here; in fact, ERLANG is "eager" by design, the rationale being that ERLANG is not intended to be used for arbitrarily large volumes of data but for the efficient and *carrier-grade* reliable processing of relatively small amounts of data.

22.4 Number representations, arithmetics, &c.

So far, PYTHON has followed C's conventions regarding numbers: All numbers are assumed to be decimal, unless prefixed with `0x` to indicate hexadecimal and `0` to indicate octal representation. The latter has sometimes been found confusing, so in PYTHON-3000 octals are prefixed with `0o`. Extending this logic, binary numbers are now prefixed `0b`.

A division of an integer by an integer will now produce a float if required. The previous truncating behaviour has been delegated to the new integer division operator `//`, e. g. $13/8 = 1.625$, $13//8 = 1$.

All integers are now internally treated as longints.

The PASCAL-style operator `<>` has now been abandonned. Use `!=` instead. Comparisons in general are now more type-sensitive.

The module `this` provides information on advanced concepts in PYTHON-3000.

22.5 Function annotations

Even though there is general agreement that PYTHON is to remain a weakly typed language, occasionally concerns have been voiced that complete freedom of types might be detrimental to stability, as it increases the likelihood of runtime errors due to incompatible data types. This is particularly liable with large teams and sparse time for writing proper documentation.

In PYTHON-3000, types for parameters and return values may be added as annotations. The core language does not check them[4], but tools are under development that shall use these annotations. Currently, they serve as an additional source of information about the parameter/return value structure of a function, as they may be inspected using the new __annotations__ attribute: __annotations__

```
ophis@aurora:~/Python-3.0a1$ ./python
Python 3.0a1 (py3k, Oct 18 2007, 15:41:32)
[GCC 4.1.2 (Ubuntu 4.1.2-0ubuntu4)] on linux2
Type "help", "copyright", "credits" or "license" for more information.
>>> def compile( ipufi: "file", opufi: "file", options: "string" ) -> "boolean": pass...
>>> compile.__annotations__
{'opufi': 'file', 'ipufi': 'file', 'return': 'boolean', 'options': 'string'}
>>>
```

22.6 String formatting and I/O

Previously, PYTHON featured two functions for accepting input: `input` and `raw_input`, the difference being that the former evaluated expressions. Thus, "3*5" was transformed to "15" when `input` was used (see the exercise on p. 125).

---

[4] There was a discussion that this might make PYTHON "too similar to JAVA".

In PYTHON-3000, `input` returns the input without evaluation, and in order to enforce evaluation, use `eval(input)`.

The `cPickle` module has vanished. Use `pickle` instead. Moreover, several previously implementation-dependent modules have been redesigned and consolidated into the general `io` module.

The `%` operator for formatting values (p. 339) is replaced by a new and more versatile function/method named `format`, which takes an arbitrary number of positional and keyword arguments, e.g: `"The story of {0}, {1}, and {c}".format(a, b, c=d)`. The following is straight from the horse's mouth[5]:

Format strings – Within a format string (i.e. the string prescribing the desired format), each positional argument is identified with a number, starting from zero, so in the above example, "a" is argument 0 and "b" is argument 1. Each keyword argument is identified by its keyword name, so in the above example, "c" is used to refer to the third argument.

Format strings consist of intermingled character data and markup.

Character data is data which is transferred unchanged from the format string to the output string; markup is not transferred from the format directly to the output, but instead is used to define "replacement fields" that describes to the format engine what should be placed in the output string in the place of the markup.

Curly braces are used to indicate a replacement field within the string: `"My name is {0}".format('Fred')`. The result of this is the string: `My name is Fred`.

Braces can be escaped by doubling: `"My name is 0 :-{{}}".format('Fred')` will produce: `"My name is Fred :-{}"`.

The element within the braces is called a "field". Fields consist of a "field name", which can either be simple or compound, and an optional "format specifier".

Simple field names are either names or numbers. If numbers, they must be valid base-10 integers; if names, they must be valid PYTHON identifiers such as variable names. A number is used to identify a positional argument, while a name is used to identify a keyword argument.

A compound field name is a combination of multiple simple field names in an expression: `"My name is {0.name}".format(file('out.txt'))`. This example shows the use of the "getattr" or "dot" operator in a field expression. The dot operator allows an attribute of an input value to be specified as the field value.

You cannot embed arbitrary expressions in format strings. This is by design – the types of expressions that you can use is deliberately limited. Only two operators are supported: the "." (getattr) operator, and the "[]" (getitem) operator. The reason for allowing these operators is that they don't normally have side effects in non-pathological code.

An example of the "getitem" syntax:
`"My name is {0[name]}".format(dict(name='Fred'))`.

It should be noted that the use of "getitem" within a format string is much more limited than its conventional usage. In the above example, the string "name" really is the literal string "name", not a variable named "name". The rules for parsing an item key are very

---

[5] This is an excellent example of good documentation and very illustrative of the way proper software development is done. Moreover, it is also a worthwhile exercise with regard to the use of objects. For these reasons, the following essentially echoes the original PEP.

Format Specifiers – Each field can also specify an optional set of "format specifiers" which can be used to adjust the format of that field. Format specifiers follow the field name, with a colon separating the two: `"My name is {0:8}".format('Fred')`. The meaning and syntax of the format specifiers depends on the type of object that is being formatted, however there is a standard set of format specifiers used for any object that does not override them.

Format Specifiers

Format specifiers can themselves contain replacement fields. For example, a field whose field width is itself a parameter could be specified via: `"{0:{1}}".format(a, b, c)`. These "internal" replacement fields can only occur in the format specifier part of the replacement field. Internal replacement fields cannot themselves have format specifiers.

Note that the doubled "}" at the end, which would normally be escaped, is not escaped in this case. The reason is because the "{{" and "}}" syntax for escapes is only applied when used *outside* of a format field. Within a format field, the brace characters always have their normal meaning. The syntax for format specifiers is open-ended, since a class can override the standard format specifiers. In such cases, the `str.format()` method merely passes all of the characters between the first colon and the matching brace to the relevant underlying formatting method.

Standard Format Specifiers – If an object does not define its own format specifiers, a standard set of format specifiers are used. These are similar in concept to the format specifiers used by the existing "%" operator, however there are also a number of differences. The general form of a standard format specifier is `[[fill]align][sign][0][width][.precision][type]`, wherein the brackets indicate an optional element and the optional align flag can be one of the following:

Standard Format Specifiers

- `<` forces the field to be left-aligned within the available space (default).
- `>` forces the field to be right-aligned within the available space.
- `=` forces the padding to be placed after the sign (if any) but before the digits. This is used for printing fields in the form "+000000120".
- `↑` forces the field to be centered within the available space.

Unless a minimum field width is defined, the field width will always be the same size as the data to fill it, so that the alignment option has no meaning in this case. The optional "fill" character defines the character to be used to pad the field to the minimum width. The fill character, if present, must be followed by an alignment flag. The "sign" option is only valid for number types, and can be one of the following:

- `+` indicates that a sign should be used for both positive as well as negative numbers
- `-` indicates that a sign should be used only for negative numbers (this is the default behavior)
- *empty space* indicates that a leading space should be used on positive numbers

"Width" is a decimal integer defining the minimum field width. If not specified, then the field width will be determined by the content. If the width field is preceded by a zero ("0") character, this enables zero-padding. This is equivalent to an alignment type of "=" and a fill character of "0".

The "precision" is a decimal number indicating how many digits should be displayed after the decimal point in a floating point conversion. For non-number types the field indicates the maximum field size – in other words, how many characters will be used from the field content. The precision is ignored for integer conversions. Finally, the "type" determines how the data should be presented. The available integer presentation types are:

- "b" – Binary. Outputs the number in base 2.
- "c" – Character. Converts the integer to the corresponding unicode character before printing.
- "d" – Decimal Integer. Outputs the number in base 10.
- "o" – Octal format. Outputs the number in base 8.
- "x" – Hex format. Outputs the number in base 16, using lower-case letters for the digits above 9.
- "X" – Hex format. Outputs the number in base 16, using upper-case letters for the digits above 9.
- "" – (None) - the same as "d"

The available floating point presentation types are:

- "e" – Exponent notation. Prints the number in scientific notation using the letter "e" to indicate the exponent.
- "E" – Exponent notation. Same as "e" except it uses an upper case "E" as the separator character.
- "f" or "F" – Fixed point. Displays the number as a fixed-point number.
- "g" – General format. This prints the number as a fixed-point number, unless the number is too large, in which case it switches to "e" exponent notation.
- "G"' – General format. Same as "g" except switches to "E" if the number gets to large.
- "n" – Number. This is the same as "g", except that it uses the current locale setting to insert the appropriate number separator characters.
- "%" – Percentage. Multiplies the number by 100 and displays in fixed ("f") format, followed by a percent sign.
- "" (None) – similar to "g", except that it prints at least one digit after the decimal point.

Objects are able to define their own format specifiers to replace the standard ones. An example is the "datetime" class, whose format specifiers might look something like the arguments to the strftime() function: `"Today is: 0:a b d H:M:S Y".format(datetime.now())`.

Explicit Conversion Flag

Explicit Conversion Flag – is used to transform the format field value before it is formatted. This can be used to override the type-specific formatting behavior, and format the value as if it were a more generic type. Currently, two explicit conversion flags are recognized:

- "!r" – convert the value to a string using *repr()*.
- "!s" – convert the value to a string using *str()*.

These flags are placed before the format specifier: `"0!r:20".format("Hello")`. Thus the string "Hello" will be printed, with quotes, in a field of at least 20 characters width.

Controlling Formatting on a Per-Type Basis

Controlling Formatting on a Per-Type Basis – Each Python type can control formatting of its instances by defining a __format__ method. The __format__ method is responsible for interpreting the format specifier, formatting the value, and returning the resulting string. The new, global built-in function "format" simply calls this special method, similar to how len() and str() simply call their respective special methods: `def format(value, format_spec): return value.__format__(format_spec)`. It is safe to call this function with a value of "None" (because the "None" value in Python is an object and can have methods.)

Several built-in types, including "str", "int", "float", and "object" define __format__ methods. This means that if you derive from any of those types, your class will know how to format itself.

The object.__format__ method is the simplest: It simply converts the object to a string. The __format__ methods for "int" and "float" will do numeric formatting based on the format specifier. Any class can override the __format__ method to provide custom formatting for that type: `class AST: def __format__(self, format_spec):` ....

Note that the "explicit conversion" flag mentioned above is not passed to the __format__ method. Rather, it is expected that the conversion specified by the flag will be performed before calling __format__.

User-Defined Formatting – There will be times when customizing the formatting of fields on a per-type basis is not enough (think a spreadsheet application, which displays hash marks when a value is too large to fit in the available space).

For more powerful and flexible formatting, access to the underlying format engine can be obtained through the "Formatter" class that lives in the "string" module. This class takes additional options which are not accessible via the normal `str.format()` method. An application can subclass the Formatter class to create their own customized formatting behavior.

Formatter Methods – The Formatter class takes no initialization arguments: `fmt = Formatter()`. The public methods of class Formatter are as follows:

- format(format_string, *args, **kwargs)
- vformat(format_string, args, kwargs)

"format" takes a format template, and an arbitrary set of positional and keyword argument. "format" is just a wrapper that calls "vformat" – the function that does the actual work of formatting. It is exposed as a separate function for cases where you want to pass in a predefined dictionary of arguments, rather than unpacking and repacking the dictionary as individual arguments using the "*args" and "**kwds" syntax. "vformat" does the work of breaking up the format template string into character data and replacement fields. It calls the "get_positional" and "get_index" methods as appropriate (described below.)

Formatter defines the following overridable methods:

- get_value(key, args, kwargs)
- check_unused_args(used_args, args, kwargs)
- format_field(value, format_spec)

"get_value" is used to retrieve a given field value. The "key" argument will be either an integer or a string. If it is an integer, it represents the index of the positional argument in "args"; if it is a string, then it represents a named argument in "kwargs". The "args" parameter is set to the list of positional arguments to "vformat", and the "kwargs" parameter is set to the dictionary of positional arguments.

For compound field names, these functions are only called for the first component of the field name; subsequent components are handled through normal attribute and indexing operations. So for example, the field expression "0.name" would cause "get_value" to be called with a "key" argument of 0. The "name" attribute will be looked up after "get_value" returns by calling the built-in "getattr" function.

"check_unused_args" is used to implement checking for unused arguments if desired. The arguments to this function is the set of all argument keys that were actually referred to in the format string (integers for positional arguments, and strings for named arguments), and a reference to the args and kwargs that was passed to vformat. The set of unused args can be calculated from these parameters. "check_unused_args" is assumed to throw an exception if the check fails.

"format_field" simply calls the global "format" built-in. The method is provided so that subclasses can override it.

22.7 Classes – as you've never seen them before

22.7.1 Metaclasses. Metaclasses are classes which upon instantiation do not yield objects but classes (and may hence comprise classes as members) – a mechanism to be found (among the "serious" languages) only in SMALLTALK, PYTHON, PERL (!) and certain flavours of LISP and PASCAL. A PYTHON-3000 *metaclass* is formally defined by inheritance from the fictitious base class type.

The new class definition allows for the inclusion of keyword arguments, among which there may be metaclasses. For example, class newclass(baseclass, other_baseclass, metaclass=strange, private=True) will specify that *newclass* is created by inheritance from both *baseclass* and *other_baseclass*; that it will be private; and that it is connected to the metaclass *strange*. Hence, *strange* will "know about" *newclass*, as about any other class connected thereto.

Metaclasses can be expected to be valuable in dynamic and reflexive programming, since they can be used to inspect the contents of classes, where these are mutable. In the end, they allow more flexibility in the creation of objects[6].

22.7.2 Abstract base classes. Occasionally it may be desirable to "inspect" an object, that is to say, to obtain information not only about its state (i. e. the current content of its data members) but rather about what kind of functionality it implements[7]. In PYTHON, asking for the presence of a method is just as easy as asking for the value of a data member; however, before doing so one must know the names of the methods which provide a particular functionality when present. Moreover, for obvious reasons related but different data types may share names of methods; for example, in PYTHON both "mappings" (such as dictionaries) and "sequences" (such as lists) do possess an attribute named __len__, and hence the presence of __len__ is not sufficient to discriminate between these two types[8].

This problem is tackled in PYTHON-3000 by instituting *abstract base classes*, abbreviated ABCs, which are essentially stencils defining the functionality

---

[6] <rant>Even the Wikipedia article on metaclasses remarks caustically: "Metaclass programming can be confusing, and it is rare in real-world Python code".</rant>

[7] Guido van Rossum and Talin themselves admit that "in classical OOP theory, invocation is the preferred usage pattern, and inspection is actively discouraged" (PEP3119) before proceeding to argue that PYTHON must be more flexible because "there is often a need to process objects in a way that wasn't anticipated by the creator of the object class".

[8] Nor is there any other good criterion, according to Guido van Rossum himself.

which the classes derived therefrom must provide (leaving the implementation to said derived classes). By merely ascertaining that any particular class is a subclass of a certain ABC, it can be made sure that each object of this class will provide the functionality characterizing the ABC. For example, the ABC named *sizeable* is characterized in that it defines a method named $\_\_len\_\_$; thus all classes derived from *sizeable* must provide an implementation of the $\_\_len\_\_$ method (multiple inheritance comes in very handy here, as this allows any class to be derived from a plethora of very elementary foundations), and it is thus guranteed that any class that has *sizeable* among its ancestors will provide $\_\_len\_\_$.

sizeable

This is related to the "type classes" of HASKELL, which yield information about whether any data type is intrinsically amenable to equality testing, ordering etc., as well as to the "interfaces" of JAVA, the "mixins" of RUBY, and the "traits" of SMALLTALK, even though the underlying intentions may be different.

In particular, PYTHON-3000 implements a hierarchy of Number → Complex → Real → Rational → Integral where A → B means "each member of B is also a member of A", and a pair of "Exact"/"Inexact" classes to capture the difference between floats and ints. This was inspired by HASKELL and SCHEME.

22.7.3 Exceptions revisited.
In PYTHON, exceptions are objects. That is to say, traditionally an exception could be of any class, such as files, lists, dictionaries...[9], and there was thus no guarantee that an exception possessed any particular functionality. Now, with the advent of ABCs, this has been solved neatly: All exceptions must be derived from `BaseException`.

In accordance with this, the syntax of exception raising has been simplified: The *raise* statement takes only one parameter, which must be an exception; further details can be specified as arguments. Thus, `raise ErrorType, ErrorValue` is now `raise ErrorType(ErrorValue)`, e. g. `raise IOError( "Out of cheese" )`.

raise

This will then be caught with an "as" syntax: `except "Out of cheese" as IOError:   ...`

Exception may now be nested or chained to process secondary errors occuring during processing of the primary exception.

22.7.4 Super – not diesel.
As we all know, a derived class may override the methods of its base class (p. 198). It is also evident that sometimes a derived class may still want to call some of the overridden base class methods – that is to say, it may be desirable to extend these methods, rather than to replace them. In PYTHON, this has always been a somewhat delicate issue when compared to, e. g., JAVA, since PYTHON allows multiple inheritance; thus there is no unambiguous definition of what the base class is.

---

[9] Why do people keep on devising the strangest things?

super

In PYTHON-3000, the function *super* yields any object's base class. To this end, it may take the name of the base class. If there is only one, this may be omitted. Thus:

```
class cologne:
    def tuennes(self): print(47)

class cologne2(cologne):
    def tuennes(self): print(11)
    def schael(self): super().tuennes()

c1, c2 = cologne(), cologne2()
c1.tuennes()
c2.tuennes()
c2.schael()
```

As will be seen, *schael* uses the *tuennes* method of the base class, rather than of its own derived class.

22.8 Class and function decorators

Decorators – similar to JAVA's "annotations" and not to be confused with the "decorator pattern" in statically typed languages such as JAVA or C^{++} – are among the rare examples of declarative programming constructs in an essentially imperative language.

decorator

A *decorator* consists of a line preceding a definition and beginning with an "@" sign, and it tells the interpreter that the following definition is to be read as the result of the decorator when applied as a function to the definition. In other words, the function which is "decorated" is passed as an argument to the decorator, and the result replaces the original function. Thus, if we have the following mechanism to display messages on screen:

```
def zenity_error_message( messageproc ):
    """Invoke X11 utility 'zenity' to display an error box, if any error occu
    message = messageproc()
    if message != None: os.system( "zenity --error --text=" + message )
```

then

```
class xfile:
    @zenity_error_message
    def __init__( self, name, mode="r" ):
        try: self.__filepointer__ = file( name, mode )
        except:
            self.__filepointer__ = None
            return "Could not open file " + name
```

is equivalent to

```
class xfile:
    def __init__( self, name, mode="r" ):
        try: self.__filepointer__ = file( name, mode )
        except:
            self.__filepointer__ = None
            return "Could not open file " + name
    self.__init__ = zenity_error_message( self.__init__ )
```

This is supposed to contribute to code reutilization and thus enhance productivity.

The same kind of thing is also possible with classes:

```
@foo
@bar
class A: pass
```

is equivalent to

```
class A: pass
A = foo( bar( A ) )
```

It should be kept in mind here that in PYTHON classes (and metaclasses) are functions which return the desired instances. So this goes well with metaclasses, too.

22.9 Miscellaneous

22.9.1 The "with" and "as" statements. The `with` statement "is used to wrap the execution of a block with methods defined by a context manager".

Let's have a closer look at this. Normally, one would simply use a variable to define things which are relevant inside a particular block:

```
for line in file( "/etc/passwd", "r" ):
    print( line.rstrip() )
```

Basically, there is a very plain and fully working mechanism for ensuring that the variable is not referred to after the block: Just don't do that. Sooner or later, the program will enter a state (e. g. by leaving the function which the block of interest belongs to) where it will be logically cogent that the variable used therein are no longer valid. Objects created locally within a block are then "finalized"; in the case of a file object, all pending read/write operations will be performed and the file will be closed before the object is vaporized (see "Extreme Unction", p. 190).

It the vast majority of cases, this is completely satisfactory. However, there may be occasions when one might want to make sure that the "finalization" is performed immediately and not some day in the hypothetical future. In the generator   present case, we might do this by writing a *generator* for accessing the contents of the file[10]:

```
def opened( filename, mode="r" ):
    f = file( filename, mode )
    try: yield f
    finally: f.close()
```

yield   Thanks to the *yield* keyword, this provides not a simple function but a generator which can be iterated over, just as the normal "file" object. This is used with the "with" and "as" statements:

```
with opened( "/etc/passwd" ) as f:
    for line in f: print line.rstrip()
print( "That's all, folks." )
```

At the end of the block, i. e. before executing the "print" line, the "finally" branch of the generator is executed, thereby closing the block.

Of course, the sense of this is not easy to see. Why should this be preferable to the following solution?

```
f = file( "/etc/passwd", "r" )
for line in f: print( line.rstrip() )
f.close()
print( "That's all, folks." )
```

This question is not naïve. In fact, this alternative solution is much more straightforward. However, it should not be forgotten that this file example is for illustrative purposes only. In more complex situations, the availability of the "with" mechanism greatly contributes to modularization and thus clarity. In particular, nested data structures such as XML and multi-threaded programs greatly benefit from this.

To sum it up, the "with... as" statement is a mechanism to tell the PYTHON interpreter that a particular object is to be used only inside the subordinate block, thereby allowing it to proceed more efficiently.

22.9.2 The "nonlocal" statement.
This allows to assign a value directly to a variable in an outer (but still non-global) score[11].

22.9.3 Extended Iterable Unpacking.
What holds true for parameter lists also holds true for plain assignments. Thus now you can write things like

---

[10] More about iterators and generators will be said in vol. 2.

[11] Whatever that may be good for. Obviously, there *must* be some use. However, the argumentation of PEP3104 is far from convincing.

a, b, *rest = some_sequence and get a sensible result[12]. And even *rest, a = stuff. The rest object is always a list; the right-hand side may be any iterable piece of data (e. g. a list or an object providing an iterator).

Today is the first day of the rest of the mess.

---

[12] Biological evolution has a general tendency to remove constraints; for example, the skeleton of a mammal is much simpler than that of a fish while at the same time providing more degrees of freedom. It seems that in programming languages in general this tendency can also be observed.

CHAPTER 23

Anna will Return

> Know the smallest things and the biggest things, the shallowest things and the deepest things. From one thing, know ten thousand things. When you attain the Way there will not be one thing you cannot see. You must study hard.
>
> – Miyamoto Musashi: *Go Rin no Sho*

Congratulations: You have made it to the very end of Volume 1! That is to say, by now you are familiar with the essentials of algorithmic thinking in general and its implementation in PYTHON in particular.

In Volume 2, we will have a look at some applications of what we have learned so far. In doing so, we will endeavour to strike the proper balance between "scientific" and "technical" issues.

Among the main topics, which we will consider again and again from different aspects, there will be bottom-up simulations, i.e., the art of simulating the behaviour of highly complex systems by registering the results of the combined activities of swarms of small independent units, which is the key to many fascinating biological problems (biological systems generally being characterized in that they are of outstanding complexity). At the same time, we will have a closer look at the way in which information can be "digested" and processed by computers; to this end, we will implement a LISP interpreter of our own.

Furthermore, we will meet the BIOPYTHON package and tools for professional development, both of which have the potential of being highly valuable to the experimenting biologist; as well as advanced technical subjects such as the basics of graphical user interfaces, server-client architectures, three-dimensional graphics and finally, as the crowning "technical" achievement of the course, a complete virtual machine with assembler and BASIC compiler included.

In addition, we will talk about real-life "scientific" applications such as some of the most commonly used algorithms in bioinformatics, e.g., sequence alignment and construction of phylogenetic trees. You will learn how to implement these on your own if desired, which will have the additional benefit of enabling you to critically appraise the respective merits of these approaches, the results of which are all too often considered as revelations that must never be questioned. Be warned, however, that the scope of this book does not allow much space

---

Bioinformatics Programming in Python. Rüdiger-Marcus Flaig
Copyright © 2008 WILEY-VCH Verlag GmbH & Co. KGaA, Weinheim
ISBN: 978-3-527-32094-3

for detail; so if you are (as I seriously hope) interested in learning more, be prepared to do some more research of your own.

And of course, there will be exemplary solutions to both parts. Do not spoil things for yourself by looking them up too early!

Thus, before you even open the second volume, I propose that you do the following:

- Turn back to the table of contents, peruse it and give a brief summary (preferably in writing) of each chapter and section listed therein.
- Scan the questions and exercises and check whether you are still able to understand all the solutions you have developed so far; if not, find out why[1].
- Look around and see whether there is any interesting real-life problem which you might be interested in solving by using the skills you have acquired, or are just about to acquire, and try to tackle this problem while beginning work on Volume 2.
- Browse the internet and find out about other programming languages. The Wikipedia (http://www.wikipedia.org) is a good starting point, and most programming languages can be downloaded for free nowadays. Try to solve some problem which you have found interesting once more, but in a different language (no matter whether it is an everyday language like JAVA or C^{++}, or an oddball like CLEAN or NEMERLE), and pay heed to the differences. Try to get a feeling for the pecularities of each language.
- Find a few bright minds and discuss things with them.

Oh, there *is* more you want to do? You would love to follow the kensei's advice and get a bit more practice? If so, then I suggest that instead of learning by heart the library functions of your favourite programming language, you ponder some of the following questions, all of which are directly linked to information and related phenomena:

SETI project

SETI@home

- The *SETI project* (Search for Extra-Terrestrial Intelligence) has recently aroused much public interest, not least because of the ingenious *SETI@home* program where millions of private computers donated idle processor time, thereby forming a cluster of what might be called "collective intelligence". From a information technologist's point of view, what is your estimate of the chances that we might ever intercept and identify as such the internal communication of an alien species? Discuss.

---

[1] If in doubt, blame it on insufficient documentation.

- Do you think there might be a multi-cellular organism comprising cells with fundamentally different genomes, i. e. cells of different species? If so, what might it look like?

- You want to model the expression pattern of a living cell. What kind of representation would you consider appropriate? Which programming language is a rational choice?

- Creationists are currently rallying under the new banner of *Intelligent Design*, pretty much to the annoyance of serious scientists. Review what is known as "collective intelligence" or "emergence" and keep them in mind while considering the arguments presented by both sides in the aforementioned quarrel. Is there any sensible reconciliation between Darwinian evolution and Intelligent Design conceivable; in other words, can you think of any theory embracing both of them? Does this theory require any supernatural influences? Explain[2]. Intelligent Design

- When you compare the immune system and the human nervous system, you will find an astonishing number of parallels, the first of which to be discovered was probably the fact that the CD4 signal transduction cell surface molecule, which is abused by the HIV virus for entering cells, is present both on immunocytes and nerve cells. Later research showed that immunocytes actually communicate by temporarily forming a structure which is so similar to the conventional synapse that it has been termed the *immunological synapse*. Now try to see all this from the point of view of IT. Are there any common goals to be solved by the design of the nervous system and the immune system, respectively? If so, what are they? In which points do they differ? Re-consider the similarities and make a statement as to whether they are really surprising. immunological synapse

- The evolution of the central nervous system of our species has long puzzled researchers (a fact which is gleefully quoted by Intelligent Design supporters, see above): From our current knowledge of molecular biology, the maximum rates of nucleic acid change and thus the amount of genetic information that can be accumulated over "deep time" can be estimated. The swiftness with which the human brain evolved from simian intelligence to its current higher[3] level thus suggests that in fact the amount of additional information (compared to the simian situation) required to form a human brain is minute. Applying all you have learned about information processing so far, do you get any idea how so much more complex a structure might be built with only a limited

---

[2] For your inspiration and relaxation, here's an exceptionally well-written thriller that might give you valuable hints: "The Swarm" by Frank Schätzing.

[3] At least in some of us.

amount of additional information? Hint: Consider the immune system, and keep in mind that *Homo sapiens* is quite unique in having by far the longest infancy of all animals[4].

- A crystal, a colloid, a gas: Which of these systems are easily described by mathematics, which are not? What does this mean for the number of *meaningful* different states in each system? How is this mimicked by modern computer technology? Hint: There are two factors to be considered, mobility and structure.

- Daneel Olivvar, the humaniform robot featured in a number of novels by Isaac Asimov, sometimes exhibits a degree of creativity that far exceeds that of his makers. Do you think that, given virtually unlimited computing power, it might one day be feasible to design a system actually capable of creative thinking? If so, what might be required? Hint: Re-think the previous question.

- Kant meticulously differentiated "understanding" and "reason", the former being the power of the mind to recognize things, and the latter, to handle abstract concepts. Is there any equivalent to this dichotomy in information technology?

- By now, science has become very good at gathering large amounts of data. The problem is that often many factors are determined together, and it is not always easy to make sense of this, i. e. to determine which combination of "input" factors is connected to any particular outcome – and how. Fuel mixtures are a textbook example: Characteristically, they comprise differing ratios of five different hydrocarbon preparations, and the resulting octane number is difficult to predict. This is called a multi-dimensional representation – in this case, a 5-dimensional one, as each mixture corresponds to a point in 5D space wherein each dimension is given by the relative amount of the respective component. How can information technology help to further our understanding of such complex systems?

Information technology *does* provide answers to all of these questions, and they will be given in the second volume. Of course, it cannot be guaranteed that our fundamental assumptions are correct, but there is an internal logic to each. By contemplating these questions, you will learn more than by "cramming" details.

---

[4] Kudos to Olaf Stapledon, who imagined the further development of our species (*Last and First Men*) and assumed that in future mankinds infancy and childhood would be extended to centuries and finally millenia.

Procrastinate later – the lemmings are gaining on you!

Glossary

address: The place in memory where a piece of information is kept. In a wider sense, any location of data.

abstract data type: A datatype based on another one, e.g., a list, an array or a stack.

AIX: →Unix.

algorithm: Mechanism for doing a calculation or other information-related operation.

array: Fixed-length group of data items of identical type.

ASCII: A standardized table for the numbers of characters (→byte).

Bayesian statistics: An unconventional approach to statistics which is suitable for biology.

bioinformatics or *computational biology*: Application of computers to biological problems.

bit: The unit and smallest piece of information. A bit may have only one of two states, ON (True) and OFF (False).

boxers, tennis players and racing-drivers: Germany's future.

BSD: →Unix.

bug: Error in a program.

byte: The smallest group of →bits forming a sensible unit. Normally a byte comprises 8 →bits and represents a single character (one out of a standardized table of 256).

closure: A →function produced by "concretization" of another function by filling in particular values.

compiler: Program that converts an →algorithm from a higher programming language into processor code.

currying: Deconstruction of a multi-parameter →function into a sequence of single-parameter →λ functions, thereby effectively generating →closures.

cytometer: →FACS

debugging: The process of removing →bugs.

Bioinformatics Programming in Python. Rüdiger-Marcus Flaig
Copyright © 2008 WILEY-VCH Verlag GmbH & Co. KGaA, Weinheim
ISBN: 978-3-527-32094-3

directory: A group of →files (and other directories) logically associated under a "company name".

disk: A mechanical (magnetic, optic or magneto-optic) →persistent storage device.

driver: (1) A piece of →software allowing software control of →hardware (printer driver, keyboard driver, etc.). (2) A person who makes several millions a month by racing in circles.

dynamic programming: A technique for dealing with situations comprising a very high number of possibilities, of which only a few are interesting.

editor: A program like a word processor used for writing the →source code.

embedded processor: A computer that is hidden within another device (a camera, a palmtop, a cellphone, an oven...)

EMBOSS: An Open Source tool collection for →bioinformatics.

FACS (fluorescence activated cell scanner): Device for raising data from single cells.

Feynman: gave a speech in 1959 which sparked off the miniaturization of computers.

file: A logically coherent block of data, generally either an executable program or a document, accessible under an individual name, on a →persistent storage device.

file system: The way in which the operating system manages the files on the →permanent storage devices of a computer. All modern file systems use →directories to create a hierarchical arrangement of files. They differ, however, in parameters such as the maximum file size and file system capacity and whether they offer failsafe facilities (e.g., "journalling", whereby any transaction is recorded, thus facilitating recovery after a hardware crash). File systems to be reckoned with are EXT3 and REISERFS (Linux), XFS and ZFS (various Unices), FAT and NTFS (Microsoft).

filtering: Selection of those components of a →list which fulfill a certain criterion.

FreeBSD: →Unix.

free software: Software distributed under a license that permits changes (thus all free software is →open-sorce too) and redistribution.

GUI: Graphical User Interface. A →shell that is not text-based but rather uses graphical signs and tokens, usually in the WIMP (Windows, Icons, Mice, Pull-down menus) form.

hannaḥ: King cobra.

hardware: The tangible parts of a computer system.

Horn clauses: Logical structures underlying declarative languages.

HUSAR: Another →bioinformatics package maintained at the DKFZ, Heidelberg.

IDE: Integrated Development Environment. A program that comprises →editor, →debugger and menu-driven control of →compiler and →linker – and/or other programs as deemed required – to facilitate writing larger programs. Especially important for languages which are inclined to excessive splitting of files (e.g., JAVA and EIFFEL and in particular C♯).

infix: Operator between operands $(a+b)$.

interface: A thing which connects other things, whether on the →hardware (printer interface etc.) or on the →software level.

interpreter: Program that executes an →algorithm written in higher programming languages.

iteration: The act of, or a strategy based on, processing data by →looping.

kernel: The innermost part of an →operating system which handles elementary things such as memory allocation and keeping track of what programs are running and what privileges they have (which makes it crucial to security).

λ *calculus*: Logical formalism underlying functional languages.

λ *function*: A function producing another function.

library: →Module containing a collection of standard functions.

linker: A program that welds together the output of the →compiler and the required →libraries to build an executable program.

list: Variable-length group of data items (usually of identical type).

list comprehension: A shorthand form for selecting items of a →list that fulfill certain criteria. Virtually equivalent to →filtering.

loop: A part of a program which is executed several times.

MacOS-X^{TM}: →Unix.

mapping: Application of a function to all members of a →list or (rarely) an →array.

Miyamoto Musashi: A samurai whose reflections on sword-fighting are still interesting today.

module: A part of a program.

NetBSD: A version of →Unix that emphasizes portability.

neur[on]al networks: A technology which tries to emulate natural nerve systems for recognition of patterns.

object: A combination of executable code and data items.

OpenBSD: A version of →Unix that emphasizes security.

open source: Software distributed under a license that requires its source code to be available.

operating system (OS): A piece of →software which does the housekeeping work of a computer (input/output, memory management, disk access, networking, etc.). Normally comprising one →kernel and lots of →drivers which are required for handling hardware, providing a common interface to different brands of chipsets. Examples for widely used OSes are (i) the →Unix-like family, comprising Linux and the genuine Unices (e.g., the BSDs including MacOS-X, Solaris and AIX); (ii) VMS; (iii) Windows^{TM}.

path: The place within a file system wherein a particular →file is kept.

PCR: Polymerase chain reaction. A technique for amplifying DNA.

persistent storage: Any device that will store data in such a form that it can exist without energy supply. Commonly used forms of persistent storage are: (i) hard magnetic disks such as common hard disks and JAZ drives; (ii) soft magnetic disks (obsolescent) such as floppy disks and ZIP disks; (iii) magnetic tapes (obsolete), (iv) optical disks (CD, DVD, Blu-Ray^{TM}); and (iv) flash memory, often found in "USB sticks".

pixel: A point on the screen. A pixel may be represented internally by a bit (black/white), a byte (256 colours or shades of grey) or a group of three bytes (256 shades of red, green and blue, resulting in a total of $256^3 = 16.7$ millions of colours).

postfix: Operator after operands, as in FORTH (*ab*+).

prefix: Operator before operands, as in LISP (+*ab*).

Python: (1) A beautiful serpent. (2) A programming language.

pointer: A piece of data that refers to another one by holding its →address.

Posix: A common standard for the "C operating system", i. e. all flavours of Windows^{TM} and all →Unix-like →operating systems.

processor: The part of the computer which does the actual work.

RAM (random access memory): The part of the computer which stores data in volatile but swiftly accessible form. Upon switching off the computer, all data in RAM will be lost. However, in contrast to →permanent storage devices, all data in RAM can be accessed directly and very swiftly by the →CPU.

read-eval-print loop: The basic design of all command line programs such as →shells and →interpreters. Basically, three steps are repeated over and over again: (1) user input is gathered from the keyboard (usually using the RETURN key as marching orders); (2) the input is evaluated, and appropriate actions are taken (e.g., a program is started, a calculation is performed,...); and (3) the results are displayed.

recursion: The act of, or a strategy based on, a function calling itself, usually with a subset of its original data to obtain a partial solution.

registers: The →processor's internal data handling facilities.

regular expression: A pattern for text matching.

ROM (read-only memory): Chips which hold "burnt-in" information; e.g., the instructions on how to load the →operating system.

rovásirás: ancient but still used Hungarian scarved glyphs (sometimes called "Hungarian runes" but evidently much older than, and possibly the ancestor to, Nordic runes); the only writing system ever developed on Earth which has, for reasons unknown to mere mortals, not been included into →Unicode. See http://www.dsuper.net/≈elehoczk/frmain.htm for a transliteration program.

shell: That which "encloses" the operating system, i.e., the software facility for user interaction. Theoretically, this definition would also comprise →GUIs but is generally restricted to text-based shells, which resemble text terminals from the 1970s. Thus, a shell can be defined as a program that manages a user's interactions with the operating system in a typical →read-eval-print loop by: (i) reading commands from the keyboard; (ii) executing those commands; or (iii) running another program; and (iv) displaying the output. On →Unix-like systems, a number of closely related shells exist, most notably *bash*, *tcsh* and *zsh*; on Windows, there is the (primitive) "DOS command prompt" which is expected to be replaced some day by the projected *Monad* shell. Sometimes the term "shell" is also used to denote the nameless scripting language in which the aforementioned →Unix shells can be programmed.

software: The immaterial parts of a working computer system.

Solaris^{TM}: →Unix.

source code: The program as it is written by the programmer (in contrast to the machine code).

Unicode: A standardized 16-bit character encoding designed to encompass every human glyph and logogram.

Unix: A name of rather ill-defined scope, originally derived (in the 1960s) from a pun on "Multics", which was the first attempt on a multi-tasking, multi-user operating system. An operating system generically known as Unix soon developed along two major lines, which were "System V" (the commercial branch) and the "Berkeley Standard Distribution", abbreviated BSD (released by the University of California as →open source), which in turn gave rise not only to OpenBSD, FreeBSD and NetBSD but also to Solaris^{TM} and MacOS-X^{TM}, all of which are thus Unices, along with the less widely used or obsolescent System V derivatives SunOS (superseded by Solaris^{TM}), IRIX, AIX, HP-UX, Ultrix, Sinix, Xenix and Monterey. The plethora of implementations led to early standardization, so a well-written program can be made to run on any of these systems simply by recompiling it – a major advantage in particular when being used with →open-source software. It should be pointed out that both Linux and Minix are *not* Unixes in the sense of being derived from the original Unix project (see the famous "SCO case"[5]) but independently implement the functionality of the general Unix standard so that they are virtually compatible with it. Adoption of Unix on early PCs was generally hampered by the fact that it was initially designed for high-range computing and thus too demanding for 20th century hardware.

---

[5] During which it was also established that the *original* Unix is copyrighted by Novell, Inc., whereas SCO own the Unix *trademark*.

UTF-8: A character encoding which is basically →ASCII with →Unicode extensions.

virtual machine (VM): A fictitious computer existing only in the form of a program running on a real computer.

VMS: A multi-tasking, multi-user →operating system whose design had some influence on that of Windows^{TM}.

Windows^{TM}: Widely used family of proprietary →operating systems for low- to mid-range computing.

wetware: Geekspeak term for people.

workstation: PC on an(n)abolics.

Hardware: the parts of a computer that can be kicked.

Index

⊕, 279
→, 364
♠, 175
  2D arrows, 271
  Adler, 307
  Airbagged integers, 319
  Area of a rectangle, 159
  Array copying, 120
  Array limits, 329
  Arrays as objects, 194
  Basics
    HASKELL, 161
    OBERON, 272
    PASCAL, 154
    PYTHON, 154
  Bitwise operations, 279
  break/continue, 135
  Breaking the code, 282
  Built-in functionality, 270
  Call-by-name, 156
  Catalogue of restriction enzymes, 97, 329
  Circular genomes, 87
  Class structure, 109
  Classification of functions and objects, 162
  Compare languages, 363
  Concatenation, 99
  Currying, 28, 81
  Dictionary, 90
  DNA information content, 301
  Double-stranded DNA, 330
  Dynamic vs static storage, 359
  Error objects, 319
  Fibonacci, 162, 164
  Find maximum, 161
  Fubaromonas annae, 323
  GenBank, 35
  Global variables, 155
  Home-made mapping function, 122

Home-made mapping function II, 123
Hydrocarbons, 177
Illegal arguments, 155
Ilustration of data type, 162
Inheritance, 197
Insertsort with functional criterion, 264
λ, 29
LISP, 29
Loop types, 135
LZW decompression, 304
LZW sample implementation, 306
Mailbomb, 307
Massive parallelization, 124
Melting temperature of DNA, 353, 356
Members of objects, 187
Object, 97
One-time pads, 284
OO and FP, 188
Palindromic DNA, 322, 323, 329, 356
Pattern matching, 140
Pattern recognition, 306
Performance of HASKELL, 266
pFOObar, 323
Pipelines and lists, 92
Pizza program, 116
Polymorphism and default parameters, 159
Postal code, 312
Primer design, 351
Programmers' work, 45
Pubert
  interactive multiplication table, 126
  multiplication table, 125
  randomized multiplication table, 125
PYTHON and ERLANG, 29
PYTHON libraries, 271
Python Pickles Module, 343
Python String Module, 344
Random DNA, 331

Recursion
    data structures, 160
    length of lists, 160
    termination, 160
REDUCE, 130
"return", 29
Riddle solver, 40
Roots, bloody roots..., 164
Scaling of quicksort algorithm, 266
Sets and bags, 241, 271
Single-pass conversions, 305
Sorting
    Bubblesort, 262
    Quicksort, 266
Stacks, 153
The C boom, 15
Transputer, 42
Trithemization of a string, 279
try/except, 114
Undefined condition, 113
White space in PYTHON, 164
Wrapper class, 194
Zlib, 307
EMACS, 65
XJED, 65
VI, 65
20–80 rule, 1

4004, 13
6502, 101, 164

A^+, 31
Abraham Lempel, 301
abstract base classes, 388
abstract data type, 92, 379, 401
abstraction, 15, 16, 29, 30, 43, 92, 146, 188, 189, 241, 288, 330, 372
abstraction level, 16
accumulator, 129
actions, 19
Ada, 20, 26, 38, 42, 47, 48, 50, 57, 80, 100, 110, 138, 155, 207, 208, 251, 254, 256, 320, 364, 368
    Ada-83, 20
    Ada-95, 20
    call-by-reference, 26, 155
    Countess Ada Lovelace, 20
    subtypes, 80
address, 12, 379, 401
Address file example, 32
Adler-32, 309
Adobe Photoshop Lightroom, 370
AIX, 379, 383, 401, 405
Alan Perlis, 62
Alec Mellor, 261

Algol, 20, 35, 83, 101, 108, 109, 130, 138, 139, 153, 206, 207, 361, 363, 364, 370, 371
    Algol-58, 20
    Algol-60, 20
    Algol-68, 20
Algol family of languages, 20
Algol-60, 48
Algol-68, 47
algorithm, 379, 401
All information can be represented as numbers, 73
Alonso Church, 25
AMD64, 13
AmigaOS, 313
ANJUTA, 65
__annotations__, 383
ANSI-C, 371
Apache, 46
Apache-2, 369
APL, 31
Apple, 13, 37, 42, 142
    IIe, 306
    Macintosh, 37
    Newton, 22, 372
AppleScript, 38
arguments, 23, 24, 155, 377, 378, 390
    exception raising in PYTHON-3000, 389
    keyword argument, 379, 384
argv, 71
arity, 256
ARM, 14, 138
array, 2, 361, 379, 401
    definition, 84
Array languages, 31, 361
Array.map, 123
artificial intelligence, 38
ASCII, 65, 75, 143, 279, 281, 337, 379, 380, 401
Assembler, 11, 14
assembler, 14, 16, 19, 363
assembly, 360
assembly language, 13–16, 22, 138, 164, 165
assumptions, 359
Atari, 42
    ABAQ ATW800, 42
    ST, 215
Athlon, 75
atoms, 78
attributes, 36
awk, 120

B, 20, 371
Babún and Rolloway, 324

Babbage, 254
Backus–Naur, 184
bash, 383, 405
Basic, 7, 13, 20, 21, 49, 57–59, 98, 99,
    105, 108, 109, 118, 138, 139, 153,
    166, 185, 234, 237, 251, 315, 341,
    377, 395
Bayesian statistics, 379, 401
BCPL, 20
Bertrand Meyer, 238
BIFs, 269
Billyboy
    coffee slave, 202
binominal nomenclature, 35
bioinformatics, 379, 401
bit, 379, 401
Bjarne Stroustrup, 372
block marks, 108
blocks
    begin and end, 108
Blu-Ray^{TM}, 287, 308
bombe machine, 283
Bonobeau and Le Mur, 333
Boole, 74
boolean, 48, 51, 52, 74, 81, 82, 109, 110,
    161, 194, 210, 279, 314
    file of file of boolean, 161
Boolean values, 75
branchings, 20
Bruin, 47
BSD, 63, 64, 307, 379, 382, 401, 404
    FreeBSD, 64, 383, 405
    NetBSD, 383, 405
    OpenBSD, 383, 405
bubblesort, 256, 262, 263
buffer overflow, 226
bug, 15, 379, 401
byte, 379, 401
bytecode, 22
bzip2, 289

C, 15, 16, 20, 23, 28, 29, 32, 35, 43–45,
    49, 50, 53, 57, 71, 76–78, 80, 81, 83,
    93, 95, 98–101, 105, 108, 110–116,
    120, 127, 130, 133, 134, 137–142,
    156, 157, 163, 166, 168, 169, 171,
    178, 181, 183, 188, 195, 202, 206,
    207, 214, 215, 217, 220, 224,
    226–228, 238, 240, 251–253, 256,
    257, 261, 265, 266, 269, 291, 308,
    313, 314, 331, 337, 339–341, 343,
    353, 361, 362, 370–372, 375, 381, 383
    C-99, 251, 270, 371, 372
C family of languages, 20

C♯, 19, 20, 56, 115, 120, 138, 217, 219,
    381, 403
C^{++}, 20, 28, 32, 38, 43, 44, 49, 51, 56,
    81, 98–101, 105, 110, 115, 120, 127,
    137, 156, 158, 188–190, 195, 200,
    214, 215, 217, 219–224, 227, 238,
    251, 256–258, 307, 308, 313, 318,
    362, 371, 372, 374, 378, 390, 396
cache, 13
Caesar cipher, 278
call stack, 153
Call-by-reference, 26, 155
CAML, 23, 29, 84, 108, 158
    CAML Light, 29
    o'CAML, 16, 29, 30, 35, 38, 43, 50, 51,
    81, 86, 93, 97, 100, 105, 107, 110,
    118, 119, 121–123, 136, 140, 142,
    155, 172, 198, 214, 216, 238, 247,
    249, 251, 269, 315, 328, 341, 382
cast, 217
censorship in a free country, 285
Chandra gupta, 323
checksum, 309
Christian Grabbe, 321
class, 385, 387, 388
    as function, 391
    base class, 389
        "type", 388
        abstract base class, 388
        methods, 389
    class decorators, 391
    datetime, 386
    definition
        in PYTHON-3000, 388
    derived class, 389
    exceptions, 389
    Formatter (PYTHON-3000), 387
    inheritance
        multiple, 389
    metaclass, 388
    superclass in PYTHON-3000, 390
classes
    description, 188
Clean, 29, 122, 172, 251, 368, 378, 382,
    396
closure, 379, 401
Cobol, 14, 19, 20, 27, 32, 50, 56, 92, 108,
    134, 137, 178
code, 15
codepages, 306
Comal, 20, 49, 108, 155, 317
Comit, 162
command line, 15, 66, 68, 70, 71, 144,
    279, 280, 346

command line arguments, 147
command line shell, 66, 70, 71
comments
    Why?, 105
Common Lisp, 53, 269, 270, 366
Compilation process, 16
compiler, 15, 16, 23, 28, 40, 48–53, 57, 78, 98, 105, 106, 110, 128, 158, 161, 162, 188, 189, 247, 249, 316, 368, 372, 379, 401
Component of structures, 35
composition operator, 151
computational biology, 379, 401
Concurrency
    Concurrency-oriented languages, 41
Concurrent ML, 29, 42
cons, 365
conserved structures, 239
constructor, 190
controlling formatting on a per-type basis, 386
CP/M, 47
CPU, 11, 251
CRC, 309
Cryptography
    project, 277
currying, 25, 28, 81, 379, 401
CVS, 258
Cygwin, 63–65
cytometer, 379, 401
cytometry, 177

Dan Browne, 283
Daneel Olivvar, 398
data aggregates
    heterogeneous, 93
    homogeneous, 84
data type, 23, 32, 33, 48, 51, 54, 57, 58, 74, 80, 84, 89, 90, 97, 98, 118, 126, 131, 161, 162, 187, 194, 214, 224, 241, 314, 315, 319, 328, 340, 372, 394, 416
data types, 73
    enum, 77
    Examples, 48
    in Ruby, 118
    primitive vs. derived, 161, 194
    range in Pascal, 126
    recursively defined, 161
    scalar, 74
    set, 90
    stack, 131
    strings, 97
    struct, record, 93
    subtypes, 80

type classes in Haskell, 80
variant record, 94
Dawkins, 176
De bello Gallico, 278
Debian, 64
debugging, 379, 401
decorator, 390
deep breath, 147
definitions, 23, 256
Delany, 17
Delphi, 20, 32, 38
Dennis Ritchie, 15
dependency, 255
Descartes, 74
design by contract, 253
destructors, 190
diamond structure, 255
dictionary, 382, 388, 389
    in PYTHON-3000, 385
Digital Fortress, 283
diglyphs, 301
Dijkstra, 5, 18, 27, 47, 363
    disliking COBOL, 27
    disliking FORTRAN, 363
    disliking PL/1, 47
Dirac, 176
directory, 380, 402
disk, 380, 402
DNA, 382
    sequence
        analysis, 382
docstrings, 106
DOS command prompt, 71
Dr. Mabuse, 106
Dr. Meerkatz, 323, 333
driver, 380, 402
DR. PYTHON, 66
dynamic memory allocation, 217
dynamic programming, 173, 380, 402
dynamic storage, 359

Easter Island, 306
EBI, 1
ECLIPSE, 65
Eden, 42
Edgar Allan Poe, 278
editor, 63, 380, 402
Edmund Spenser, 311
Edward Gibbons, 227
Eiffel, 38, 120, 138, 194, 200, 217, 238, 251, 253, 254, 258, 312, 375, 381, 403
Eiji Yoshikawa, 7, 358
Einstein, 67
ellipsis, 35

Ellipsis (..), 35
EMACS, 65
emacs, 237
embedded devices, 363
embedded processor, 380, 402
embedding, 362
EMBOSS, 2, 3, 352, 380, 402
endianness, 225
ENIAC, 15
Enigma, 279, 282, 283
ensure, 253
ERIC, 66
Erlang, 15, 23, 28, 29, 34, 40, 41, 44, 53, 57, 78–80, 86, 87, 97–99, 105, 110, 111, 116, 121, 122, 136, 140, 142, 171, 172, 208, 234, 251, 252, 256, 315, 316, 318, 341, 362, 365, 368, 382
Error, 311
   BASIC, 315
   in ERLANG, 316
   signalling by return value, 313
error object, 128
Examples
   HASKELL, 244
exception, 389
   chains in PYTHON-3000, 389
   raising in PYTHON-3000, 389
Exceptions
   Raising of, 344
Explicit Conversion Flag, 386
extending, 362
Extending and embedding PYTHON, 43

F♯, 29
FACS, 380, 402
factorial, 140
Feynman, 380, 402
Fibonacci, 162, 164, 168
file, 15, 146, 379–381, 389, 391, 402
   access
      generator, 392
   as object, 391, 392
   binary, 381
   GenBank, 382
   name, 381
   text, 381
   unicode, 381
file system, 31, 380, 402
filtering, 380, 402
flag, 74, 110, 138, 147
   error flag bit in o'CAML, 314
   processor flag in the Acorn ARM, 138
Flight 501, 319
flow charts, 20

FooBar, 14, 17, 142, 360
Foobar
   "throwing" and "catching" of
      exceptions, 316
   Docstrings in PYTHON, 106
   invention, 341
   Lambdoid nature of PYTHON
      functions, 158
   Pattern matching in ERLANG, 140
   Type classes in HASKELL, 50
   Types in PYTHON, 53
Foobasic, 185, 251
for, 146, 280
FORALL, 361
Formac, 162
Format Specifiers, 385
Format strings, 384
Formatter Methods, 387
Forth, 20, 22, 99, 181, 207, 372, 382, 404
Fortran, 14, 19–21, 29, 30, 35, 49, 50, 58, 84, 99, 108, 110, 118, 121, 134, 135, 139, 154, 252, 256, 361, 363, 366
   Fortran-2003, 20
   Fortran-2008, 20, 71, 361, 363
   Fortran-66, 20, 21
   Fortran-77, 20, 134, 220, 363
   Fortran-90, 20, 108, 251, 361
   Fortran-95, 20, 361, 363
   Fortran-I, 47, 162, 363, 366
   Fortran-IV, 20, 166
Frank Schätzing, 397
free software, 380, 402
FreeBSD, 64, 380, 402
Freemasons' ritual, 61
fubarase, 323, 333, 334, 347, 353, 357
   sequence, 324
Fubaromonas
   F. annae, 323
fun, 59
function, 23
Functional languages, 23
Functional parameters
   Benefits, 52

Gaius Iulius Caesar, 278
Gaius Iulius Octavianus, 277
garbage collection, 217
gcc, 16, 372
GEDIT, 65
generator, 392
Gentoo, 64
Georg Büchner, 233
George Orwell, 105
GIF, 303
glossary, 6

GNU, 257
  Richard M. Stallman, 257, 370
  the Ada compiler, 250
  the GNU C compiler, 16, 70, 250, 257, 363
  the GNU Pascal compiler, 250, 251
  the GNU project, 257
GNU C compiler (gcc), 16, 70
Go Rin no Sho, 5, 395
Goethe, 1, 201, 287
Gollum, 277
Gont, 139
graphical user interfaces, 37
guards, 368
GUI, 204, 205, 380, 402
Guido van Rossum, 57, 342, 343, 388
guru
  and λ functions, 158
  Dijkstra, 4, 27, 47, 363
  meditation error on AmigaOS, 313
  scanner, parser and back end, 254
GVIM, 65
gzip, 289
Gøteborg ML, 29

hardware, 380, 402
Haskell, 16, 23, 24, 26, 29, 35, 40, 50–52, 54, 56, 57, 80, 81, 84, 86, 87, 98–100, 105, 106, 109, 110, 119, 121, 122, 125, 126, 129, 136, 140, 142, 151, 155, 158, 160, 161, 207, 214, 216, 224, 230, 244, 245, 251, 261, 262, 265, 266, 279, 311, 361, 368, 382, 389
  Haskell-98, 29, 86, 251
Haskell B. Curry, 25
head→tail, 34, 96, 156, 160, 230, 362
header file, 220, 221, 252
header files, 257
heap, 218
heapsort, 267
hexcode, 83
high-level, 31
higher programming languages, 14
Hint
  "code", 15
  Advantage of interpreter, 16
  Agarose gels, 333
  by Dijkstra, 4
  Characteristics of functional programming, 27
  Complex numbers, 107
  CVS, 258
  Danger of espionage, 285
  Electrophoresis
    Agarose gels, 355
    Blots, 353
    Low MW bands, 356
  GenBank reader, 328
  Green fluorescent protein (GFP), 347
  Importance of comments, 105
  Irregular DNA, 353
  Meaning of OO, 38
  Modules *sys* and *file*, 280
  Natural amino acids, 354
  Objects and inheritance, 241
  PCR methods, 351
  PCR troubles, 355
  Prettyprinting, 332
  Static *vs* dynamic structures, 331
  Strength of typing, 48
  Types and variables, 189
  Understanding algorithms, 303
  White space in PYTHON, 164
Hoare, 41
Homunculus, 201
Horn clauses, 380, 402
HP-UX, 383, 405
HTML, 45, 46, 105
Hungarian notation, 35
HUSAR, 380, 402
Husar, 3
hypothesis generator, 288

IBM, 13
IDE, 250, 381, 403
identifiers, 30
IDLE, 65
iff, 147
immunological synapse, 397
impact factor, 143
implementation, 239, 256
import, 146
Incident
  Saturn V, 58
  Flight 501, 319
#include, 220
inference and polymorphisms, 48
infix, 381, 403
inheritance
  multiple, 389
initializer, 190
Inmos
  transputer, 42, 372
Innenwelt, 67
Input, 11
input, 379, 383, 384
  complex factors, 398
  with and without evaluation, 383
INRIA, 51

insert sort, 148
Insertsort, 263
insertsort, 263
InstallShield^{TM}, 64
instantiation, 188
instruction codes, 12
integers
    unsigned, 76
integrated development environment, 65
Intel
    8086, 14
    8088, 14
    80286, 14
    x86, 13, 14, 101, 225
        assembly language, 138
Intelligent Design, 397
interface, 239, 256, 389
interpreter, 16, 39, 90, 96, 106, 107, 124,
        136, 162, 190, 191, 195, 270, 313,
        315, 318, 337, 369, 381, 403
IRIX, 383, 405
Isaac Asimov, 398
Isaac Newton, 313
isomorphic, 123
Itanium, 13
iteration, 381, 403
Iverson notation, 31

J, 31
J. R. R. Tolkien, 277
Jacob Ziv, 301
Java, 19, 20, 23, 32, 38, 44, 49, 51, 54,
        56–58, 82, 98, 99, 101, 105, 109, 110,
        115, 120, 127–130, 137–139, 142,
        156, 158, 163, 188–191, 194, 195,
        197, 199, 200, 203, 205, 206, 208,
        214, 215, 217, 219–221, 223, 224,
        228, 230, 238, 239, 251–254, 256,
        257, 261, 296, 315, 316, 338, 362,
        363, 374, 375, 378, 381, 383, 389,
        390, 396, 403
JavaScript, 45, 55
Joanne Rowlings, 63
John Backus, 29, 363
jumps
    "go to", 138
    "gosub", 153

K, 31
K&R, 120, 269, 371
Kafka, 359
Kamasutram, 282
Kant, 67, 254, 398
Kary Mullis, 348
KATE, 65

KDE, 56, 65, 345, 362
KEDIT, 65
kernel, 381, 403
key, 31, 384, 385
keyword parameters, 377
keyword-only parameters, 378
Klingon computers, 254
Ksi, 251
Kyte–Doolittle, 334

λ calculus, 23–25, 381, 403
λ function, 24, 381, 403
Larry Wall, 342
lazy evaluation, 382
least significant digit, 83
Lenin
    Wladimir Uljanowitsch, 131
Leuchttonne, 282
Levenshtein, 24, 229, 230
library, 17, 381, 403
    linking together with, 17, 256
library files, 256
library function, 381, 396
    don't learn by heart, 396
    strcpy(), 226
library functions, 3, 43, 44, 58, 208, 261,
        269–271, 308
    longjmp(), 140
    os.path.walk(), 205
    strcmp(), 110
Line feed and carriage return, 337
linker, 16, 221, 252, 256, 258, 269, 381,
        403
Linné, 35
Linus Torvalds, 372
Linux, 63–65, 266, 307, 345, 382, 383,
        404, 405
    Black Lab, 63
    Debian, 64
    Gentoo, 64
    Red Hat, 63, 64
    SuSE, 64
    Ubuntu, 64, 66
    Yellow Dog, 63
Lisp, 7, 14, 28, 29, 46, 47, 53, 59, 78, 85,
        86, 93, 94, 99, 100, 102, 103, 105,
        107, 110, 113, 118, 138, 151, 162,
        181, 197, 237, 251, 255, 341,
        361–363, 365–368, 372, 377, 382,
        388, 395, 404
list, 381, 382, 387–389, 403
    being replaced by generators, 382
    comprehension, 382
    of parameters, 378, 392
list comprehension, 128, 381, 403

List.map, 123
lists
    as objects, 230
    head–tail, 86
local variables, 19
Logo, 58, 118, 363
loop, 381, 403
    as alias for mapping in PYTHON, 124
    combined, 137
    "for"–"next", 118
    generalized "for", 120
    instead of mappings in PYTHON-3000, 382
    non-rejecting, 132
loops, 20
low-level, 31
Lua, 20, 30, 44, 45, 119, 362, 370, 371
LZ78, 301
LZW, 300–303, 306, 307, 340

Machine control languages, 42
machine language, 13
machismo, 361
MacOS-X^{TM}, 63, 381, 383, 403, 405
mailbomb, 300
make, 257
making tracks through possibility space, 176
malloc, 178
Malthus, 283, 287
MAP, 116, 122–124, 130, 164, 281
Maple, 108
mapping, 381, 388, 403
Mark Adler, 309
Mary Queen of Scots, 278
masterhood, 1, 6
math, 346
Matz, 369
memory, 382
memory leak, 156, 219
Mercury, 40, 41, 172, 251, 363
Merkmale, 67
metaclass, 388
methods, 36
Micro-Willy, 321, 323, 324, 326, 332, 333, 347, 351, 354–357
Microsoft, 381
Microsoft Office^{TM}, 362
Middle Egyptian, 301, 302
Milner, 29
Minix, 383, 405
MIPS, 372
Miranda, 29, 121, 368, 382
mixin, 389
Miyamoto Musashi, 5, 381, 395, 403

mnemonics, 13
modifiers, 114
Modula-2, 20, 100, 108, 195, 207, 253, 364
Modula-3, 20
module, 381, 403
module hierarchy, 255, 257
Monad, 383, 405
monomorphic, 51
Monterey, 383, 405
Moscow ML, 29
Moses Schönfinkel, 25
Motorola, 13
    68000, 14
    68020, 14
    68030, 14
mouse
    and SMALLTALK, 37
    input device, 11, 12
MS-DOS^{TM}, 13, 337
multi-dimensional representation, 398
multiprocessor, 31, 39, 42, 117, 123, 206, 290, 361, 363
Mycoplasma genitalium, 326

name server, 31
Naval Enigma, 283
NEDIT, 65
Nemerle, 378, 396
NetBSD, 381, 403
neur[on]al networks, 381, 403
NeXT, 372
Nikolaus Wirth, 28, 154, 251, 361, 364
non scholae sed vitae, 4
None, 142, 233, 329, 330
numbers used as references, 73
numbers with intrinsic meaning, 73

Oak, 20
Oberon, 20, 32, 38, 57, 100, 135, 139, 197, 207, 217, 247, 251, 253, 258, 272, 361, 364, 365
object, 381, 403
    exception, 389
    finalization, 391
        now or never in PYTHON-3000, 391
object code file, 16, 256
Object-oriented languages, 32
Objective C, 20, 120, 195, 215, 217, 222, 372
Objective Haskell, 188
objects
    banking object, 205
    explicit destructors, 190
    generatio spontanea, 213

range in RUBY, 126
Shelf object, 343
Occam, 42, 109, 116, 363, 372
octane number, 398
Olaf Stapledon, 67, 398
one-time pad, 284
open source, 381, 403
open-source software, 63
OpenBSD, 381, 403
OpenOffice, 47, 236, 251, 362
operating system, 13, 15, 23, 59, 68, 71, 153, 234, 269, 346, 359, 360, 371, 372, 382, 404
  CP/M, 47
  FreeBSD, 64, 383, 405
  Linux, 63–65, 266, 307, 345, 382, 383, 404, 405
  MacOS-X^{TM}, 63, 381, 383, 403, 405
  Minix, 383, 405
  MS-DOS^{TM}, 13
  NetBSD, 383, 405
  Newton, 22
  NeXTStep, 372
  OpenBSD, 383, 405
  ProDOS^{TM}, 13
  RTOS, 42, 137
  Solaris^{TM}, 63–65, 307, 383, 405
  Unix, 15, 371
  Windows^{TM}, 23, 64, 65, 67, 226, 307, 337, 372, 382, 384, 404, 406
Operators
  Type inference in the ML languages, 50
Orca, 108, 271
Orthogonality, 100
os, 346
os.system, 68, 346, 390
Output, 11
output, 379, 384

Paṇini, 184
package management, 64
Pan troglodytes, 323
parameter, 377, 378, 383, 385, 388, 389, 392
  args, 387
  "ragman", 378
  keyword, 377, 378
  types in PYTHON-3000, 383
Parameters
  explicit fetching, 25
parameters, 24
partial evaluation, 25
Pascal, 17, 20, 23, 28, 32, 35, 50, 59, 63, 77, 78, 82, 84, 92–94, 100, 105, 108, 110, 113–116, 118, 126, 127, 132, 133, 138, 139, 144, 145, 154–157, 161, 163, 168, 188, 207, 215, 217, 251, 252, 256, 261, 271, 312, 313, 361, 364, 370, 383, 388
  Delphi, 20, 32, 38
  GNU Pascal, 251
  TurboPascal, 20, 252
  UCSD-Pascal, 362
path, 382, 404
pattern matching, 121, 140, 183
PCR, 2, 3, 323, 334, 347–351, 353–357, 382, 404
  hotstart, 351
  touchdown, 351
PDF, 303
PDP-11, 101
Pearl, 42, 137
people need to understand the software, 106
PEP, 384, 388, 392
PEPs, 377
Perihelion, 42
peripherals, 11
Perl, 7, 25, 30, 38, 42, 44, 45, 49, 56, 57, 59, 87, 97, 99, 100, 108, 114, 137, 168, 171, 253, 342, 362, 369, 388
persistent storage, 382, 404
PHP, 44, 56, 89, 120
π calculus, 39
Pierre Weis, 51
pipeline, 69
pixel, 382, 404
PL/1, 20, 47
PL/M, 47
Plum, 47
PNG, 303
pointer, 364, 382, 404
polymorphism, 51, 158
poplib, 346
Posix, 372, 382, 404
postfix, 382, 404
PostScript, 22, 23
PPC, 12–14, 142, 225, 372
  family of processors, 13
  G5, 75
prefix, 382, 404
preprocessor, 220, 221
Prettyprinting
  C, 339
  Python, 337
private, 36
procedures, 19, 153
Processing, 11

processor, 382, 404
ProDOS™, 13
Productivity of languages, 44
Prof. Snape, 40
programming in bioinformatics, 3, 4
Programming multiprocessor systems, 42
Project Venona, 284
projection, 123
Projects
  strategy, 334
Prolog, 40, 41
pseudovirions, 18, 67
Psyco, 58
public, 36
PURE, 361
.pyc, 22
PyPy, 58
Python, 4, 6, 7, 16, 17, 22, 24, 26, 28, 29,
    32, 35, 37, 38, 43–45, 47–49, 53–59,
    63–66, 68, 71, 77, 81–83, 86–90,
    92–94, 96–100, 102, 105–107,
    109–116, 118, 121, 123–132, 134,
    136, 138, 141, 142, 144–148, 151,
    154, 155, 157–159, 161, 163–165,
    171, 173, 174, 178, 181, 183,
    188–191, 193–202, 205, 208, 210,
    214–217, 219, 221, 222, 224,
    228–230, 232, 238, 240, 241, 244,
    245, 253–255, 261–266, 269–271,
    280–282, 290, 291, 296, 304,
    307–309, 312, 313, 315, 318, 319,
    329, 332, 337–346, 361, 363,
    369–371, 377, 379, 381–384, 388,
    389, 391, 392, 395, 404
  advanced techniques, 337
  Compiler, 49
  prettyprinting, 337
Python Enhancement Proposals, 377
Python-3000, 377, 378, 381, 383, 384,
    388–390

Quicksort
  algorithm, 264
  in C, 265
  in HASKELL, 265
quicksort, 256, 266, 267

raise, 389
RAM, 12, 13, 22, 75, 315, 382, 404
range(), 382
read-eval-print loop, 382, 404
recursion, 382, 404
  dissection of type declarations, 48
  efficiency, 162
  Fibonacci's function, 162

finding prime numbers, 244
in O'CAML, 110
in compilers, 162
in declarative languages, 39
infinite
  stack overflow, 171
mathematically, 26, 159
processing large quantities of data, 160
requiem for a pug, 136
sets and bags, 161
tail optimization, 136, 159
terminal condition
  comparison to loops, 137
the *map* function, 123
recursive data structure, 109
recursive descent, 109, 185, 206, 257
Red Hat, 64
redirection, 68
REDUCE, 130
register, 382, 404
register machine, 13
registers, 12
regular expression, 382, 404
Regular expressions
  scanf, 341
repetition
  in LOGO, 118
  in REXX, 118
  in RUBY, 118
Reptiles
  Anna the hannah, 8, 39, 59, 62, 72, 88,
    97, 103, 126, 142, 151, 164, 179, 185,
    241, 245, 260, 267, 273, 282, 285,
    309, 320, 324, 331, 334, 340, 346,
    358, 376, 378, 384, 393, 398, 406
  Bodo the Boa, 340
  Is Anna a lizard?, 39
  newts are no –, 80
  Sammy the slowworm, 340
requires, 253
restriction endonucleases, 322
results, 23, 24
reverse Polish notation, 22
Rexx, 25, 44, 118, 135, 137, 167, 171
RISC, 101
ROM, 383, 405
rongorongo, 306
Ronnie James Dio, 243
→, 35, 218
rovásirás, 306, 383, 405
RTOS, 137
Ruby, 17, 23, 28, 35, 38, 44, 45, 49, 108,
    109, 113, 114, 118, 126–128, 188,
    191, 193, 194, 200, 208, 210, 214,

215, 217, 221, 238, 316, 361, 362, 369–371, 389
Ruby Gems, 362
Ruby on Rails, 362
runtime system, 16, 71

S-Lang, 28, 44, 45, 55
Scheme, 29, 136, 172, 270, 366, 389
Schopenhauer, 61
SCITE, 65
sed, 120
Serialization, 54
SETI project, 289, 300, 396
sets and bags
  as recursive data types, 161
  definition, 90
  functionality, 271
SGI, 42
  MIPS, 14
Shell, 45, 108
shell, 15, 383, 405
shellsort, 267
significant whitespace, 108
Simula, 37, 38
singularity of choice, 36
Sinix, 383, 405
Sisal, 32
sizeable, 389
slice, 361
slice operator, 146
SmallEiffel, 251
Smalltalk, 23, 37, 38, 107, 118, 138, 194, 195, 200–202, 208, 217, 227, 232, 291, 362, 369, 375, 388, 389
smtplib, 346
software, 383, 405
Solaris^{TM}, 63–65, 307, 383, 405
sorting
  bubblesort, 256, 262, 263
  burstsort, 267
  heapsort, 267
  insertsort, 263
  quicksort, 256, 266, 267
  shellsort, 267
source code, 383, 405
spaghetti code, 139
SPARC, 14, 372
stack, 22
  definition, 90
standard input, 67, 779
standard error, 379
Standard Format Specifiers, 385
Standard ML, 29
standard output, 68, 379
static storage, 359

StdErr, 379
stdin, 67
stdout, 67
Storage, 11
string, 15, 146, 381, 384–387
  constant, 380, 381
  format string, 384, 385, 387
  module in PYTHON, 381, 387
  unicode in PYTHON-3000, 381
string.find(), 147
strings
  as arrays, 87
  as objects, 224
Strong typing
  by postfix, 49
  in ADA, 50
subroutine, 153
Sun, 14, 42, 205, 374
  Fire 15K, 124
  SPARC, 12, 14, 170
  SPARC64, 75
SunOS, 383, 405
super, 390
super-encryption, 282
SuSE, 64
SV40, 176
symbol, 31
symbol table, 30, 49
sys, 71, 146
sys.argv, 72, 147, 280
sys.stderr, 379

tail→head, 34, 96
tarball, 64
Tcl, 44
Tcl/Tk, 205, 370
tcsh, 383, 405
Templates, 51
tensor, 289
Terry Pratchett, 247, 269
The Book of the Five Rings, 6
The Gold Bug, 278
The Number of the Beast, 74, 82, 309
time, 346
trait, 389
transputer, 42
trash, 89
Trees, 80, 92
triglyphs, 302
Trithemius, 279
true functional modularity, 106
tuple, 24, 96, 97, 111, 141, 146–148, 156, 159, 161, 183, 216, 316, 318, 329–331, 340, 357
TurboPascal, 20, 252

TurboProlog, 40
type classes, 389

Ubuntu, 64
UCSD-Pascal, 362
Uexküll, 18, 67
Ultrix, 383, 405
Umwelt, 67
Unicode, 281, 306, 383, 405
unicode, 380, 381
Unix, 15, 23, 43, 45, 251, 332, 337, 345, 371, 372, 383, 405
UNIX kernel, 14
unixish, 381
Use OO to avoid repetitions, 239
user interface, 15
User-Defined Formatting, 387
UTF-8, 281, 306, 380, 381, 384, 406

VAX, 12
version control system, 257, 258
VI, 65
Vigenère Cipher, 282
virtual machine, 384, 406
Visual Basic, 38, 54, 56, 237, 375
VM, 13, 141, 172, 252, 256, 384, 406
   Erlang, 142, 368
   Foobar
      assembly language, 172
   Java, 142, 205, 375
VMS, 384, 406
von Neumann, 38

walk down a list, 121
Wanna be an IT macho?, 55
Weak typing
   Functional languages, 53
Website
   caml.inria.fr/FAQ/general-eng.html, 51

caml.inria.fr/ocaml/htmlman/manual005.html, 198
sun.java.com, 205
www.dsuper.net/≈elehoczk/frmain.htm, 383, 405
www.ebi.ac.uk/2can, 1
www.emboss.org, 352
www.python.org, 4, 59, 64, 377
www.sunfreeware.com, 64
WHERE, 361
Whorf's theorem, 3
Why PYTHON?, 57
Windows^{TM}, 23, 64, 65, 67, 226, 307, 337, 372, 382, 384, 404, 406
Wirkmale, 67
Wirth, 287, 372
WordStar, 234
workstation, 384, 406
World Of Warcraft, 370

X11, 345
x86, 372
x[:b], 146
x[a:], 146
x[a:b], 146
XEMACS, 65
Xenix, 383, 405
XML, 47, 92, 236, 346, 392
xml*, 346
xrange(), 382

yield, 392

Z80, 101, 164, 234
zlib, 307, 309, 346
zsh, 383, 405
zychology
   and witchcraft, 200